BRAUN

BRAUN

Designed to Keep

Klaus Klemp
with Oliver Grabes, Nina Sagra,
Renke Thye and Benjamin Wilson

Part 1

History

The Braun Story

The Braun Story

The Braun brand has a firm place in design history. Products by the household electrical appliances company that was founded in Frankfurt in 1921 can be found in the collections of nearly every design museum in the world. Indeed, Braun's long-standing chief designer Dieter Rams is regarded as one of the most influential champions of twentieth-century industrial design. Accordingly, much has been written and published on the company. However, the focus has often been on specific themes or products and a complete overview of the 'Braun phenomenon' has not, until now, appeared. For over a century now, Braun has without a doubt been a truly remarkable phenomenon. The company represents an extraordinary synthesis of business ethics, high-quality technology and durable, user-orientated and innovative design. In this way, Braun exemplifies a unique corporate philosophy that is still highly relevant today.

This book reconstructs the company's history, keeping as close to the facts as possible. At the same time, it paints a portrait of the bigger picture while highlighting many of the smaller stories. It explores what is special about the brand, what has been at its core for over a century and what continues to be essential to it today. These 'core identities' also demonstrate how Braun may inspire the global and sustainable design of tomorrow. Braun's 'core identities' together constitute a stance that has helped shape modern product design worldwide as is clearly realized by hundreds of Braun products. Simplicity, usefulness and longevity are key concepts when it comes to understanding the 'Braun phenomenon'. The Braun understanding of products today is more topical and timelier than ever – and it can certainly serve as a role model for the future of product development.

Braun has also been shaped by the individuals who have helped build this company and its products. Some of the numerous people who have helped to create this 'Braun phenomenon' are introduced in short biographies in the third section of this book. Without question they are figures who were especially significant to the emergence and development of Braun Design and its successes, yet without the support of those working in the background they would be nothing. To honour all Braun employees would fill another book, so the biographies here should be seen as representative of the many who go unmentioned. These biographies offer a new perspective on the company and show clearly that without the creativity and courage of many individuals, it would have achieved little as a brand.

100+ Years of Braun

On the surface, this book covers the hundred-year history of a company primarily known for its design. What design expert or enthusiast is not aware of Braun products? Consider the works produced during the Arts and Crafts movement or Viennese Modernism around 1900, Peter Behrens's pioneering German corporate design for the AEG electrical company, early twentieth-century Bauhaus designs, the works of the Eameses, Olivetti, the Hochschule für Gestaltung (HfG) Ulm and the creations of Apple in the early twenty-first century – Braun's products without question join the ranks of the most well-known objects of design modernism to date.

New design always stands on the shoulders of its predecessors and, in the best cases, learns not only from insights gleaned during the creative process but also from wrong turns taken. Only the gods can perform 'creatio ex nihilio', or produce something from nothing. And while people who establish corporations, develop technology and design objects do create something, they are certainly not gods. This is true not only for the project of modernism in architecture and design since the early twentieth century but also for Braun, which is why it is worth looking back in order to prepare for the demands of tomorrow. Therefore, this book does not set out to be simply a historical account but rather a guide to the future.

The evolution of the Braun company is unquestionably a success story. From a four-man operation in Frankfurt in 1921 to a medium-sized enterprise and then a global player with almost 4,000[1] employees in Germany, the company has not only had a long journey but followed a continuous and winding path. En route there have been setbacks and problems, which are discussed in this book alongside the successes. The Braun story shows that reinvention is necessary, that challenges are inherently a part of development and that things often go forwards but sometimes also backwards. Business success is dynamic and volatile. It is not a linear process, though many company histories might like to make it out that way.

The Braun company has not always invented truly new things. It often optimized existing products in such a way that they could be meaningfully and easily used. Those who believe this to be of secondary importance are mistaken. After all, as the philosopher Martin Heidegger put it, the 'readiness to hand' of items and their appropriate use are truly what add value to a product. It is no wonder, then, that Apple's founder Steve Jobs and its long-standing head designer Jonathan Ive referred to Braun and its legendary innovator Dieter Rams as role models.

Design History

As early as 1.75 million years ago and up until about 40,000 years ago, our human ancestors developed and used flintstone hand axes. The hand axe, then, makes for a very long product cycle. Paleoanthropology and neuroscience have revealed how using a tool such as a hand axe slowly yet steadily changed the neuronal networking and abilities of its users. In fact, the tool already fulfilled some of the demands laid out for the first time by Dieter Rams in the 1970s in his Ten Principles for Good Design.[2] The hand axe was innovative, usable, self-explanatory, honest, durable, coherent, environmentally friendly and involved very little design at all.

Over centuries of use, hardly any formal improvements were made to it. What followed, thanks to constantly expanding human brain power, was a flood of design innovations from the earliest advanced cultures to present day. Design history ultimately starts well before the nineteenth century. And the beginning of Braun Design has a similarly long prehistory.

Longevity Yesterday, Today and Tomorrow

Humans shape things and things shape humans. The design of our world of things thus entails much more than just manufacturing 'stability, utility and beauty', which are the architectural properties the Roman architect and theorist Vitruvius called for back in the first century BCE. These three qualities are by no means obsolete and, in terms of today's product design, could be translated to 'durability, usability and beauty'.

Nowadays, the durability of products is a priority. The longer a product can be used, the smaller its ecological 'footprint' over its lifecycle. Unfortunately, business today is not concerned with durability but a technical and, most notably, aesthetic obsolescence, resulting in an object's premature visual ageing. Predetermined breaking points that technically phase the item out of use may be intentional or simply the result of a lack of engineering or production expertise. Visual obsolescence happens only when products follow fashion trends that fade quickly. By definition, fashion is the perpetually new and, especially with clothing, visual longevity tends to be the exception rather than the rule. Visual longevity should, however, be a fundamental criterion when it comes to technically durable consumer goods. Architect and design theorist Adolf Loos recognized this back in the early twentieth century. A product, he insisted, should remain visually 'tolerable for as long as it was physically durable,' meaning a ball gown's life should be shorter than a bureau desk. Loos saw this primarily from an economical viewpoint; today we must see it from an ecological perspective.

Visual longevity has played a special role in Braun devices since the mid-1950s. You do not become weary of these objects or tired of looking at them because their design is restrained and yet remains beautiful and desirable. This has always been and continues to be a central aspect of Braun Design.

Details Are Not Details

'God is in the detail' as the German art historian Aby Warburg put it. And the writer Antoine de Saint-Exupéry captured the essence of reduction in his famous quote: 'in anything at all, perfection is finally attained not when there is no longer anything to add, but when there is no longer anything to take away.'[3] The kind of design that obeys such principles is not developed through grand concepts but devotion to the details.

Unlike the rather cold colour schemes generally used for technical devices, Braun products feature a warm palette, with hues such as yellow, red or green as feature points on light grey or black casings. This subtle introduction of emotion into technical household devices changes our perception of them and is at the heart of our fascination with Braun products. Details are not details, rather 'they make the design' as US designer Charles Eames put it.[4]

People at Braun

Do the clarity, comprehensibility and functionality of Braun products reflect the clear, comprehensible and functional conditions in which they are developed? Most definitely not. Instead, several outstanding minds, who were extremely different and sometimes a challenge to work with, created countless products of far-reaching significance in a way that seemingly goes against everything textbooks advise. How was this possible? It was not a straight path but often a curvy and very rocky one.

When Max Braun founded his company in 1921, World War I had just ended and Germany faced a period of great social, political and economic upheaval. The debate over what constituted cutting-edge design was already in full swing. In Germany, the Deutscher Werkbund, the Bauhaus, the New Frankfurt movement, architects, designers and companies were all involved in driving this animated debate. Philosophers such as Friedrich Nietzsche or Reformers from the late nineteenth century onwards called for a 'new human' in response to the transformations brought about by industrialization, which also made new designs for their changing environment possible. Max Braun was impervious to all the theory involved. He was a pragmatic engineer who wanted to construct and manufacture meaningful appliances. Yet this was precisely what allowed him to contribute to a new industrial world of products that were manageable and usable for consumers.

Things were different when it came to his son and company heir Erwin Braun. Without a doubt, Erwin played a special role in the emergence and advancement of Braun Design. It was he who, in the 1950s, gathered a group of like-minded colleagues to produce appliances 'in the style of our times'. Essential to all of this (and this continues to be the case today) was an intensive culture of discussion and constructive debate, fostered most notably by Fritz Eichler, the first member of the board to bear responsibility for the design department.

Such a culture has brought the company sustainable success. Otl Aicher and Hans Gugelot of the HfG Ulm must also be mentioned in this context, not just for their important design work but precisely because of their zest for debate. When it came to creating this book, the author and current Braun designers have remained true to this principle of lively discourse.

Erwin Braun was a seeker, not someone who already knew how everything could be. He was not a visionary but a networker who recognized which paths were emerging and successfully went further down them. In 1961, he wrote:

> What is hoped for in all of this: we assume the basic fact that people are different to be true. From this point of view, the individual can only feel good (be happy) to a high degree if his life circumstances are adapted to his peculiarity. The individual has just a small chance of finding his ideal life circumstances. He must live according to given circumstances ... But it may be possible that he essentially has to only get accustomed to an international 'technical basic culture'. This could connect most of the individual people socially and create a condition for them – in groups or individually – to live according to their individuality better than they do today.[5]

His younger brother, Artur, should also not go unmentioned in this context. He was a man of few words but had a pragmatic technical mind and gathered a group of equally gifted engineers around him.

Likewise the quite extraordinary history of Braun does not end after the company was sold to Gillette in 1967. It continues through its incarnations first as a Gillette, and then a Procter & Gamble, subsidiary. This new constellation brought about entirely new conditions and opportunities, new successes as well as challenges, all of which are discussed in this book.

In a sense, writing the history of Braun is like composing a guide to the future. There were and are decisions, attitudes and principles that can still serve as models for good business and design today, not as a blueprint but as inspiration for how a successful company can gear up for the rest of the twenty-first century.

Global Design

Germany's industrial rise in the nineteenth century was based on a long global tradition of making copies of existing products and processes (especially of English ones) before going on to improve them. This is certainly true for the production processes for steel or for the knives produced in the Bergisches Land mines, which took their cue from the UK's coalfields in Sheffield. In an attempt to halt such copying activities, the British parliament introduced the 'Made in Germany' label, attaching negative connotations to any product with German origin. This label, however, has ultimately evolved from being a stigma to a seal of quality. Even today, few other countries have a better image in terms of quality products than Germany. This image is based on many technical improvements and the country's intense design and architecture discourse, which has been ongoing since the early twentieth century.

Yet 'global design' does not refer solely to the adoption and improvement of things that originated somewhere else. For example, from the late 1950s onwards, Braun products have been 'culturally compatible', meaning they blend in with many cultures due to their visual restraint. This was the case, for instance, with the formally reduced Thonet chair, masterminded more than 150 years ago. However, this is different from the 'design colonialism' exhibited by American car design in the 1950s and early 1960s, for example, with its tail fins and chrome trim that were adopted by Mercedes in Germany as well as in the Soviet Union for its government limousines.

Braun has established a place for itself in design history as a pioneer of modern design for household appliances. With its consistent focus on how a product functions and how to make those functions as visually simple and comprehensible as possible, Braun is unique and has achieved international acclaim and relevance.

Today, numerous global brands aspire to adopt this successful basic understanding of modern appliance design. Braun's restrained creative approach as part of a larger 'international design' fits well with the concept of a global economic system where the efficient, quality product is the one that is simple to use and relevant to as many worldwide users as possible. However, the results are often of questionable quality – simplicity is not easy to replicate.

To be successful and persuasive, design must develop a clear stance of its own. In addition to attentively grasping the situation of the consumers who want to use and experience a product, this also requires reflecting on your history and values – and then looking ahead. The future of product design relies on experience, innovation and the courage to do something new, something that is truly your own and better than what was there before.

Braun Design was not and is not about quick market turnaround or immediate profit generation. Instead, it offers an economically and ecologically mindful perspective and presents realistic solutions for the future of our material environment.

How to Read This Book

This book is, for the most part, chronologically structured and divided into chapters and subchapters. It does not necessarily have to be read from start to finish but invites you to turn the pages and enjoy discovering the history and philosophy of Braun.

The texts on Braun's company history in the first part of the book are framed by others on the historical context throughout, which provide a snapshot of the social, cultural and economic conditions of the respective time.

What has distinguished Braun over the decades is summarized in eight separate chapters as 'core identities' in the second part of the book. These 'core identities' describe Braun's 'design DNA'. The book's third part contains short yet illuminating biographies of the creatives and leaders who helped shape Braun's history. In both of these sections, potential company founders, design lovers and other kinds of readers may find a few stimulating ideas, rather than recipes that can simply be followed. This also holds true of Dieter Rams's famous Ten Principles for Good Design, which should always be considered and advanced forward.

After all, thinking is the quintessence of design.

01
A Start-Up in Frankfurt

In 1921, engineer Max Braun set up shop in a Frankfurt courtyard with just four employees and only one product, which immediately emerged as a bestseller. A native of East Prussia, he came to Hesse in Central Germany because he had fallen in love, moving to the cosmopolitan city of Frankfurt with his new wife. The emerging medium of radio soon became the focus of what swiftly evolved into an international business.

By 1931, only ten years later, Braun's payroll had risen from 4 to 1,000 employees.

'Trumpf' Belt Connector 1921

'Trumpf' Drum Detector 1923

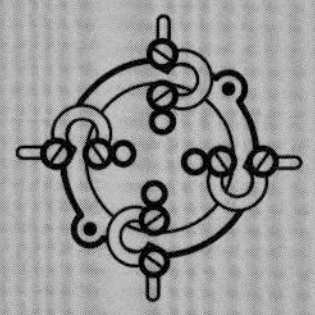

Injection-Moulded Parts 1925

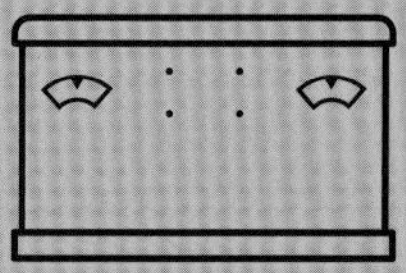

NE 3 1929

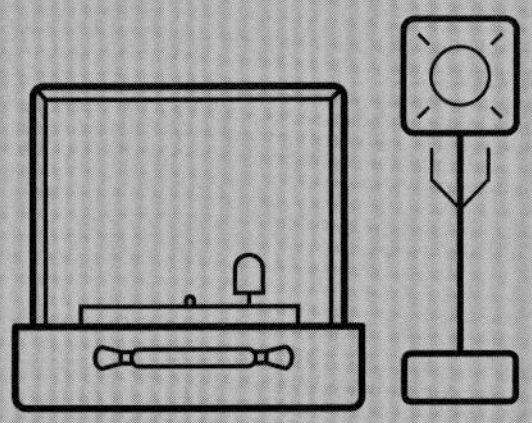

Home Recorder 1935
BMF 2020 1932

Cosmophon 333 1932

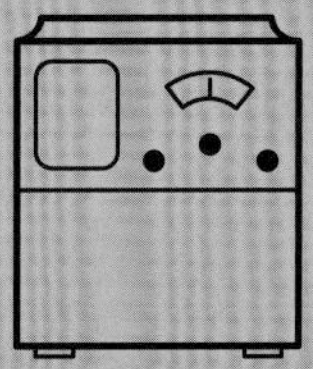

Cosmophon 777 1933

A Start-Up in Frankfurt

In the Silicon Valley of the recent past, many start-ups' origin stories began in garages and parents' basements. Max Braun similarly set up his company in a nondescript building in a rear courtyard in Frankfurt, working from the basement with family support. Braun's origins, then, make for a founding history that fits well into the scrappy and successful start-up tradition.

Wilhelm Max Braun came from deep in the German countryside, from a province close to the Russian border. He was born the sixth of seven sons to farmer and sailor Friedrich August Braun and his wife Karoline on 25 October 1890 in Schillgallen, a village with three hundred inhabitants in former East Prussia, now Russia. Friedrich was raised by a Protestant family that hailed from the Austrian state of Salzburg, from where they had been exiled in 1731 due to Catholic intolerance. Prussian King Wilhelm I accepted them into his Protestant country and entrusted them with a stretch of land in the easternmost part of the state, situated by the mouth of the river Memel. The family found their new village to still be overwhelmingly Catholic. Though religion dominated the culture and way of life well into the twentieth century, the more liberal environment of Prussia in which Max Braun grew up without a doubt influenced his later cosmopolitan attitudes.

The Braun family farm was quite impressive in size and employed several agricultural workers. Yet Max Braun's father was primarily a sailor and often away at sea for long periods. In the year Max was born, his father returned sick from a voyage and died soon afterwards, at the age of only forty-six. In 1895, his mother married the marine soldier David Koschubs, and together they expanded and modernized the farm, especially through the use of new machines, which may have had an influence on Max's affinity for technology. Nevertheless, in 1919 the couple divorced, the farm was sold and the assets were completely lost in the German hyperinflation crisis before the end of 1923. There was nothing left for Max to inherit.

After World War I, Max ended up in the Rhine-Main area from Berlin essentially by chance. At his brother's wedding in nearby Wiesbaden, he had met and fallen in love with Mathilde Göttelmann. To marry her, he had bowed to his future father-in-law's demand that he relocate to southern Germany from Berlin, since Mathilde's father wanted to protect her from the dangerous world of the capital. Max took the relocation after the betrothal in early 1920 as an opportunity to get off the AEG payroll in Berlin and go it alone. His wealthy father-in-law supported his endeavour, providing considerable sums of money as funding. This allowed Max to buy into Marnet & Co. in June 1920, a Frankfurt-based manufacturer of lathes. However, the company soon turned out to be unsound, and he parted with it quickly to salvage his investment.

Production in the Courtyard

In October 1920, Max Braun and his wife bought a house in Frankfurt. In February 1921, he founded his own company with just four employees and started work in premises encompassing 200 sq. m on the third floor of a building in a rear courtyard at 12 Jordanstrasse in the city's Bockenheim district. The new firm was also able to use the basement rooms.

The product that was to lay the foundations for the company was a now legendary belt connector, a device that fixed broken transmission belts. To grasp the importance of this small device, one should bear in mind that transmission belts were crucial to early-twentieth-century industry. Decentralized electric engines were still the exception; many machines such as lathes, drills or punches were driven by a central engine, with the drive motion transmitted to the workstation by means of rods, shafts, pulleys and belts. The leather belts occasionally broke in the process and had to be shortened or lengthened. Stitching them back together was an onerous matter, and while there were already other connecting systems, Max Braun's invention was more efficient, safer and cheaper. All the belt-fixing operations – the smooth cutting of the belt ends, the perforating and joining of the ends with a steel wire – could now be performed with a single device. About 36,000 of the connectors were produced at a price of 36.50 Reichsmarks each. This corresponded to sales of about 1.3 million Reichsmarks[1] – not bad for a start-up. Fittingly, it was given the name 'Trumpf', which means 'trump card' in German.

While Max Braun may have developed the product during his time with Marnet & Co., he had already been aware of the problem with torn belts while working in factories in Hamburg and Berlin.

|↑→
Max Braun set up his first workshop in a rear building at 12 Jordanstrasse in Frankfurt. His company had just four employees.

↑
Grandfather Peter Göttelmann with his daughters Annie (left) and Mathilde (right), Max Braun (rear) and the young Erwin Braun on his knee.

↑
The wedding of Max Braun and Mathilde Göttelmann at the bride's parents' house in Armsheim in the southern Rhine-Main region, 6 November 1920.

↗
Max Braun, 1921.

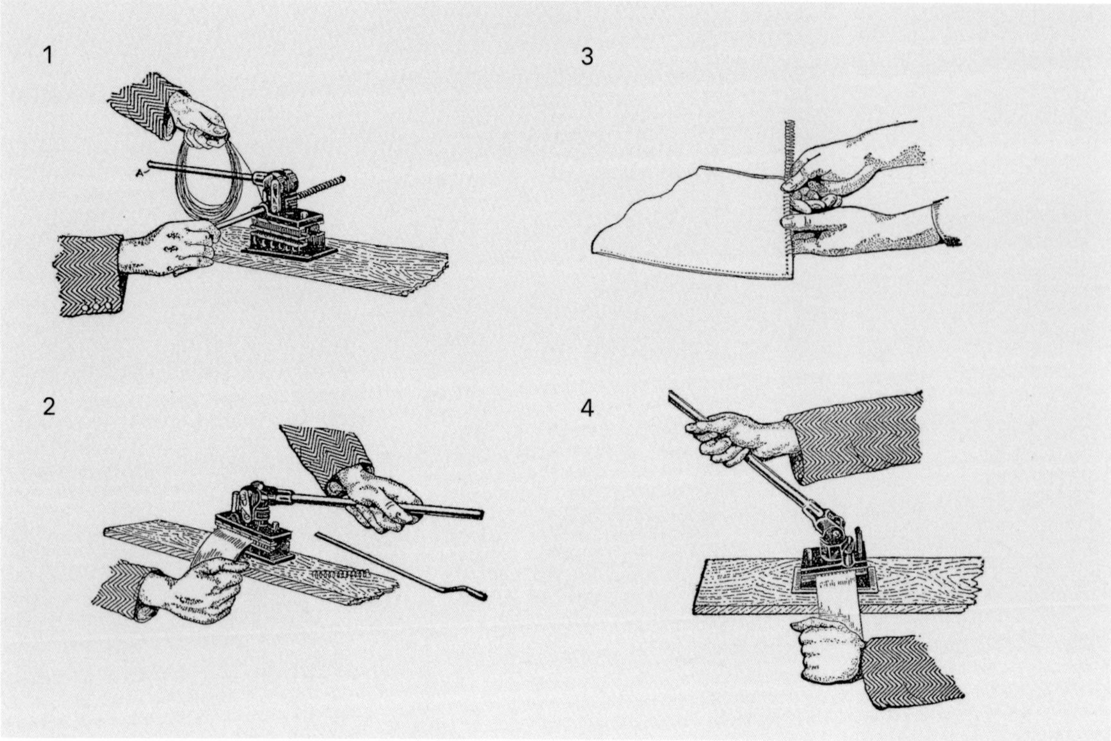

Perhaps his idea for a possible solution had been germinating in his mind for a long time. Either way, he was focusing on a topic that was of great relevance to the corporate world of his day.

A Problem-Solver

The belt connector was a true problem-solver – not fundamentally new, but simpler and better than the competition and, most importantly, cheaper. This first product by Braun thus embodied a creed that was to become characteristic of the company: make the better product at the better price. Throughout the company's history, Braun has not so much created new product types but, above all, optimized existing appliance concepts in technical or combination terms and, since 1955, in their designs. This is true not just of the belt connector, but the radio detector, the combination of record player and radio, the dynamo flashlight, shavers, kitchen appliances, Hi-Fi systems, hair dryers, epilators, electric toothbrushes and blood-pressure devices. In all cases, the optimization was often more challenging than the basic idea. After all, it required a high degree of inventiveness, both in technical and design terms, down to the very last detail. Braun was always a company that optimized products and made them accessible to the masses, with a view to technology, design and price.

Perhaps just as importantly, Max Braun began early on to think beyond Germany's borders. At the Frankfurt Autumn Fair of 1921, he booked half a stand to present his belt connectors, and it was there that he piqued the interest of an English manufacturer. Camp & Watson, which had already constructed a belt connector in England, became Braun's trading partner and successfully sold the device in the United Kingdom. Perhaps most importantly, they paid in pound sterling, which was a considerable help during the days of rampant inflation in Germany.

Mathilde Braun also played a strong role in the company. Not only was she present at the first fair events, but she also worked in the company's administration in the early days. Mathilde was keenly aware of her role as the company's co-financer. Her sister Annie also acquired a stake in Radio Braun GmbH in the late 1920s. For Max Braun, marrying into the Göttelmanns, a family of wealthy merchants, undoubtedly created a launchpad for his successful start in business.

↖
The company made its success with a device for fixing transmission belts that was inexpensive, efficient and easy to use. It was aptly named 'Trumpf' (Trump card) and immediately became a bestseller.

→|
Around 1960, Julius Calzaferri demonstrates the function of the belt connector to master machine builder Alfred Rada and Artur Braun (right). The device had been found during clean-up work in the factory, and thus a piece of company history had been rediscovered.

The Inter-War Years

The period in history when Max Braun set up his company was anything but easy for new businesses. Confronted by food shortages, rising prices, reparation payments and violent clashes, Germany faced great challenges after the end of World War I. While the nation did become a parliamentary democracy following elections to the new National Assembly in early 1919, neither the extreme right nor the extreme left were content to leave things with debates in the Reichstag and instead engaged in violent battle on the streets of Berlin. As the capital city thus appeared too dangerous a location for its parliament, the Reichstag convened in the small provincial city of Weimar. It was the city where Goethe had lived and worked and the seat of the new Bauhaus design school. The immediate postwar period in this 'Weimar Republic' was characterized by attempted coups and political murders. The years 1919–24 saw more than 400 politically motivated murders, with Communist Party founders Rosa Luxemburg and Karl Liebknecht, the first Minister President of the Free State of Bavaria Kurt Eisner and Reich Foreign Minister Walther Rathenau, son of AEG-founder Emil Rathenau, among the most prominent victims. Inflation also soared in the early 1920s, morphing into hyperinflation and destroying all monetary assets. In October 1923, a ticket for the tram cost 20 million Marks (so-called Papiermarks).

It was, however, also a time of new opportunities. The end of the German Reich unleashed energies that had already laid the foundations for a modern era – and for modern design – before World War I. The beginnings of modern public services and the service industry also occurred during this period.

There was a period of relative stability and economic recovery between 1924 and the Great Depression triggered by the New York stock market crash of 1929 – and those five or six years certainly offered favourable conditions for the development of Braun's company. In this context, with his involvement in radio and phonograph technology and strong focus on exports, Max Braun had bet on the right horse and entered a profitable field of business. From 1921 to 1933, the company grew from a payroll of initially 4 to some 1,000 staff members.

The years before the Nazis seized power were again characterized by political instability, and when the Nazi party gained its parliamentary majority in 1933, it marked the end of a democratic system. The first six years under the new rulers were defined by a fascist dictatorship increasingly showing its true inhuman face, combined with an economic recovery, mainly due to major state investments in armaments.

The politically and economically unstable inter-war years were certainly no simple birthplace for the young Braun company, but the period's modernist thrust in terms of technology and culture offered considerable opportunities.

Post-War Modernisms

In large parts of Europe, the period immediately after World War I saw a cultural renewal involving music, literature, painting, sculpture, theatre, dance, architecture and design – a renaissance that had its roots in the pre-war period. At the turn of the century, Vienna had become a hotbed of this early modernism, and the German Werkbund had been founded in Munich as an association of companies and creatives to improve the quality of German products, especially for export purposes. From 1916 onwards the Dada movement fundamentally challenged art. Marcel Duchamp declared it impossible to complete an artwork and, when it came to his 'readymades', ceded the interpretative authority to the viewers alone. The visual arts and architecture saw a suffusion of new 'isms': Cubism, Dadaism, Expressionism, Surrealism, Futurism, Constructivism, Functionalism and more.

Setting Out into a New World

The horrors of the war had shattered many traditional conventions of a personal, religious, social or cultural nature. Furthermore, the economic challenges of the post-war period led to a rural exodus to the major cities, which resulted in both a considerable housing shortage and a change in urban societies. The flourishing media of radio and cinema created a new public sphere and gave broader swaths of society access to culture.

The 1920s were a time of crises but also of the beginnings of a new world. The United States emerged as a major political and economic power. The ten-millionth Ford car rolled off the belts in 1924 and in 1927 Charles Lindbergh crossed the Atlantic Ocean in an aeroplane. In the 1930s a new medium took shape primarily in Great Britain and Germany that was later to have a fundamental impact: television.

The Bauhaus

In 1919, founding director Walter Gropius established the Bauhaus in Weimar as a modernist college with the goal of overcoming the differences between art, on the one hand, and architecture and design, on the other. All creative disciplines were to be devoted to a new architecture:

> The final goal of all creative activity is construction! Decorating housing was once the noblest task of the fine arts, they were inextricable components of the great art of building. Nowadays they stand in self-sufficient singularity, from which they may only be liberated through conscious cooperation and interaction between all workmen.

Such reads Gropius's first Bauhaus manifesto of 1919. He goes on:

> Architects, sculptors, painters, we all must go back to craftsmanship! For there is no 'art by profession'. There is no difference in nature between the artist and the craftsman. The artist is a progression of the craftsman.

Written while the nineteenth-century English Arts and Crafts movement was still at its zenith, this ideology did not work quite so well in practice. The crafts mentality no longer corresponded with the reality of highly industrialized societies, and the Bauhaus masters tended to address their own respective genres in art. In 1921, painter Paul Klee was appointed at the school and, in 1922, the Triadic Ballet by Oskar Schlemmer premiered there. In 1923, László Moholy-Nagy joined the Bauhaus as perhaps its most experimental artist and that same year Gropius corrected his demand from 'art and craft' to 'art and technology – a new union'. It was only from 1927 onwards, after the move to Dessau, that training in architecture was provided at the Bauhaus under Hannes Meyer.

Design Modernism in Other Places

While with its new approach to teaching, prominent lecturers and experimental approach the Bauhaus was without question the most important German design college of the 1920s, it was far from being the only place where modernist design was taught and implemented. New Objectivity in architecture was to be found first and foremost in Berlin, Stuttgart, Magdeburg, Breslau and Frankfurt. And modern industrial design was likewise not only taught at the Bauhaus.

Radio Time

In the mid-1920s, radio was the talk of the town in Frankfurt. And for Max Braun radio offered the prospect of an important new field of business.

The number of fee-paying broadcast subscribers was growing fast. In Frankfurt, the total rose from just 36 on 1 January 1924 to no less than 33,212 only six months later.[2] Twelve months on, there were about half a million registered and paying listeners in Germany.

Max Braun no doubt quickly realized that this was the shape of things to come. Together with his friend Rudolf Ackerle, he had been attending evening classes on radio technology at the Senckenberg Museum in Frankfurt since 1923.

The 'Trumpf' Drum Detector

In the cellar of 12 Jordanstrasse, Braun and Ackerle experimented with lead crystal detectors, the most important building block of early radio sets which were used to set transmission frequencies. 'Max Braun has succeeded – whether he is the first or not remains unclear – in replacing these crystals with artificial, self-fired stones, which makes production considerably cheaper.'[3] One can imagine that this process of invention spanned months and involved painstaking experimentation, as described by Manuel Aicher in his biography on Max Braun. The 'Trumpf' drum detector, with its artificial stone, was to be a crucial development for radio. It was more efficient, precise and simpler to adjust than traditional detectors with their lead crystals. Once again, Max Braun had found the solution for a problem and opened a new, lucrative area of business. At the Second Grand German Radio Exhibition of 1925 in Berlin, his invention was correspondingly well received.

Radio Parts

The next step was obvious: to produce entire radio sets. However, due to the patent situation this was not yet possible for Braun. The only patent for the manufacture of radio sets in Germany was held by the Telefunken company, which was very restrictive in granting licences. And so the company initially shifted its focus to the production of accessories.

Early radios with detectors worked only through headphones because of weak output signals and they also did not need electric current. The second generation, the so-called Audion-sets, were operated using mains power, tubes and loudspeakers. In 1926, Loewe Radio GmbH introduced such a device to the German market for the first time. However, this required a 'power amplifier'. Braun's first complete device would later be such an amplifier.

Alongside the detector, Braun initially produced new injection-moulded plastic parts for radios, which at the time were truly cutting-edge. In 1925, Max Braun came across a new plastic powder called Trolitul at the Frankfurt Spring Fair. He devised special machines that were able to handle such thermoset plastic at lower

→|
At the end of the 1920s, Braun sold numerous radio components, especially tube sockets and connecting systems. Simplicity and elegance were already part of the company's advertising.

|↙↓
A resilient tube socket that acted as a shock absorber for the sensitive tubes was particularly successful. Only the product and product name were featured on the packaging.

|↘↓
In 1928, Max Braun had an ultra-modern factory built that was designed in the New Frankfurt architectural style and had 5,000 sq. m of floor space. However, just one year later, Ernst May, then head of the City Planning Office, had the remaining space around the factory developed for the Hellerhof housing estate. This meant the factory could not be expanded and must have dampened Max Braun's enthusiasm for the urban modernization project.

RADIO Braun ELECTRO

Mit BRAUN-Teilen einfachste Montage

Nr. 8 Experimentierfassung für Zweifachröhre, mit 6 Anschlüssen im Kreis, beide Anoden rot RM 0.80
Gewicht 14 g Tel.-Wort: sixi

Nr. 9 Versenkte Fassung für Zweifachröhre mit Befestigungsgewindering. 6 Anschlüsse im Kreis
Gewicht 26 g RM 1.30
Telegr.-Wort: sixitop

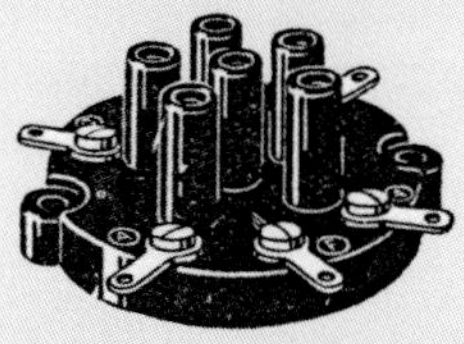

Nr. 10 Experimentierfassung für Zweifachröhre 5 Anschlüsse im Kreis, 1 Anschluß in der Mitte RM 0.80
Gewicht 14 g Tel.-Wort: sixo

Nr. 11 Versenkte Fassung für Zweifachröhre 5 Anschlüsse im Kreis, 1 Anschluß in der Mitte. Ausführung wie Nr. 9 RM 1.30
Gewicht 26 g Tel.-Wort: sixotop

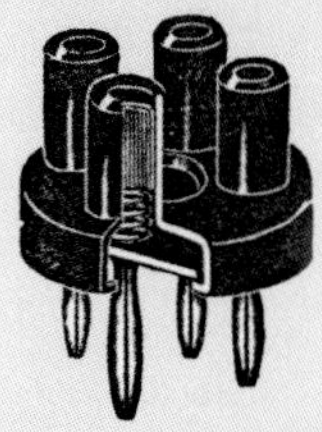

Nr. 6 Federnd. Zwischenstecker für Europasockel, DRGM-DRP. a. Zur Umwandlung normaler Fassungen in federnde, gegen Beschädigungen gesichert RM 1.—
Gew. 15 g Tel.-Wort: simplex

Nr. 6w wie Nr. 6 für Wechselstrom Röhren mit Mittelkontakt. RM. 1.15
Gewicht 17 g Tel.-Wort simplow

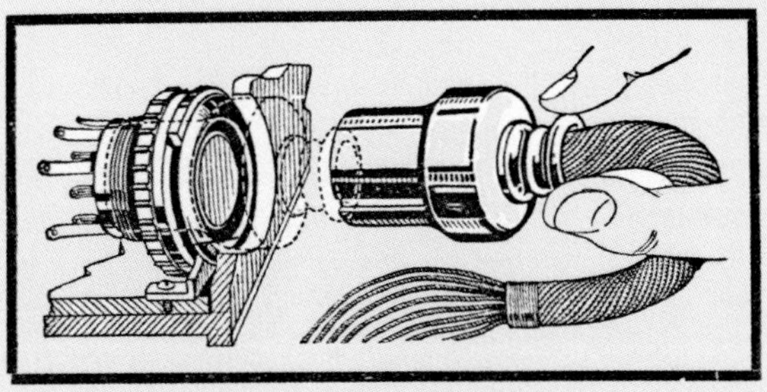

Nr. 31 Batterieanschluß »Kontakt«
Neuer kurzschlußsicherer Batteriestecker nebst zugehöriger Fassung mit Befestigungsgewindering. Einfachste Montage. An- und Abschalten aller Batterien mit einem Griff. Mit 7 Anschlüssen ohne Batterieschnüre. RM 2.80
Gewicht 32 g Telegr.-Wort: contact

Nr. 33 wie Nr. 31, jedoch mit Rändelmuttern einschl. Winkel und Befestigungsschrauben RM 3.60
Gewicht: 54g Telegr.-Wort: contactor

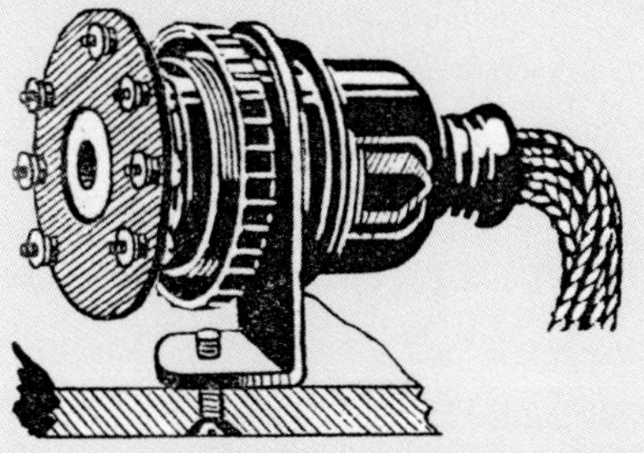

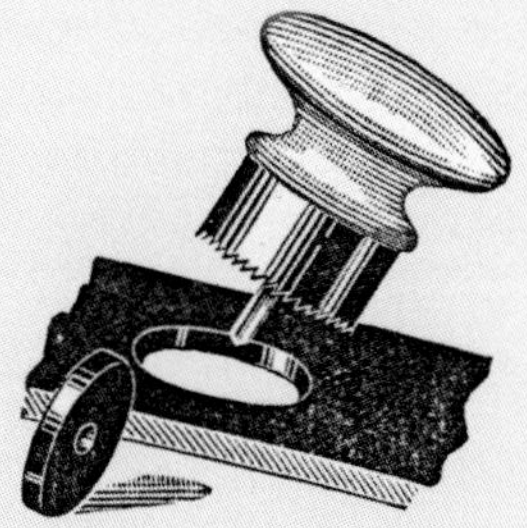

Nr. 51 Kreisschneider für Topffassung Nr. 3, 4, 7, 9 und 11. Man schneidet in ca. einer Minute ein kreisförmiges Loch aus Holz, Hartgummi und dergl. RM 1.20
Gewicht 80 g Telegr.-Wort: cirkel

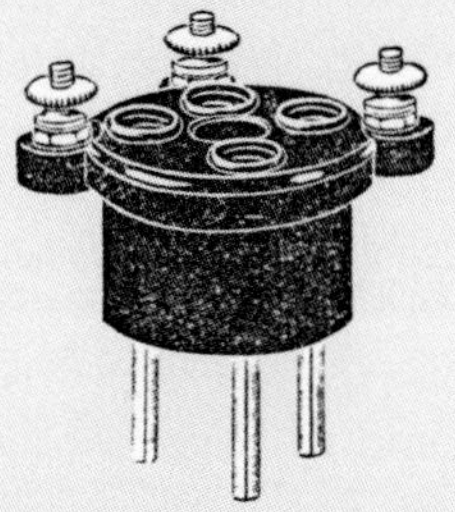

Nr. 20 Anschlußsockel zum Anschluß einer Elektro-Grammophon-Dose (pick-up). Die Audion-Röhre des Empfängers wird in den Anschlußsockel gesteckt und dient als Vorverstärker. Diese Schaltung arbeitet sehr rein und lautstark.
Gewicht 17 g RM 1.80
Telegr.-Wort: picksoc

Nr. 45

Nr. 45 Klinken-Stecker. Zweckmäßige Konstruktion. Keine blanken Teile, keine Schrauben (daher auch keine elektr. Schläge). Für Bananenstecker und Kopfhöreranschlußnadeln in gleicher Weise verwendbar.
Gewicht 18 g RM 1.20
Telegr.-Wort: plug

Nr. 32

Nr. 40

Batterie-Anschluß »Kontakt« mit Schnüren

Nr. **34**	wie 31	mit Schnüren	7/8	fach	RM 5.20
» **35**	» 33	» »	7/8	»	» 5.80
» **36**	» 31	» »	6/7	»	» 5.—
» **37**	» 33	» »	6/7	»	» 5.60
» **38**	» 31	» »	5/6	»	» 4.80
» **39**	» 33	» »	5/6	»	» 5.40

Nr. 32 Befestigungswinkel für »Braun-Kontakt« (Nr. 31) RM 0.40
Gewicht 25 g
Telegr.-Wort: standard

Nr. 40 Montagewinkel vernickelt
Gewicht 15 g RM 0.20
Telegr.-Wort: fix

Ein Gerät aus BRAUN-Teilen ist elegant

prices than the competition. On 19 June 1925, he filed a patent for such a heatable press with the German Reich Patent Office.[4] 'He operates, as it were (and he succeeds in this again and again) on the interface of a completely new branch of production and an equally new material.'[5] Braun thus sold both the products and the machines. This strategy was the real breakthrough to success and led to growth and a now much larger company.

Another factor was added: the radio valves used at the time were sensitive to vibration during operation. A spring-loaded base developed by Max Braun fixed this problem. It was met with great demand from the radio industry and, moreover, meant the tubes could now be used in portable devices, which soon became a specialty of the company. In addition, there were further developments for which utility model protection was registered. This paved the way for 'Radio Braun'.

Radio Braun

In 1928, the company built its own factory in Frankfurt at 91 Idsteiner Strasse; the premises boasted 5,000 sq. m of floor space. After only six months of construction work, 400 employees started work at the factory on 1 October 1928. The company was primarily a supplier to the phono and radio industry, building pick-up arms, phono drives and chassis. Since the construction of entire radios was protected by the patent held by Telefunken, Radio Braun collaborated from 1929 onwards with the Höchst-based manufacturer Carl Sevecke, who held such a licence. Max Braun subsequently bought up Sevecke's company to use its licence under his own brand. In 1932, Braun launched the first complete radio with receiver, amplifier and loudspeaker combined in a single unit: the Cosmophon 333W. Until 1931, the name Sevecke was used on all nameplates and in 1932 'Max Braun, Frankfurt a. M.' was added.[6] The radio set retailed at 119 Reichsmarks. This was equivalent to almost two months' rent in the new terraced houses of the famed housing estates designed by Ernst May. Radio was thus still a relatively expensive medium but became ever more popular.

Braun also offered an attachment set that made it possible to receive stations across Europe until listening to foreign radio stations was made illegal during World War II. Even so, many still used this device during the war – daring to listen to the BBC only when in bed and hiding under the blankets.

Record Player and Radio-Phonograph Sets

Alongside radios, Braun also built electric record players and the two together gave rise to a new product: combined radio-phonograph sets. The first such unit by Braun was presented at the Leipzig Spring Fair of 1930 under the name Rectofar.

Long-playing records were considerably higher fidelity than medium-wave broadcasters. This was no less than a revolution for middle-class living rooms. Families used to hear music in the home only if they played it on instruments themselves. With the new combination sets, it was possible to experience concerts and operas, as well as news and spoken-word programmes, from the comfort of your armchair. This was a giant leap in media history, and one that would only be matched in terms of significance by television and the internet. Technology and conceptual design came together here, which shows just how strongly this 'pair of siblings' changed twentieth-century society.

At the beginning of the 'radio era', when Max Braun became involved in the field, radios were black metal boxes whose outside showed little of the mysterious inner technology that made them function. (Similarly today we encounter many new-fangled black boxes in the shape of smartphones or smart speakers with even more mysterious inner lives and almost no buttons, increasingly operated using voice control.) Only two adjustment scales and turning wheels attached to the black box defined the first radio sets. Six screw heads pointed to the fact that something had been fastened inside. In the following decades, the inner functions increasingly moved to the devices' exteriors, and Braun pioneered the design of these new radios and listening devices.

↗
An advert by Werner Epstein, a student of Hans Leistikow at the Frankfurter Kunstschule, probably for the Südwestdeutsche broadcasting service.

|↓↓
Electronics in generous wooden housing. Audio furniture production at Braun's Idsteiner Strasse factory, about 1934–35.

Sevecke NE 3

Radio receiver, 1929,
Max Braun

The first radio by Max Braun went by the name of the cooperating firm, Sevecke. It was a kind of black tin box with a minimalist design, weighing 4.5 kg and costing 132.50 Reichsmarks.

REN 904

Radio: A New Medium

The social impact of radio broadcasting in the early twentieth century is certainly comparable to that of the internet today – even if unlike today's World Wide Web the older medium's effect was one-dimensional, meaning a one-way street from sender to receiver. Radio broadcasting decisively changed information sharing and communication in society, and time and space gained a new meaning. Even later, during the Third Reich, radio also became the medium of choice for spreading what we would now call 'fake news' or untruths.

On 29 October 1923, the first regular radio broadcast in Germany, Berlin's Vox-Funk-Stunde, went on air and could be received on longwave throughout the country. To begin with, the makers of the show were content to play live music from the studio, but new formats soon followed. The Südwestdeutsche Rundfunkdienst AG (SÜWRAG) broadcasting service, founded in late 1923 by a factory owner's son Dr Carl Adolf Schleussner and his friend Dr Hans Flesch, began broadcasting a regular programme on 1 April 1924. Alongside Schleussner, Frankfurt city council and Fritz von Opel also invested in this novel project:

> Even before the Frankfurt broadcasting station was founded, the first radio sets were presented at the Frankfurt Autumn Trade Fair in 1923 in the context of a small radio exhibition. The Municipal Director for Fairs subsequently initiated a separate 'radio fair' being established and this became a great hit with the public from April 6–18, 1924.[1]

Hans Flesch on Air

SÜWRAG's broadcast was the most experimental to date. This was primarily thanks to the founding director Hans Flesch, a medical graduate. During his studies Flesch had gained initial insights not only into X-ray technology but also into the field of the electromagnetic waves on which they are based. He also had a great interest in culture. When it came to his radio programme, he was not simply content to broadcast dance music but also produced the first German radio play and initiated live broadcasts from downtown. Ambitious lectures were broadcast in the Gedanken zur Zeit (Thoughts on the Times) series, which featured Theodor W. Adorno and Le Corbusier, among others. New music also had its place in the programme. Furthermore, Flesch experimented with different listening experiences by broadcasting the same piece of music with different microphone constellations. This provided practical evidence for media theories of the time. The station also collaborated with Hanns Eisler, Anton von Webern and even Bertolt Brecht. Hans Flesch became director in Berlin in 1929, in 1933 he was driven out by the Nazis, in 1942 the Wehrmacht conscripted him as a military hospital doctor and after the end of the war in 1945 he was considered missing in action – a forgotten pioneer.

Radio for All

In 1920s Frankfurt the radio project met with the approval of the Frankfurt city council and especially of the lord mayor, Ludwig Landmann, because it aligned with the city's drive toward radical modernization. The council's resolution no. 645 of 22 June 1925 gave two reasons for the technical expansion of the radio facilities and further support for the Frankfurt broadcaster:

> in order to propagandistically ... show the importance of Frankfurt as a central place of West German culture ... In the occupied zone in particular it must now become possible to listen to the radio using cheap apparatuses and indoor aerials, for it is only in this way that a broad dissemination among wide circles of the middle and working classes, who cannot afford expensive devices, is ensured.[2]

The municipal building councillor Ernst May had a cable radio of sorts built into the new housing estates constructed in Frankfurt, since the individual radio licence fee of 2 Reichsmarks per month was too high for many. While this was not particularly of interest to radio manufacturers initially, it did make the medium popular and thereafter convinced many inhabitants of housing estates to buy their own radio sets in order to be able to receive all stations.

The New Frankfurt Movement

In the period 1925–30 a comprehensive public housing project was launched in Frankfurt under the slogan 'The New Frankfurt'. The scheme was much more than just a programme of architecture. Parallel to the Bauhaus – but involving a decentralized network of institutions and individuals – it strove to enact fundamental social reforms by designing new and different worlds of objects and communication.

It was Ludwig Landmann, elected lord mayor of Frankfurt on 2 October 1924, who coined the programmatic term 'The New Frankfurt'. In the first edition of the magazine of the same name, he referenced Friedrich Nietzsche's cultural criticism and called for the 'establishment of new emotional and aesthetic linkages based on a new sense of time, a will to time, and the conditions of the time'. He did not mean this to constitute a rejection of the past, 'but a new generation, a new time, must work out in its inner and outer shape, the content that corresponds with its own striving for happiness and its perceptions of the ideal.'[3]

The 'new man' propagated by Friedrich Nietzsche and others was to be made possible here. After World War II, in a Braun advertisement this idea morphed into 'for the people of today' or 'in the style of our day'.

More than a Housing Construction Project

The plan was to create no fewer than 10,000 new homes for the many people seeking accommodation – within the space of only five years and in the context of a large-scale housing construction project. On 9 June 1925, architect Ernst May was elected as head of the municipal urban and housing development department by a large majority. Design here became more than a peripheral aspect. The idea was to take a holistic approach, in which not only the volume of the new homes but also their features were to be redefined – with furniture that was as small and adaptable as possible.

The Frankfurt Art Academy

In 1923, the city of Frankfurt took over the Kunstgewerbeschule (School of Applied Arts) and the private Academy of Arts Städelschule and merged them to form the Kunstschule für Freie und Angewandte Kunst (Art School for Fine and Applied Arts), which became one of New Frankfurt's central institutions. Art historian and director Fritz Wichert attempted to woo individual Bauhaus masters to work at the new school in Frankfurt. In 1925, he succeeded in appointing the master craftsman in charge of the Bauhaus metal workshop, Christian Dell, as well as Bauhaus teachers Josef Hartwig and Karl Peter Röhl. Moreover, in 1925–26 Paul Renner relocated from Munich to Frankfurt to lecture in advertising graphics and typography. In 1927 architect Franz Schuster arrived from Vienna to teach interior design and in 1928 artist and advertising designer Willi Baumeister moved up from Stuttgart. There was now ample design potential to be found in the city.

Frankfurt Opts for Standardization

A central topic in Frankfurt was the standardization of as many details in the home as possible. 'Frankfurt norms' were created for doors and door handles, ovens, light fittings, sanitary ceramics and furniture. Even the fitted unit kitchen designed by Margarete Lihotzky for all the homes was standardized and consisted of modules that could be combined in different ways – it was not until the 1950s that both Hans Gugelot and Dieter Rams revisited this notion of modular design in furniture and radio sets.

Numerous Frankfurt companies embraced the new design trend. In the 1920s, the city was what one would now call a 'design hotspot'. This was the environment where Max Braun's rational understanding of construction came to the fore, even if no direct links to Frankfurt's reform designers can be traced and no collaboration occurred. Perhaps this was a missed opportunity. A radio set by Mart Stam, Franz Schuster, Ferdinand Kramer or Walter Gropius, all of whom were working in the city, would have been a design milestone, and would certainly have been an icon for the Braun company.

Frankfurt and Beyond

Max Braun saw himself as a machinist and mechanical engineer, which of course he was in terms of the work he had done in Hamburg and Berlin. His second product, the drum detector for frequency tuning in early radio sets, sold incredibly well, as had his belt connector, and together the two products made for a substantial income. Yet Max was acutely aware of the fact that this would not last. Sooner or later, technological progress would render both products obsolete. And so, at an early date, the detector drew its inventor's attention to the field of radio technology, a step which, in a matter of years, would turn the small company into a formidable operation. Max, who as a farmer's son had worked his way up the ladder first from an industrial worker to machinist, from there to design engineer and finally onwards to entrepreneur, was not one to rest on his laurels. He was always looking for the next big thing.

Exactly five years after its entry in the commercial register on 1 June 1921, the company and a payroll of what was now some 180 members of staff relocated into a newly acquired building on neighbouring 41 Kiesstrasse (formerly a lathe factory owned by Lorch & Schmidt). Transformers, capacitors, blocking circuits, graduated dials and plugs, but especially valve sockets, were produced here, which the radio industry bought up in large numbers. A sales hit was the flexible socket, which was better able to protect delicate radio valves from vibrations and became a bestseller at the Leipzig Spring Fair. Braun's British business partners, Camp and Watson, also bought large numbers of these sockets. Max Braun sent a sample machine to the United Kingdom for inspection, yet this led to English sales of the socket going down – the partners had simply copied the machine and were now producing their own sockets. On a trip to London, Max Braun saw the many illegal 'offspring' of his machine and promptly sued Camp and Watson – and won. This put an end to the business relationship, but it also prompted Max Braun to think for the first time about producing abroad and opening his own sales branches outside of Germany.

Braun in Paris

In 1928, Max Braun journeyed to Paris with his wife to negotiate with a possible importer. The couple travelled in their own car, an American Whippet limousine, down a number of country roads. A deal was soon done, and the new partner in Paris ordered large quantities, but then fell behind with payments and began demanding more and more credit, claiming that business development was still difficult. In the summer of 1931, during a visit to Paris, Max Braun found out that the partner's coffers were empty. The director had increasingly resorted to 'private withdrawals', or so the accountant revealed. An auditor dispatched from Frankfurt specifically to keep an eye on things also turned out to be the wrong choice – he too partook in the 'withdrawals'. Eventually, the German-speaking Swiss accountant Christian Gmachel, who had uncovered the swindle, became a partner to Max Braun in a full-fledged Parisian subsidiary agency Max Braun & Cie., founded on 26 October 1931. This resulted in sound cooperation. Max (with a stake of 10,000 francs) and his sister-in-law, Annie Schnell (who put up 40,000 francs), became partners. Braun became widely known in France as a phonograph manufacturer on the back of his growing staff of sales representatives, active participation in trade fairs and much advertising. Braun would later be called 'Le roi du pick-up' and France did indeed turn into the most important export country for the new electric record player.

Crises and Opportunities

Thanks to extensive exports and the consequent switch in production to radios and combined radio-phonographs, Braun weathered the Great Depression from 1929 to 1932 relatively smoothly. Of the twenty-eight German producers of radio sets, Braun was initially one of the smallest – in 1939 before war broke out, the company was placed seventh in the rankings. Business was so good it was even possible for the Braun family to build their own and larger residential house at 17 Cretzschmarstrasse as early as 1930.

From 1930 onwards, the company established Braun representative offices, warehouses and assembly plants in numerous other countries. The first Braun factory outside of Germany opened in 1930 in Brussels, assembling radios and phonograph sets.

↑
Max and Mathilde Braun in front of a LUFT HANSA Dornier Komet III plane in 1926, probably before a flight to Geneva.

→
Max Braun exhibition stand in Paris, 1934; Braun Cosmophon BMF 2020 microphones can be seen in the background.

COSMOCOR
COSMOCORD

THIS
FACTORY
AND LAND
FOR·SALE
CHAMBERLAIN
&
WILLOWS
23 MOORGATE E·C·2

AYS

However, the management there was not very successful, and the factory was liquidated just two years later. Instead, a talented sales representative was subsequently found in Belgium, who also worked throughout the north of France.

The project involving the greatest commitment was launched in England in 1930. It was here that Max Braun built a factory in Enfield in north London. The new edifice looked like a mix between Ernst May's 'New Frankfurt' project and the prevalent Art Deco style.

Braun now had three hubs in Europe: in Belgium, France and Great Britain. Other branches in the Netherlands, Switzerland, Spain, Tunisia and Morocco followed. Most of the necessary contacts had been made at the Leipzig Fair. From an early date, Max Braun was already thinking internationally. Just nine years after launching the company in a Frankfurt courtyard, such expansion was more than remarkable.

On Construction

Max Braun remained a technophile design engineer throughout his life. From 1931 to 1937 he produced a record device, which he marketed as the 'Heim-Ton-Schallplatten-Selbstaufnahme-Apparat', meaning literally a device for recording records at home. The technology had been borrowed from a foreign model and it was possible to create 'spoken letters' this way. 'The "home recorder" was one of Max Braun's favourite ideas. He wanted to be the first to bring a record-cutting device for the home to market ... What was lacking was a cheap device that was easy to use ... that private individuals could use to record at home ... The "home recorder" turned into an obligatory touchstone for every new design engineer at Braun. Half a dozen engineers tried their hand at this idea, one after the other.'[7]

Yet it proved truly impossible to solve all the many technical problems. Furthermore, blank records were expensive and such a recording device therefore remained a marginal item in Braun's portfolio. A principal idea had been identified, but the right technology was just not available to produce it at the time. Braun had his on his hands a 'Leonardo problem' (referring to Leonardo da Vinci's idea for a helicopter that was only realized several centuries after he first conceived of it). Braun's portable TP 1 record player of 1959 was, likewise, missing a crucial technical element to make it practicable, since the small record only had a running time of two to three minutes. Sony's Walkman, designed to play the compact cassette, was the right medium for listening to music on the go, and first hit the streets two decades later.

A different device by Braun became much more successful after 1933. The Cosmophon 777W, which combined a radio and record player all in one, was the first of its kind and would remain a Braun specialty for many years. It was followed by the SK 4 in 1956 and by audio systems in the 1960s and 1970s.

The Boss

The story of the Braun company between 1921 and 1933 can without a doubt be seen as one of success. In only twelve years it had evolved into a major corporation with a payroll of around 1,000 – a company that centred around the dominant figure of Max Braun. He was a down-to-earth patriarch who was often described as tough, disciplined and hard on his staff (and himself) but also as a man who treated his employees justly. Nationalist thought was foreign to him, he was an entrepreneur who early on took an international approach and not just in a business sense – born in a backwater, he had become a man of the world.

Max Braun was thrifty: a carelessly dropped screw or paper clip on the floor would make him angry. That said, he could also be generous. He would invite business partners to expensive Parisian restaurants but would not order a starter or dessert. In 1923, he treated himself to an expensive flight to London from Frankfurt because he was interested in the new technology.[8] He was always keen on finding out about technological innovations, and his special interest lay in how to make them more manageable. He was not primarily concerned with culture or the design of the items he produced. For him, what was most important was their functionality, value for money and, most of all, the simple construction required to achieve these goals. As Leonardo da Vinci once quipped: 'Simplicity is the highest form of sophistication.'

|↑↑
The Braun factory in Enfield, London. Production happened there under the name of COSMOCORD.

|↖
Pick-up arm assembly in the Idsteiner Strasse factory, November 1931.

|←
Max Braun on the airfield in front of the British Imperial Airway's long-haul passenger plane, the Handley Page HP 45, Paris, 1932.

Home Recorder, BMF 2020

Recorder and microphone, 1935 and 1932, Max Braun

For recording on records, complete with microphone. This was Max Braun's attempt to design an inexpensive recorder for private use. However, none of the engineers at Braun were able to solve the device's many technical problems.

02
Dark Times, New Hope

When the Nazis came to power in Germany in 1933, arms manufacturing increased at an exponential rate. Radio, still a young mass medium, was also immediately used for propaganda purposes. Max Braun was by no means a sympathizer of the Nazi regime, quite the contrary, but his company now had to produce electronic equipment for the army. Like all manufacturers, Braun was obliged by the new government to produce a simple radio receiver, known as a Volksempfänger or 'People's Radio'. But the company also produced a special attachment for the radio that enabled strong reception of other European broadcasts, which was forbidden by the Nazi regime.

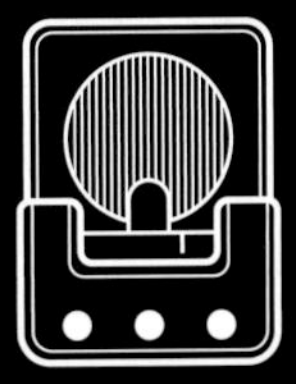

Volksempfänger VE 301 1933
+ Braun-VE-Vorsatz-Gerät 1935

BSK 36 1936

Ventilator BV 37 1936

BSK 238 1938

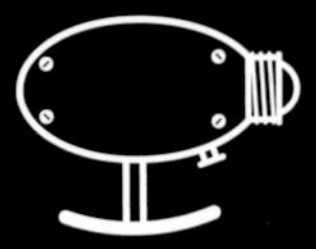

Manulux 1940

S 50 1950

Multimix 1950

The Third Reich and Wartime Production

In his biography on Max Braun,[1] Manuel Aicher concluded that Max completely rejected National Socialism:

> Rather unusually for a company that was so important for armaments and thus for the Nazi state, he never became a member of the NSDAP, the Nazi party, nor any of its branches. He never wore a party badge, but, in deliberate contrast, often wore a fresh flower, usually a carnation, in his buttonhole. In company meetings or in public he greeted the attending celebrities not with party titles, but by their names.

The following episode also made his attitude clear:

> In February 1933, a sales representative visited and was visibly proud of the party dress uniform he was wearing, adorned with colourful ribbons on the chest. After knocking on the door, he stepped into Max Braun's office. 'Heil Hitler, Herr Braun.' The latter mustered him silently. The sales rep turned red. 'I just came to ...' he started. Max Braun walked over to the telephone as though there was no one in the room and called his sales manager to the office. Until the manager turned up, Braun stood and looked out the window, while the man in the uniform stuttered 'Herr Braun ... Herr Braun.' Once the sales manager arrived, Braun turned to him and asked: 'Do you recognize this item of clothing?' The sales manager, smelling trouble, awkwardly replied: 'Yes, Herr Braun.' – 'Well, why don't you enlighten me!' After the manager had explained that this was a Nazi party uniform, Max Braun looked at him serenely and casually exclaimed: 'Please let this gentleman know that I do not wish to see such items of clothing in my building.' He then left the office and softly closed the door behind himself.[2]

We know of numerous instances in which Max Braun expressed his disapproval of the new rulers. There were frequent conflicts with regional party officials, yet it seems that Max had contacts in Berlin whom he could call upon if things got serious. This would presumably have included electrical wholesaler Friedrich Wilhelm Liebig, who had been a World War I fighter pilot in the famed Richthofen squadron. He contacted Max in the second half of the 1920s and in 1932 took over representation as Braun's first general agent. 'Liebig saw himself as the Braun company's "governor" in Berlin,' as Hans Wichmann wrote in his biography of Erwin Braun.[3] A friendship developed that would later also include Erwin.[4]

Attempt to Oust Max Braun from His Company

In 1938, the district leadership of the Nazi party tried to strip Max Braun of his powers and remove him from his position in the company. Through his loyal purchasing manager, he contacted the Wehrmacht High Command, which appreciated Braun for its quick delivery of special equipment. Through this intermediary, Max ensured that he was questioned by top-level officers, and subsequently was exonerated. Such an interrogation duly took place in the Reich Security Main Office, during the course of which it was concluded that Max was indeed a troublemaker, but a harmless one. The subordinate Frankfurt authorities were thereupon instructed accordingly to leave Max in his leadership position.

In 1943, when it became more likely that the Nazi regime would fall, a second interrogation followed in the Reich Security Main Office. Again, the Frankfurt offices had collected evidence against Braun. Now it was a matter of his exhibiting fundamental hostility to the Nazi party, intercourse with 'non-Aryans' abroad, daily eavesdropping on enemy broadcasts, sabotaging the recommendations of the Reich's economic leaders, snubbing the governors and showing signs of defeatism, or lack of faith in the Third Reich's 'final victory'.[5] What saved him this time was his role in supplying the Wehrmacht with special equipment that other factories, with a few exceptions, could not offer.

Wilhelm Wiegand

While domestic transactions were going badly in 1934, the loyal representation in Paris continued to increase record-player sales. Up to 350 record players were shipped to Paris every day and helped Braun weather an

→
The Volksempfänger VE 301, or 'People's Radio', was one of the most important propaganda instruments of the Nazi regime. All German radio manufacturers had to sell it for 76 Reichsmarks if they wished to continue receiving material. Braun produced it in 1933 and 1935. The radio, its attachment (Vorsatzgerät) and turntable (Volkslaufwerk) were designed by Otto Grissing and Walter Maria Kersting with input from Max Braun.

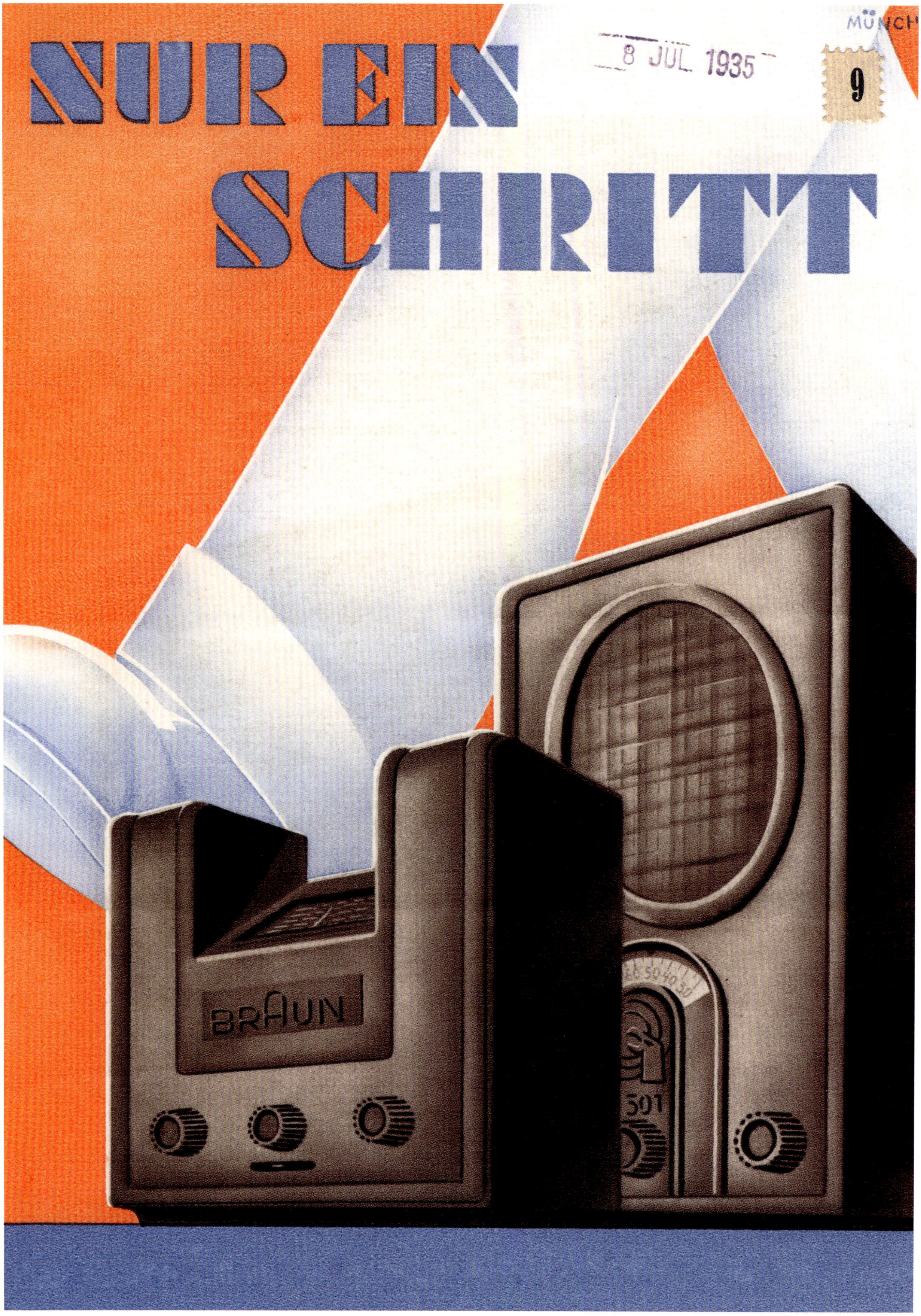
NUR EIN
SCHRITT
8 JUL. 1935
9
BRAUN
501

|←
After the mid-1930s, Braun produced an attachment for the VE 301. The radio had poor reception, but thanks to the attachment, it could receive European stations. There is a subtle ambiguity to the 1935 advertising poster by graphic designer Will Münch: there was definitely a lot of marching done in Germany then, but not to hear European voices. Braun also boasts about the excellent European reception in other contemporary ads.

↑
In mid-1941, the Nazi Block Wardens were required to go to every flat and affix this card to any radio receivers.

↗
Braun exhibition stand at the Great German Radio Exhibition in Berlin, which ran from 28 August to 6 September 1936.

eighteen-month-long financial crisis in the wake of the Great Depression. On 1 January 1935, Wilhelm Wiegand was employed as sole signatory at the Frankfurt head office. The talented businessman with a knack for organization and negotiation successfully established a network of sales outlets in Germany and abroad. Right up until Max Braun's death in 1951, Wiegand would be his most important and trusted staff member. In the year 1935, Braun also acquired a licence to construct its own radios from the Sevecke company. From 1933 onwards, Braun, like all other German radio manufacturers, produced the Volksempfänger VE 301 radio set which, in combination with the radio stations that had been forced to toe the Nazis' ideological line, was the party's most important propaganda instrument.[6]

Whether this was an obligation that the company had to fulfil is unclear. Ernst Gerhard, who at one point held power of attorney at Braun and from the early 1960s onwards was a member of the Frankfurt City Council, later reported that the companies had applied for contracts to build the VE 301. They had probably done so because companies were only allocated material for the VE 301, and not other products, so in this way they had an order book that was full enough to ensure their workforce was not conscripted into military service. All that changed during the war years. In 2021, after analysing all available archives for Braun, a team of expert historians determined:

In 1939, Braun employed 910 people in the two Frankfurt factories on Idsteiner Strasse and Mainzer Landstrasse. A few months after the start of the war, the newly established 'Reichsministerium für Bewaffnung und Munition' [Reich Ministry for Armament and Ammunition] under the leadership of Minister Fritz Todt pushed ahead with the war economy orientation and centralisation of the entire German industry in order to prepare the transition to a complete war economy. In 1940, Max Braun was also forced to reorganise production to meet the needs of the Wehrmacht in order to be able to maintain operations. By March 1943, 180 of the regular workforce had been drafted into the Wehrmacht. 945 people now worked for Braun, including 269 foreigners and nine employees registered as Jews.

Regarding Braun's potential employment of forced labour, the historians stated:

> During the Second World War, Braun employed foreign forced labourers and also prisoners of war. The so-called 'deployment of foreigners' in the German war economy was organised via the labour offices. As far as we know, forced labourers from concentration camps, who were made available to many industrial companies important to the war effort by the SS-Wirtschaftsverwaltungshauptamt [SS Economic Administration Main Office]

from 1942 onwards, were not employed at Braun.[7]

'Aryan' and 'Non-Aryan' Radios

The Volksempfänger was a single-circuit receiver with relatively weak reception. Its two units retailed respectively at 65 and 76 Reichsmarks. Braun offered an attachment set that provided a more powerful dual-circuit receiver, and later a record player, the Braun-Volks-Laufwerk, was added to the range. In the mid-1930s, the Braun logo with the larger letter A, the brainchild of graphic designer Will Münch, appeared on devices and in brochures for the first time. The oldest brochure featuring the enlarged letter A dates from the year 1933, like the shape of the Cosmophon 333 casing, which perhaps inspired it.[8]

A ban on working with 'non-Aryan' partners abroad (in Braun's case this applied to its representative offices in Naples and Belgium) led to new sales representatives having to be hired there. However, Max Braun made it a requirement – and the instruction was made only verbally – that the old partners would receive five per cent commission 'under the table' for every order. This arrangement was strictly adhered to until 1945.

Access to the French market, which was so important to Braun, became increasingly difficult. This in part resulted from a temporary import ban on complete sets, something the company got around by opting for final assembly of the sets in France. This was also in part the product of a waning acceptance of German goods in the neighbouring country. From 1936 onwards, Braun also produced table fans to generate additional turnover. In 1936, the first battery-powered portable receiver, the BSK 36, was brought to market. While this product in and of itself was not new, it was nevertheless better than the competition since it was a very handy and compact device. It was promptly awarded the Great Prize at the Thessaloniki International Exhibition, and by March 1938, 11,000 models had been sold. After this, a cheaper version with an imitation leather cover was also produced.

Small Radios and Smaller Premises

Space was getting tight at Idsteiner Strasse, even though the company now had more than 5,400 sq. m of production and office premises thanks to extensions. The subsidiary in Paris moved to a larger building in 1937. From 1 August 1938 onwards the company rented two floors offering a total of 5,200 sq. m on the former Peters Union grounds on Frankfurt's Mainzer Landstrasse and, in the following year, a third floor with a further 5,000 sq. m of space was acquired. Braun production facilities now totalled 15,000 sq. m, meaning that manufacturing premises had tripled in a short period of time. In 1939, before the start of World War II, Braun had 910 employees and a turnover of 9.1 million Reichsmarks. Max Braun's son Erwin confirmed to biographer Manuel Aicher that

↖
After 1940, Braun had to produce various devices for the military: radio transmitters and radio control units, including remote controls for small tanks and mine detectors.

→|
Produced initially for the military, and then after the war for civilians roaming around the ravaged, unlit cites at night, the dynamo-operated Manulux flashlight provided light without batteries.

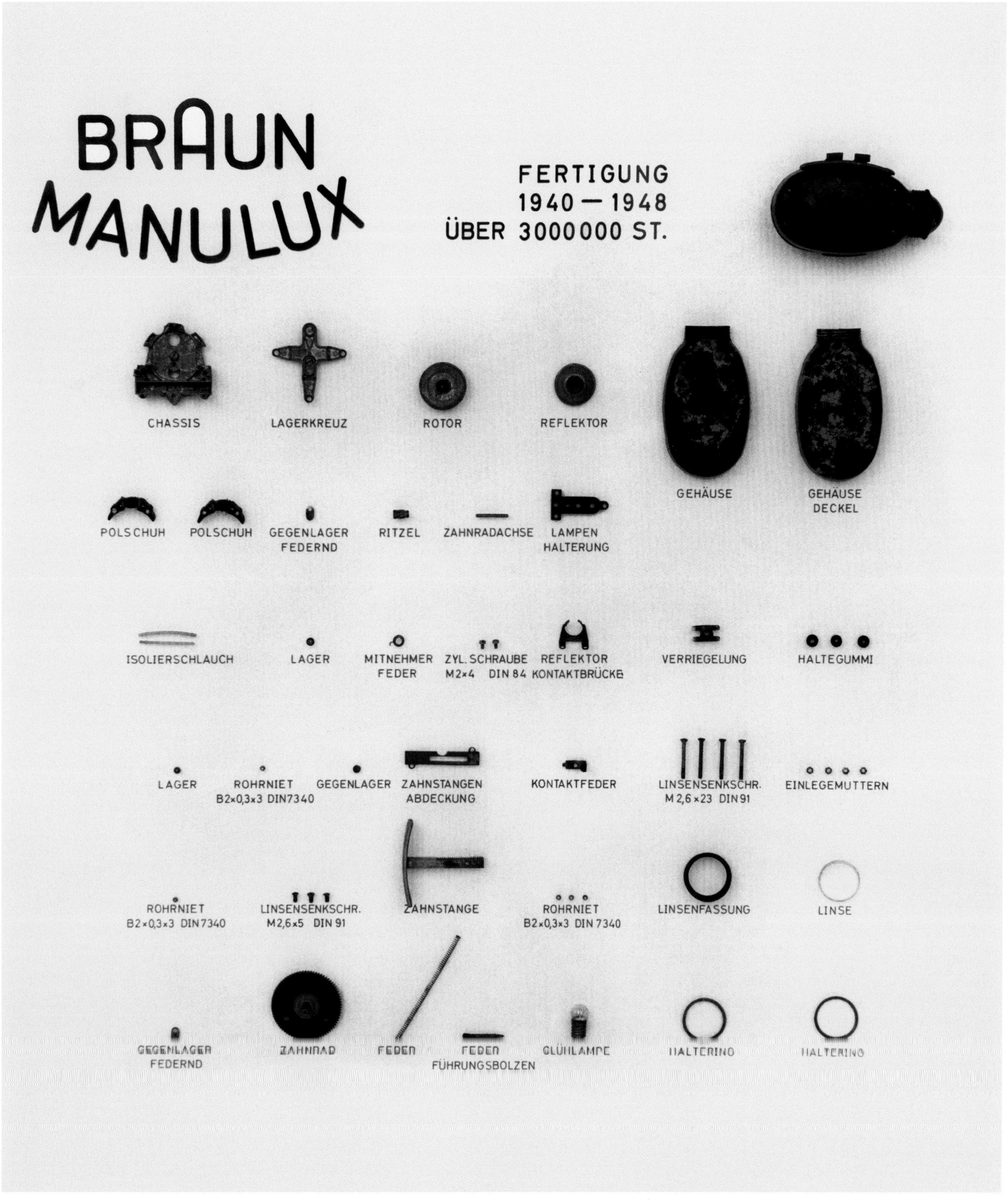

BRAUN
MANULUX
FERTIGUNG
1940 — 1948
ÜBER 3000000 ST.
CHASSIS
LAGERKREUZ
ROTOR
REFLEKTOR
GEHÄUSE
GEHÄUSE DECKEL
POLSCHUH
POLSCHUH
GEGENLAGER FEDERND
RITZEL
ZAHNRADACHSE
LAMPEN HALTERUNG
ISOLIERSCHLAUCH
LAGER
MITNEHMER FEDER
ZYL. SCHRAUBE M2×4 DIN 84
REFLEKTOR KONTAKTBRÜCKE
VERRIEGELUNG
HALTEGUMMI
LAGER
ROHRNIET B2×0,3×3 DIN7340
GEGENLAGER
ZAHNSTANGEN ABDECKUNG
KONTAKTFEDER
LINSENSENKSCHR. M 2,6×23 DIN 91
EINLEGEMUTTERN
ROHRNIET B2×0,3×3 DIN7340
LINSENSENKSCHR. M2,6×5 DIN 91
ZAHNSTANGE
ROHRNIET B2×0,3×3 DIN 7340
LINSENFASSUNG
LINSE
GEGENLAGER FEDERND
ZAHNRAD
FEDER
FEDER FÜHRUNGSBOLZEN
GLÜHLAMPE
HALTERING
HALTERING

Manulux

Dynamo-operated hand torch, 1938 (construction/ design), 1940–49 (production), Max Braun

his father opposed wars. A telling incident occurred in relation to this:

> A week after the war had broken out, Max Braun met with his foreign representative and the Berlin representative in a café on Kurfürstendamm in the morning. 'Order a bottle of sparkling wine will you,' he asked his surprised companions. 'Gentlemen, this is our farewell bubbly, for who knows when the time will come when we won't be able to buy Champagne anymore. You know what's happening. And you also know it's going to go wrong. I have a globe at home. In order to cover up Germany on it, all I need is my thumb. I can hardly hold the rest with both of my hands. Since when is the thumb stronger than an entire hand?'[9]

War Apparatus

Up until 1940, Max Braun did not accept commissions from the Wehrmacht. The last radios of that year were sold under the slogan 'Der edle Ton – Braun Tradition' (The Braun Tradition – A Classy Sound) and with a signet showing the company name printed on a coat of arms with a musical note. Had Braun wanted to continue producing radio sets, this would have required obtaining procurement permits for the raw materials, which were only issued for products important to wartime efforts. Therefore, production in Frankfurt had to switch to walkie-talkies and radio-control units and remote controls for small tanks, as well as mine detectors and dynamo flashlights. All in all, 1.3 million Manulux flashlights were produced by 1949. The Wehrmacht received the largest portion of these. Radios were now only produced in the factories in France, Belgium and the Netherlands.

During the heavy air raid on Frankfurt on 22 March 1944, thirty-five per cent of the premises on Idsteiner Strasse were destroyed, and in December, eighty per cent of those on Mainzer Landstrasse. Forever pragmatic, on 1 October 1945, Max Braun immediately set about rebuilding, with some 150 to 200 staff at his side. With a little luck and cunning, he managed to convince the Allied occupying forces' dismantling commission of the 'worthlessness' of his machines.

|↑↑
Partly destroyed by an air raid in 1945, the factory on the Mainzer Landstrasse was rebuilt by employees.

↖
More of a ruin than a factory: work was resumed on Idsteiner Strasse in a building without windows that showed signs of fire damage.

↑
Before It was damaged,
the Braun factory was
an exemplary workplace
suffused with light.

The Third Reich and World War II

When Adolf Hitler seized power on 30 January 1933, it marked the beginning of a twelve-year dictatorship in Germany that would lead the previously industrially and culturally developed nation to completely self-destruct. The already latent anti-Semitism suddenly became a national goal and led at first to forced displacements and, later, during World War II, to the murder of millions of Jews both inside Germany and in the foreign countries it occupied.

The cultural avant-garde in literature, music, theatre and art was outlawed and soon subject to persecution. Numerous Germans of Jewish origin, many of whom were often outstanding representatives of science and culture, were silenced and/or fled the country. The formerly international character of the social, cultural and economic life in the country from 1918 to 1933 gave way to an inward-looking emphasis and the spread of Germanization.

Design and Architecture in the Third Reich

The Nazis regarded the 'white modernism' of the New Frankfurt movement, the Bauhaus or other progressive architectural trends with suspicion because of the socialist leanings of those movements' leaders. Furthermore, given that the Nazis took their cue from rural nineteenth-century life, they were unable to fathom the modernists' white cubes.

Some middle-class architects and designers, such as Walter Gropius, Ludwig Mies van der Rohe or Ferdinand Kramer, at first believed the new political conditions would be short-lived. Gropius continued to participate in architectural competitions after 1933 and became a member of the Reich Chamber of Culture. However, in 1935, he explored opportunities to work in England while initially keeping his options for returning to Germany open. It was not until 1937 that he permanently emigrated to the United States. The fact that he was able to take all his household effects with him and was not forced to pay tax under the new regulations for emigrating from the Reich shows that the German authorities did not consider him to be an opponent of the regime.[1]

Moderate representatives of modernist design and architecture who had not been ostracized or expelled from the country were often instrumentalized for the fascist ideology. Former Bauhaus students thus also worked in Nazi Germany. The Bauhaus and the Werkbund were disbanded in 1933, but many of their design methods were permitted by the new holders of power, such as in the luminaires of Bauhaus master craftsman Christian Dell, who had been working at the Frankfurter Kunstschule from 1925 onwards, or the 639 porcelain tableware set by Wilhelm Wagenfeld.

The Nazi rulers in Germany especially welcomed all technological innovations, including, for example, Ferdinand Porsche's Volkswagen or the radio set designed by Walter Maria Kersting in 1929, which was rebranded as the 'radio receiver of the people' – more than twelve million units had been produced by 1939 thanks to government subsidies. Kersting had submitted it as a competition entry in 1933, and later was to distance himself from the revisions made to the design.

German Designers in the United States

Nevertheless, modernist design was not advanced in Nazi Germany – instead, life was defined by the dull recycling of modernism's residues, while the foundations and prerequisites for any advances had been lost. Europeans in exile found the United States to be a new, more productive field in which to work – both in terms of the willingness of American consumers to engage with modernist design and architecture and as a fertile intellectual context.

Reconstruction and Expansion

When Max Braun reopened his company on 1 October 1945 with a staff of around 150 to 200, Artur and Erwin were also there. After returning from doing active service, the sons, at just twenty and twenty-four, were simply integrated into the company by their father without much ado. Artur completed an apprenticeship as an electrical engineer in his father's company, Erwin did a commercial apprenticeship and, at the same time, started a course in business administration at Goethe University in 1946. This did not necessarily correspond to his own outlook on life since he would have preferred to study medicine.

Resuming production turned out to be difficult initially since the suppliers for radio apparatus parts were only able to deliver at a limited capacity. Furthermore, there were restrictions on deliveries from other occupied zones. In a letter dated 7 December 1945, director Wilhelm Wiegand wrote:

> What we are concerned with is essentially the repair of broadcasting devices as well as currently with the production of portable dynamo flashlights, as thank God we still have materials in stock for them, meaning that this is in fact the only item still being made. However, this item is subject to control by American Occupying Forces, meaning that from time to time only a certain quantity is approved for civil purposes, namely for important official and governmental services.[10]

Handheld Electricity and Other Post-War Improvisations

The Manulux dynamo flashlight was nevertheless to fill the company treasury, since in war-ravaged Germany the battery-free flashlight was an important aid when it came to people navigating unlit city streets on the way home after dark. Braun employees received two flashlights as part of their monthly salaries, which they could easily swap for food items on the black markets, much like the 'cigarette currency' that came in the form of American tobacco.

The radio sets the company repaired often included devices belonging to GIs, since American radios were built for 110 volts and were often simply plugged into German 220-volt power grid. To this end, the transformer had to be replaced, or an attachment set was required, which Braun was able to supply in a simple wooden box. Asked in a 2011 interview whether this also led to the company becoming acquainted with American radios, Artur Braun replied:

> The contact with the GIs was always very lively, and America had a pronounced pioneering role any way. Of course, they had dry shavers, they had radio sets, they had all kinds of things available at the time, and everyone liked to get a glimpse of what was being made in the US. After all, development essentially consists of looking at what everyone else is doing and then figuring out whether you can do it better – sometimes that's a very simple process.[11]

These contacts emphatically influenced the early post-war design of Braun's food processors, radio cases and first shaver, the S 50. Artur Braun remembers: 'Seen alongside the American models, our shaver looked a little shabby. I asked Schiebelhuth, a cabinetmaker, for some lime wood and built two wooden models that emulated the chic style.'[12] When it came to the table radios, production of which was resumed in 1947, the old shapes and techniques were largely retained.

Artur Braun was in charge of the new food processor segment. During that period, Max Braun devoted almost all his time to the idea of an electrical shaver, which he had already experimented with during the war. Manuel Aicher renders Erwin Braun's description of his father's relationship to the shaver as follows:

> Actually, Father was never particularly interested in broadcasting. He had always been annoyed with the difficulties and animosities with the patent and licence holders. According to him it happened more often than not that other patent holders butted in rather than he himself managing to register a patent. Once he had the idea to develop the shaver, he left others to get on with the whole radio issue, that just ran alongside. As Braun was the biggest company producing radio sets in Germany in 1945, it was expected

Braun's post-war product, the S 50 shaver, went on display in stores such as Storck in Hildesheim.

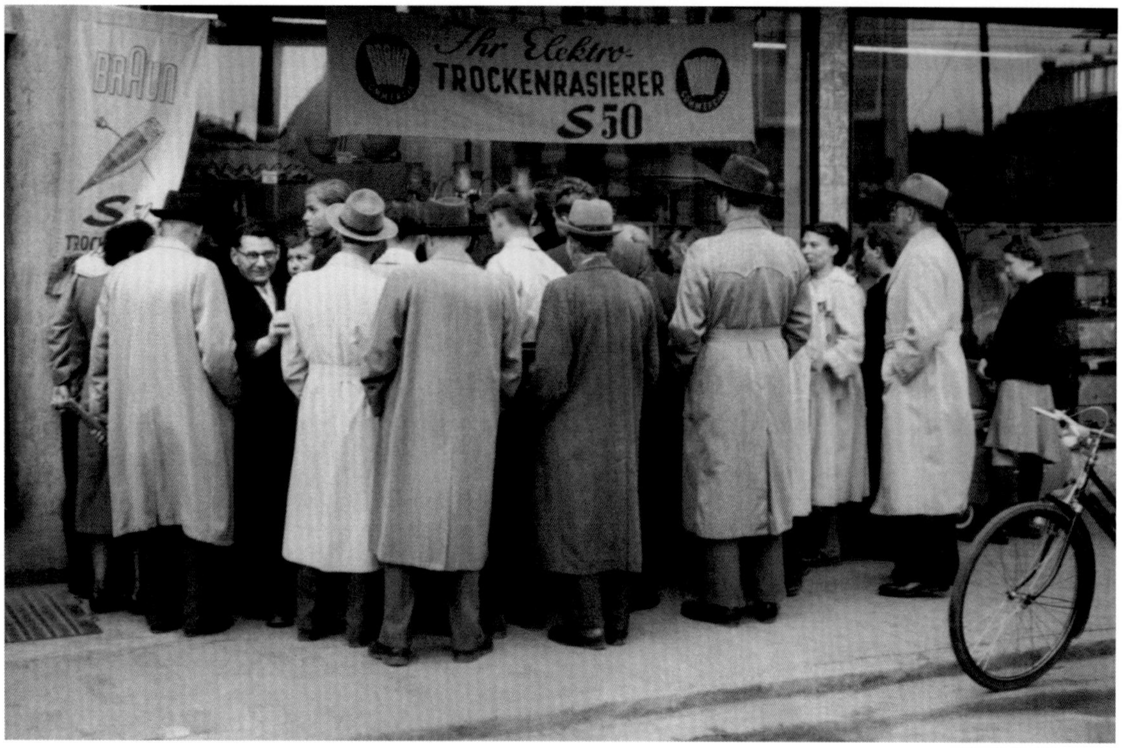

that radios would bring in the money required for the shaver project. His core interest lay in the shaver at this time though. He put all his energy – and the money – into it.[13]

An Innovation in Shaver Development

Max Braun achieved a crucial improvement with the mounting of the shaving head. The Sunbeam company had been producing a shaving foil system in the United States since 1934. The shaving head moved back and forth across a fixed perforated foil. Braun used a flexible mounting for the shaving foil that allowed it to be pressed directly onto the cutter block when shaving, which made for a markedly better result. Artur Braun and Bodo Fütterer provided crucial input to the development of the shaver.

On 21 December 1950, Max Braun received a patent for his dry shaver, effective 30 January 1945.[14] This was important, since only in this way was it possible to later sell a licence to the American company Ronson. In 1949, Max Braun took his shaver with him on his trip to the United States and had business cards featuring an image of the S 50 printed for the occasion.[15]

The entire Braun family pulled together when it came to presenting the S 50 and the Multimix food processor to the public for the first time at the Frankfurt Fair. The Piccolo portable radio, with its undeniable American influence, was also launched in 1950. That same year the Braun brothers founded the company Braun Commercial GmbH in order to step up sales of shavers and food processors. Wilhelm Wiegand now also had a share in the company's nominal capital of 50,000 Deutschmarks.

A 'Perfect Dicer for Foodstuffs'

While Max Braun had conceived of the shaver during the war, the development of the food processor required a great deal less time. In principle it was a recreation of the Starmix device available on the German market, which Max Braun had brought home with him from a fair. The main difference was that the Braun product retailed more cheaply. Surprisingly, he was nevertheless granted the patent for the 'perfect dicer for foodstuffs' for which he applied on 16 November 1950.

In line with his desire to practice medicine, Erwin Braun was particularly concerned with healthy eating, which the new food processors were supposed to promise prophylactically. The Multimix and Multipress came with index cards illustrated on brochures that featured recipes and health tips. 'Multimix gives you joie de vivre!' was to be the resulting motto. Around that time Dr Fritz Bode was also recruited as a medical consultant, thus laying the foundation for the later in-house 'health service', which, in contrast to 'traditional' company medical care, was primarily preventive in nature.

↖
Another view of the store Storck in Hildesheim. There were so many people it was sometimes hard to see anything for sale.

↑| and ↓|
Max Braun on a fact-finding trip to the US in 1949: in Philadelphia on the Delaware River Bridge, later renamed the Benjamin Franklin Bridge, and at Niagara Falls.

S 50

Electric shaver, 1950,
Max and Artur Braun

Shavers were a completely new business segment and Max Braun's passion since the war. The firm's own electric shaver became its most profitable product. And that is still the case for its shavers today.

←
After 1950, the Multimix food processor became one of the most important post-war products for the company alongside shavers. Braun did not reinvent the machine but made it better and less expensive than competing products.

↑| and ↓|
The Multimix presented to passers-by in the window of an appliance store and at a kind of trade fair for household appliances, 1953.

BRAUN Multimix
Max Braun Frankfurt a/Main

↑
The company logo, updated by Wolfgang Schmittel in 1953, installed on the new factory on Rüsselsheimer Strasse.

Rüsselsheimer Strasse

The number of employees at Max Braun oHG had risen from 240 to 392 between June and December 1948 following the German currency reform. The premises rented from Continental at 196–212 Mainzer Landstrasse, most of which had been destroyed during the war, were not rebuilt. Braun consequently acquired a plot comprising 24,000 sq. m on neighbouring Rüsselsheimer Strasse, where, in the years that followed, premises straddling a total of 4,600 sq. m of usable space were constructed. The first building was completed in 1950. For the first time, electric S 50 dry shavers and Multimix food processors were now produced alongside radios and record players. In 1952, the first Multipress centrifugal juicer followed; it relied on the same motor block as the food processor. In the two years before Max Braun's untimely death on 6 November 1951, foreign sales sharply increased. From late 1949 to the end of 1950, they rose fourteenfold, and the subsequent year by a further 274 per cent. No less than thirty per cent of sales were now booked with exports, well above the average of eleven per cent in the German electrical industry. The company was certainly on an upward trajectory.

↗
When the presentation of shavers and kitchen appliances becomes something special, for want of other sensational events in the 1950s, Braun employed numerous women presenters who toured West Germany in a Messerschmitt Cabin Scooter to show off its shavers.

↑
The Braun factory II on Rüsselsheimer Strasse in the early 1950s.

STREBELWERK
BRAUN

03

A New Company

In 1951, Max Braun passed away at sixty-one due to a heart attack. His sons, Artur and Erwin Braun, took over the company. Following the architectural reconstruction of the company's buildings after the war, they prioritized the well-being of their employees, introducing new ways of working and new departments. They also embarked on developing innovative products such as kitchen appliances, electric shavers and flashlights.

Hobby 1952

combi 1955

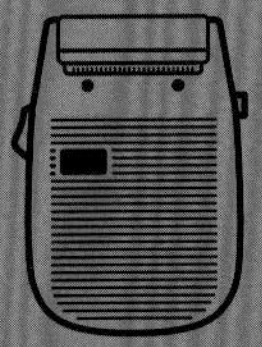

combi DL 5 1957

A New Company

Morning Contemplation on 6 November 1951

Standing on your toes
You don't have a solid stand
Walking with your legs spread wide
You won't move forward
Those who want to shine in person
Won't be enlightened
Those who want to be something special
Remain unimportant
Those who boast
Won't succeed
Those who show off
Will remain petty people.

– Erwin Braun referring to the day of his father's death[1]

Max Braun suffered his first heart attack in August 1948. Several others followed, one of which occurred during his trip to the United States in 1949. Yet he ignored his wife's plea to take thing more slowly since he felt it was an undue obstruction to his activities. On 6 November 1951, he suffered another heart attack at his desk in the Idsteiner Strasse plant, and this time it was fatal.

After his premature death, Max Braun's sons had to step up virtually overnight and assume responsibility for a company that was only just beginning to find its feet again. After 7 November 1951, the company was run as a general partnership (Max Braun oHG) with Erwin and Artur Braun as the lead associates. New areas of business – the electric shaver and food processor product lines – had already been staked out. The days of exclusively making radios and record players now belonged to the past. Over the coming decades, 'Radio Braun' would turn into a diversified producer of smaller electrical appliances.

In 1951, the company's annual turnover was 14 million Deutschmarks. The company assets Max Braun left behind totalled 4.9 million Deutschmarks and his personal wealth comprised 60,550 Deutschmarks. After inheritance tax, the sum was 3.3 million Deutschmarks (equivalent to approximately 9 million Deutschmarks or €4.5 million today), which constituted a strong financial launchpad for his sons.

Networks

Within a short period of time, Erwin Braun expanded the company's senior management by drawing on his personal network of friends from before and during the war. For example, his childhood friend Gerhard Lander, who had studied physics at Frankfurt's Goethe University while Erwin read business administration, joined the company as a freelancer in 1951. Another friend from Erwin's school days, Georg Hohm, began working for Braun in a commercial role in 1950, and Werner Apel-Dube, a further schoolmate, was responsible for setting up the photo technology department from 1952 onwards. Hagen Gross, whom Erwin Braun had met in a wartime sanatorium, joined Braun in 1955. Erwin had met Wiesbaden-based print shop owner and publisher Dieter G. Mundas at the military intelligence college in Halle and, after 1953, the latter worked for Braun on a freelance basis, mostly as a recruitment consultant. War comrade Richard Rohlf oversaw setting up the shaver factory in Walldürn from 1954 onwards, while another former fellow soldier, Rudi Kurz, who would later teach at the Frankfurt College of Engineering, recommended Bodo Fütterer, who, after 1950, became one of the Braun company's most important design engineers. Yet the most impactful hire from among his circle of friends was certainly Dr Fritz Eichler, with whom Erwin Braun shared a friendship since their time together in the Panzer Signal Corps and who was to make a major contribution to organizational reform in the company from 1954 onwards. Every morning at 10 a.m. Erwin Braun and Fritz Eichler would take breakfast together at a review meeting.

Erwin, who was in his early thirties, had thus established a circle of close confidants within the company, men who not only supported his plans for restructuring but actively helped move them forward. Needless to say, some of the old management had rather critical views of the new owners' activities.

'Honesty and Humanity'

Big new competitors, such as Max Grundig, and a price collapse had made the radio market much tougher than before. Erwin Braun knew that the company would need to add new product segments to its portfolio. Together with

↗
Artur and Erwin Braun, c. 1954.

→
In 1954, Braun signed a franchise agreement with American manufacturer Ronson for electric shavers to the tune of $10 million the single largest export transaction in post-war Germany at the time. Around the table for the signing of the contract (left to right): Artur, Erwin and L. V. Aronson, President of Ronson. A framed photograph of Max Braun hangs near the window.

Gerhard Lander in 1952, he developed the idea for the inexpensive and light Hobby flash unit for the then newly emerging market of amateur photography – a device that would be marketed efficiently and successfully. Once again, this was a Braun success story that started not with an invention, but with a considerable improvement on an existing product that was then made more affordable for the broader public.

Yet despite such successes it was still not what Erwin had in mind for the company. He imagined something much more holistic: 'Is it also possible to be successful in business through more honesty and humanity?'[2] The world of tomorrow would be different from that of the past. Every opportunity that followed such a 'new path' was pursued and often the company was lucky to find the right partners for its endeavours.

This approach fit in with the liberal-minded ethics of the post-war era, as evidenced by the West German president Theodor Heuss's remark that:

> Objects should become so neat that they are no longer admired as art but are simply regarded as a matter of course. Our era should in particular comprehend one thing: the law of poverty, to which we are subject, may blend well with the law of grace, if it is imbued with the breath of human emotion.[3]

The post-war era, then, was defined by reconstruction efforts, as well as attempts to bridge the gap between contemporary culture and the modernist designs of the 1920s. Heuss, a liberal economist, journalist and politician, had been the full-time managing director, then a member of the board of the Deutscher Werkbund, since 1918. He served as president of West Germany from 1949 to 1959 and championed the promotion of industrial design and initiated state funding for design during his term. In 1953, the German Design Council was established as an independent, yet publicly financed, foundation based in Darmstadt by resolution of the German Bundestag. Its director Mia Seeger curated the first German contribution to the 1954 Triennale Milano, the theme of which was Prefabrication and Industrial Design.

Patrons of Design

One year before the foundation of the German Design Council, the city of Darmstadt and Louis, Prince of Hesse and by Rhine, a supporter of modern design, had already founded the Institute for New Technical Form. The idea was sparked by the city's Darmstädter Gespräche Mensch und Technik (Darmstadt Discussions on Man and Technology) in 1952, which called for a new approach to design and new forms of industry as well as corresponding institutions. The patron of these talks was none other than Theodor Heuss. The Darmstadt discussions received a great deal of media attention nationwide, and Erwin Braun and Fritz Eichler would

↖
Wilhelm Wagenfeld (left), who worked as the first professional designer for Braun after 1954, during a visit to Frankfurt.

→|
The combi, a combined radio and record player designed by Wilhelm Wagenfeld in 1955, photographed together with the Diamond Chair by Harry Bertoia. The German American furniture manufacturer Knoll International became an important image partner for Braun.

CRONIN
DIE ZITADELLE

presumably have been aware of them as well. Ludwig Erhard, who was Federal German Minister of Economic Affairs from 1949 to 1963, also promoted modern architecture and design. In 1955, he acted as patron of the Grosse Deutsche Rundfunk, Fernseh und Phonoausstellung Düsseldorf (Düsseldorf Consumer Electronics Exhibition).

Two years earlier, Raymond Loewy's book Never Leave Well Enough Alone: The Personal Record of an Industrial Designer was published in German with the somewhat inaccurate and sensationalist title Hässlichkeit verkauft sich schlecht. Die Erlebnisse des erfolgreichsten Formgestalters unserer Zeit (Ugliness Sells Badly: The Experiences of the Most Successful Designer of Our Time).[4] The book had a great impact, and not only did Erwin Braun read it himself but distribute it to senior colleagues at the company. Bundesverband der Deutschen Industrie (BDI, the Federation of German Industries), founded in 1949, also addressed the subject of industrial design for a time through its non-profit organization Kulturkreis der deutschen Wirtschaft im BDI e. V. (Cultural Society of the German Economy in the BDI) and the Arbeitskreis für industrielle Formgebung (Working Party on Industrial Design).

The Verein für Industrieform (Association for Industrial Form), founded in 1955 primarily by industrialists from the Ruhr region, instituted a permanent public exhibition on industrial design in the Krupp Villa Hügel in Essen; the show was curated by a jury of Werkbund members. Together with the German Design Council and the Arbeitskreis für industrielle Formgebung, by 1955 there were thus four new entities under private law busy promoting design in Germany. The consumer goods fairs likewise addressed the topic. From 1952 onwards, the Spring Fair in Frankfurt showcased a Sonderschau formschöner Industrieerzeugnisse (Special Exhibition of Beautifully Designed Industrial Products), which in 1956 and 1960 was also featured at the Autumn Fair in Cologne.[5] In 1953, the Sonderschau formgerechter Industrieerzeugnisse (Special Show of Well-Designed Industrial Products) went on display at the Hanover Industrial Fair, partly on the initiative of Philip Rosenthal and the BDI. Wilhelm Wagenfeld gave the opening speech. In 1959, the show was renamed Die gute Industrieform (Good Industrial Form). After 1954, a competition was also held in which the iF International Design Forum GmbH awarded the iF Design Award to designers, which continues to be one of the three major design prizes in Germany.

From Theory to Practice

The evolution of a new product design at Braun coincided with an economic upturn, the promise of a 'social market economy' and an institutional debate on contemporary design. A lecture Wilhelm Wagenfeld gave to art teachers in Darmstadt on 18 September 1954, which the Braun brothers attended, can be viewed as a synthesis of these business and design developments. The lecture emphasized the importance of art instruction as taste education, called for meaningful

↖
General Manager Wilhelm Wiegand presents the first cheque from Ronson to the Braun brothers.

↗
The first shaver for Ronson from the Braun factory in Walldürn.

→|
23 November 1955: Braun audio devices stand packed and ready on the factory grounds destined for Chicago. General Manager Wiegand and plant manager Calzaferri are visible in the photo.

↓↓|
Braun factory II, 1960. On the triangular site there is a separate building for the firm's health service (seen in the upper right of the image), to which a gym and small park would later be added. The large building in the foreground houses the canteen serving healthy food from a menu devised by Erwin Braun.

TOP
DIC 83
N=3
MADE IN WESTERN GERMANY
CHICAGO
VIA NEW YORK
This Side Up
TOP
DIC 83
N=31
MADE IN WESTERN GERMANY
CHICAGO
VIA NEW YORK
This Side Up
BRAUN
Type MM 2
Fragile

product worlds and advocated for the abandonment of tawdry design fads. Wagenfeld also hinted at how companies needed to structure their design process: 'In any case, the better design certainly requires ... intelligent producers who carefully think through each product in term of its purpose, utility, and longevity.'[6] The Brauns had found a suitable partner and just two days later Erwin Braun asked Wagenfeld if he might 'name the conditions under which your institute will work for industry'.[7] Erwin Braun was no doubt also familiar with Wagenfeld's 1948 book Wesen und Gestalt der Dinge um uns (The Essence and Form of Things Around Us).[8] Much of what the author called for may well have guided Erwin's hand when reorganizing the company. However, Wagenfeld only provided consultation for Braun's first practical approach towards implementing a new design in 1955 for the combi, a small portable radio-phonograph. The device, which marketed at 223 Deutschmarks, was not a commercial success and was only produced in a small series. Therefore, it is now a highly sought after and expensive collector's item.

Experiments

Radios and phonographs were initially intended as a field of experimentation since sales of traditional devices were no longer performing well. Financial resources for the experiments were available because Wilhelm Wiegand had negotiated a licensing agreement with US corporation Ronson for Braun's patented shaving foils system in 1953. The terms were for a volume worth $10 million and the agreement was signed on 22 February 1954, covering supplies and royalties. Braun was thus more than solvent. The press celebrated the arrangement as the greatest German foreign transaction of the post-war era. Nevertheless, this should not be seen just as an exceptional entrepreneurial feat but also in the context of the European Recovery Program, or so-called Marshall Plan, which was implemented to 'promote the export capability of Western European countries to the Dollar area'.[9] The programme intended to facilitate access to the American market for German companies interested in exporting. As of 1951, West Germany continuously posted export surpluses[10] and an above-average growth in gross domestic product. This grew by around 10.5 per cent annually between 1951 and 1963.[11]

Erwin Braun later described the licensing deal with Ronson as his greatest youthful folly: when it was sold to Gillette, both the American and entire British Commonwealth markets remained closed to the Braun brand until 1975.

At the first German export fair held by West German industry in New York in 1949, which Max Braun also attended, German products were seen as hopelessly out of date. In the United States, modern consumer-focused styling was considered the benchmark, and the German products were comparatively almost 'ready for a museum', as one observer remarked.[12]

↖
Artur and Erwin Braun inspect the new canteen with the head of the health service, Werner Kuprian.

→
Diet meals tastefully presented. Even when photographing seemingly unimportant things like healthy dishes, everything at Braun was designed down to the last detail, demonstrating the new standard that Braun sought to achieve through its communications.

↑
Gym and exercise room – part of the health services offered to Braun employees.

Healthcare and More

From the start, Erwin Braun was not only interested in creating useful products but in running a strong company. He viewed his workers not just as productive inputs but as humans who spent a large part of their time at the company. Erwin pursued a shift away from Max Braun's paternalistic and thus hierarchical management style, which was nevertheless also orientated towards the well-being of his workforce, and in the direction of self-motivation and personal responsibility. He was convinced that this would be the only way Braun could evolve into a large corporation. The focus was on fair wages and salaries, a reduction in working hours and a company pension. Profit sharing was introduced in 1958 and timecard control, considered degrading, was abolished the year before. Staff turnover at Braun was low and, over the years, its workers came to have an extraordinary sense of company loyalty, identifying with 'their' company. Employees saw themselves as 'Braunians' and the company as an extended family.

Erwin Braun's greatest motivation lay in setting up an in-house healthcare service that was preventive above all else – after all, his great interest in questions of health had already shown itself in his desire to study medicine. In 1957, the company medical service was expanded to include a 'dental station' with preventive check-ups and even denture services.[13] As early as 1954, physiotherapies, saunas and gymnastic courses were on offer. Physiotherapist and sports teacher Werner Kuprian was appointed to head the newly established healthcare service. By 1959, he had developed a wide range of services in the plants, including therapeutic horseriding, hydrotherapy and physiotherapy in the company's own gymnasium and saunas.

In September 1960, a second company canteen was established in the Rüsselsheimer Strasse plant as a diet cafeteria that served lacto-vegetarian food and was used daily by 300 employees. As early as 1957, only organically produced ingredients were used. The company magazine Betriebsspiegel stated that same year: 'It has long since been one of our company's principles to promulgate a healthy and sensible lifestyle with regards to nutrition, too; we also advocate this to the outside world, for example with a view to our food processor.'[14] And Fritz Eichler later remarked: 'Design and health were not separate processes – they formed an organic part of a concept that spanned the entire company.'[15]

For Erwin Braun the healthcare service was far more than a run-of-the-mill social benefit provided by a company; it was the vision of a different life. While in Germany the so-called 'gluttony wave' was reaching its pinnacle, with people indulging in high-calorie, heavy food, Erwin promoted a vegetarian diet and an active lifestyle – an understanding of health that would only become widespread in Germany after the turn of the millennium.

↗
Employees exercise using hoops in front of the gym.

combi

Combined radio and record player, 1955, Wilhelm Wagenfeld

The combi was only produced in small numbers and is a highly valuable collector's item today.

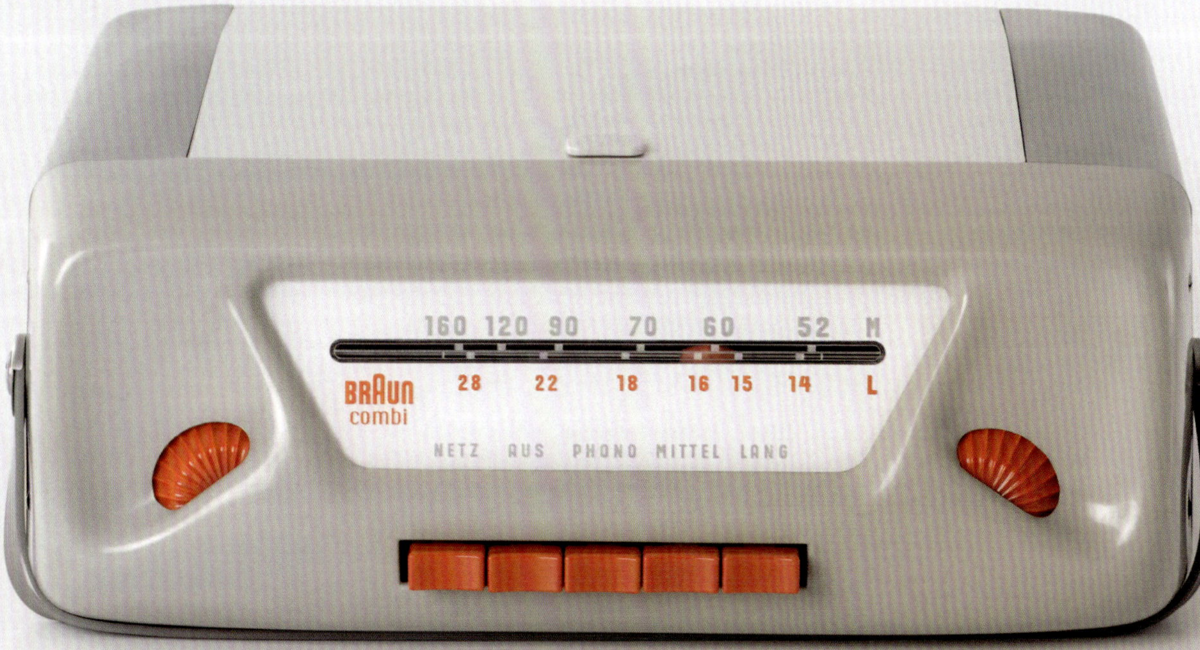

The Early Post-War Period: Middle-Class Mediocrity and New Conflicts

After World War II, the average German's initial focus was on sustaining basic needs, such as food security and ensuring a roof over one's head. Steel helmets were reworked into cooking pots. There was no demand for new forms – people were happy if they could salvage old things or were able to continue producing old models.

Nothing New in Sight (Yet)

In the first years following the war, the world of German products and everyday life was still a mixture of disillusionment and helpless continuation of what was assumed to be a culture with strong local roots. German girls still wore pigtails and German boys mainly wore lederhosen. Living rooms were dominated by a style as mundane as it was middle-class, with heavy cabinets, bulging and rounded upholstered furniture and abundant ornamentation.[1] Large radio sets, with their gold trim and fabric fronts, still looked exactly as they had in 1935. Bookshelves, if there were any, featured Goethe and Schiller in the better cases, or more commonly novels set in some nostalgic rural Germany.

At the 1949 German Industry Show held at New York's Rockefeller Center, German consumer products received a crushing verdict: they were condemned as 'the epitome of Germany lagging behind American product design developments'[2] – this created a sense of déjà vu, mirroring the situation before World War I, when German products had taken a similar battering at the 1904 St Louis World's Fair.

War After the War

The period after World War II was anything but peaceful. The Korean War, which raged from 1950 to 1953, saw once again the brutal use of multiple nations' full range of military hardware, resulting in the near total destruction of the Korean peninsula. The Korean War was one of the largest satellite conflicts during the new Cold War, with the North supported by China and the South supported by the United States, reflecting the larger global split between communism and capitalism. On the one hand, this conflict led to larger fears of a World War III, especially in Germany, but, on the other, it also had considerable positive economic effects. The so-called Korean boom led to an increase in foreign demand for consumer and investment goods from Germany and became the main driver of the German 'Economic Miracle'. This also gave Braun, a traditionally export-orientated company, an economic boost and confirmed that an internationally compatible design was then a good choice.

There was also more positive news on the home front: Christmas 1952 saw Nordwestdeutscher Rundfunk launch Germany's first regular post-war television programme.

A New Marketing

In the years after the war there was no need for product marketing. After all, when there is a shortage of almost everything, people will buy almost anything. Nevertheless, companies only began to generate sales again after the currency reform of June 1948. Braun still relied on its old radios at first, but it soon became clear that a new product was needed – and this was to be the first dry shaver, which Max Braun had been experimenting with since the war. However, the American shavers that the GIs brought to Germany and that Braun became aware of were also key stimulants. The company's own system with an oscillating cutterhead was perfected to create a marketable device. In a moulded box with a transparent plastic lid (similar to the later acrylic cover of the SK 4 radio-phono-device), the shaver came complete with an instruction manual and a packaging design created by Will Münch; it certainly was an attractive product. Soon the company was rolling out 200 shavers a day, with 25,000 devices completed in time for the Frankfurt Spring Fair of 1950. 'It sold splendidly,' wrote Artur Braun. The shaver was also available in black and it was not until the launch of the sixtant in the early 1960s that another Braun product would be offered in that colour.[16]

What the Consumer Accepts

The first new product created with the Braun brothers at the helm, the Hobby flash unit, showed that the company was adopting a new strategic outlook. Along with working for his father's company, Erwin Braun had studied business administration at Goethe University and possessed completely different business management skills compared to his father, who was after all a self-made man in commercial terms. When Erwin took over, the company began to rely on market analysis and systematic marketing strategies. Braun's success during the 1950s was thus no coincidence, but rather based on up-to-date market insights and creative new ideas.

'Strategies for breakthrough are not armchair decisions – they're much more likely to crop up on a walk in the forest or on dusty streets,' declared Albrecht Schultz in the opening sentences of his memoirs. Schultz, who rose from market researcher to member of the executive board, was referring to Erwin Braun's tendency to discuss his ideas with employees on long walks through the woods in the Taunus hills, and to Schultz's own visits to specialist retailers throughout West Germany.

The opinion pollsters at the Institut für Demoskopie in Allensbach published a study under the title Der Wohnstil (The Living Style) in 1954 that showed the benefits of modern forms of home living with light and bright furniture and a corresponding ambience. The EMNID institute for consumer research in Bielefeld came to similar conclusions the following year, especially with a view to radio sets.[17]

Nevertheless, the company did not rely solely on the opinion surveys of the corresponding institutions. Instead, it developed its own method of covert inquiry:

→|
The Multimix in a Berlin TV studio around 1953. The new medium of television was used at an early stage to market Braun products.

> The initial idea: a seeming end user, an interested potential customer like many others, appears and asks his questions. In the ensuing sales talk, he makes all his observations – on the layout of the store, the decoration, the placement of the products and so on. He asks all the pertinent questions in order to find out, what we are interested in ... He must remember everything down to the last detail and write it down after leaving the shop ... In point of fact, Erwin Braun and Schorsch Hohm had little scruples when it came to sounding out the retail trade like this ... The questioner was put into action in spring 1953, equipped with a VW and a typewriter, to scout out the situation from Flensburg to Garmisch and Trier to Hanover and to provide account of events on a daily basis ... The probably most important result was the insight gained that one could not expect salesmen to have any more knowledge of the product than one had given them. Consequently, Schorsch Hohm hired two product consultants ... who offered both presentations for end consumers in retail stores, as well as training the salesmen and most notably their wives, who tended to be the ones working in the stores.[18]

Erwin Braun was thus an early advocate for comprehensive market research and marketing strategy, which were at the time still new concepts in Germany. All findings were saved in a punch-card file system and condensed in a 'master retailer system'. The company knew where it stood and, in this respect, success was not coincidental.

In 1952, Erwin Braun wrote in a sales portfolio for the traders of the new flash units: 'On the following page we present our plan.' The portfolio goes on to outline a situation analysis, address the question of price in detail, advertising and customer service and provide instructions for actions on marketing.[19]

Direct Sales: The Hobby Flash Unit

In 1952, engineer Gerhard Lander developed Braun's flash unit for amateur photographers that had only a quarter of the volume and a fifth of the weight of competitors' devices and cost only a fifth of the usual sales price, too: the Hobby. Specialist photography equipment retailers sold the established brands from Agfa to Zeiss – nobody knew Braun here. The company thus developed a rather idiosyncratic, but ultimately quite successful, strategy: they wrote 'flash messages' on a postcard to eighty of the most important German photo retailers. Werner Apel-Dube, responsible for the new photo trade, remembers:

> There was no written order, no order confirmation, no delivery bill, no invoice, no reminder. The client could order from us three, six or twelve devices, and they could do so only through an advance check with a predetermined amount, where a hefty discount had already been deducted. It was a simple cash transaction over what today

↖
A 1953 promotional photo demonstrates that the Hobby flash unit BH 55 is truly shatterproof.

↗
In 1962, the Braun advertising department used this press photo from Toronto showing Winston Churchill after his recovery from an illness to demonstrate the wide use of the Hobby flash units with a round reflector. However, other manufacturers are also featured in the image and [illegible] likely not all by Braun.

→
After 1952, the Hobby flash unit, designed by Gerhard Lander / Braun Technology, became one of Braun's most important products. It was the first device of its kind for photography amateurs and unbeatable in price, performance and weight. In 1957, It was already for sale in 112 countries.

would be called a virtual counter. There was only one type and one colour – and that was black. There was no paper. We explained that we had no invoice or reminder department. The money thus saved benefited our prices, and by extension the client's business. The checks were delivered to Peter the accountant in the morning in a shoe box. They were always covered. The devices then went out the same day, always per express ... The photo retailers were electrified. This was unprecedented.[20]

At the end of the first business year, the photo retailers were told in a letter that Braun had managed to turn a better-than-expected profit, which the company now wanted to split half-half with the clients. A corresponding cheque was enclosed in the letter. It was hardly surprising that enthusiastic partnerships were forged in this way, which would later also pay off when Nizo film cameras came to market. The flash units were not provided to wholesalers of photographic equipment or mail order companies.

Fixed Prices Everywhere

For the first time Erwin Braun no longer went through wholesale traders but directly approached retailers, who he sought to establish as 'partners'. This 'philosophy of partnership with the specialized trade'[21] was still relatively unique at the time and formed one of the cornerstones for the company's rapid commercial success. This was also applied to shavers, household appliances and radio sets. Braun's strategy also included Erwin's championing of fixed prices, which ultimately made the existence of small, specialized retail outlets possible, since a product cost no more or less there than it did in large shops. Even Braun employees did not get a staff discount and had to buy products from specialized retailers like everyone else – these prices were even harder to stomach for female employees, or the 'women working the assembly line', since their hourly wages were only between 1 and 1.14 Deutschmarks an hour, as set by collective bargaining in Hesse.[22]

The later removal of set prices led to more retail competition and often lower consumer prices, and spelled the end for traditional retailers with their repair services and customer advice. Only a few such retailers still exist in Germany, with the market largely dominated by major chains partaking in price wars.

Erwin Braun set up an 'information service for prime specialized retailers', so-called postal retail dispatches. In March 1955, this took the form of a seventeen-page letter entitled 5 Minuten vor 12 (It's High Time), that warned retailers not to undermine fixed pricing. It cited the United States, with its variable prices, as a cautionary tale. At the same time the recipients were asked to provide their 'opinion on the problems' on an attached postcard.[23]

Retailers and wholesalers were obliged by means of a written declaration (on the reverse side of the postcard) to adhere to the fixed price. This 'second-hand price maintenance', pertaining to retailers, was intended to successfully prevent a 'grey market' for Braun products. To achieve this, it was necessary to ensure the fixed prices were followed by all, which could only be achieved by meticulous insistence (if necessary, by legal means) that retailers acted accordingly. A 'Braun fiduciary agency' that was separate from the company itself was established to this end. However, the internationalization of sales and the increasing glut in the markets made this system more and more complicated. In the early 1970s, sales of price-fixed goods amounted to only five per cent of retail turnover. Finally, in 1973, the possibility of price fixing in West Germany was abolished by the legislature.

The more than two decades of fixed prices, including at 'prime specialist retailers' who themselves then offered repair services and advice, contributed significantly to building the Braun brand as a special promise of quality. There was also the option of paying in installments: 'On July 11, 1950, the Teilzahlungs-Organisations-Gesellschaft mbH (Organization for Installment Payments), which is affiliated with the Braun company, was entered in the Commercial Register at the Frankfurt District Court. This company has the task of collecting hire-purchase installments on Braun equipment sold by the radio dealers under installment plans.'[24]

Multimix and combi DL 5

Alongside the electric shaver, the company's strategy was driven first and foremost by the Braun Multimix blender which was launched in 1950 and came with a price tag only slightly more than half that of the then sole competitor's product on the German market, the Elektrostar Starmix. While it formally still stood entirely within the tradition of machine aesthetics, the blender brought American flair to German kitchens, since juices and shakes were both healthy and chic. The only actions the machine could not perform yet were stirring and kneading.

→|
The 1950 recipe book for the first Multimix features a raucously colourful cover.

BRAUN
MULTIMIX
REZEPTBUCH

Raw vegetables and fruit juices were promoted in the United States by Gayelord Hauser (born Helmut Eugen Benjamin Gellert Hause in Tübingen in 1895), among others. The charismatic and shrewd entrepreneur for health food and raw vegetable diets had been a successful author in the United States since the 1930s and provided nutritional advice to Hollywood greats such as Marlene Dietrich, Gloria Swanson and Greta Garbo and, later, Grace Kelly and Ingrid Bergman. In 1950 he published his bestseller Look Younger, Live Longer. Medical professionals had a somewhat sceptical view of his activities: 'a mix of common sense, questionable dietary advice, name dropping, continental charm and bashful garrulousness'.[25] Others saw in him a far-sighted pioneer of natural nutrition. Braun, for one, engaged him for its advertising and enclosed corresponding recipes with its devices. Manuel Aicher rightly considers this evidence of a new marketing at Braun: 'For the first time in the history of the Braun company what we discern here – if only rudimentarily – is an attempt to achieve something that would later make Braun very successful: namely not just selling a product but embedding it in a certain attitude to life, a particular lifestyle.'[26]

Omnipresent on television today, cooking programmes were already around in the 1950s. Braun was also in touch with television chef Horst Scharfenberg from Frankfurt, who became a household name as a journalist for the channel Südwestfunk and as a producer of entertainment programmes, as well as through lectures held at universities. In 1962, he founded the Koch Club on TV, which he presented until 1970. Braun also published cookbooks as supplements to their products to corner this burgeoning market.

Starting in 1954, Braun set up a system of initially twelve female 'customer advisors', who held talks and presentations on modern kitchen management and healthy eating in the retailers' shops, at trade fairs and markets, in municipal authorities and schools – demonstrated using new Braun kitchen devices, naturally. The women drove around the entire country from one presentation to another in a Messerschmitt Kabinenroller, the legendary smallest car in post-war Germany. Additional sales staff were trained in a special teaching kitchen in Frankfurt. American role models also played a part here: since 1948, Earl Silas Tupper had presented his plastic household devices at 'Tupperware parties' directly in customers' homes.

The KM 3 food processor, launched in April 1957, introduced a completely unique appliance aesthetic to the market. Together with the SK 4 radio-phonograph released six months earlier, it paved the way for future Braun design. That same year, the combi DL 5 shaver hit the market, already bearing the signature style of Dieter Rams and Gerd A. Müller. It also included a long hair trimmer and, at 58 Deutschmarks, cost no more than the previous model. Braun was even prepared to take back the stocks held by retailers of the old model and swap these for the new one. Given that there were 20,000 wholesalers and retailers, this was no small feat in terms of logistics – and one that needed to be completed in a single week. Nevertheless, the campaign no doubt truly boosted dealers' confidence.[27]

Advertising through the Ages

In the early post-war period, promotion in ads and on posters still took its cue from the 1930s. Graphic designer Will Münch, who had created the logo with the larger letter A, continued to work for Braun well into the 1950s. Advertisements for traditional radio sets showed the product with the model's name written in a flourish of script and often included extensive explanations on three-dimensionally displayed sheets that looked more like title deeds. The brand slogan 'Der edle Ton – Braun Tradition' (The Braun Tradition – A Classy Sound) had been used for a long time. The advertising for the new portable appliances already had a distinct cosmopolitan ring to it and again featured the Braun globe. The ads for the Multimix and Multipress were primarily based on the notion of preparing healthy food, while the early shaver ads for the S 50 used storytelling with a dash of humour. Here, the initial aim was to make the public familiar with electric shaving in the first place, which was why the fundamental advantages were communicated through stories that were often witty in tone. 'Of the shaving devil and its guiles' was one such series, in which the connecting cable of the S 50 was depicted as a lasso, wrapping around the text and image of the potential adversities of a wet shave and the dry shaver is presented in the proper light as the source of salvation.

Braun advertisements of the late 1940s and early 1950s were as conventional as the products of that time. An early ad by Wolfgang Schmittel for the Hobby flash unit was, however, strikingly different in a positive way: it consisted of a collage in blue and black reminiscent of pre-war modernism, with which the student of Frankfurt's Städelshule would certainly have been familiar. By comparison, other works by Schmittel created before the middle of the century were somewhat more conservative in terms of typography. What was convincing, however, was his 1953 revision of the Braun logo, which was subsequently used for many decades. A 1957 ad for the launch of the

↗
Like a pop star, Gayelord Hauser, the German American author of the 1950 bestseller Look Younger, Live Longer, celebrates fruit juice made by Mrs [illegible] in front of a large audience in Frankfurt's exhibition centre. The self-appointed nutritionist was not without his detractors but was popular particularly among female Hollywood stars at the time.

↓
[illegible] visiting Frankfurt's Lord Mayor Dr Walter Kolb (left), early 1950s.

combi DL 5 shaver is also worthy of note. Here, too, a story is told. A rather sober typography and illustration poses the question: 'What does your subconscious know about shaving?' The answer: 'Practically nothing.' In a humorous-cum-ironic vein, the advert alluded to the intellect of the men to whom it was pitching the Braun shaver: 'You'll only know whether it shaves and how well when you try it and other devices out. Maybe you'll find another one that does a better job – if so, feel free to buy that one.'[28] This was now a far cry from the stories of the 'shaving devil'.

In 1979, Erwin Braun reminisced about other advertising strategies: 'Remington hit the streets with a campaign where they shaved a peach, which we countered with one that showed a gooseberry being shaved, and Philips were not particularly happy when our answer to their "Why not take Philips" was "You're better off with Braun".'[29]

Cooperative Thought and Action

Erwin Braun nevertheless also sought to cooperate with comparable corporations from other sectors also looking to create new designs. He initiated the Verbundkreis für Industrieform (Network for Industrial Form), a loose association of nine design-orientated companies geared at joint public relations and advertising activities. The Verbundkreis was founded in 1955 by Dieter D. Mundas, Erwin Braun's HR consultant. Alongside Braun, it initially included the companies Bremer Tauwerk-Fabrik (carpets), Knoll International (furniture), G. M. Pfaff (sewing machines), Gebr. Rasch & Co. (wallpapers), H. Römmler (Resopal), Rosenthal (porcelain), WMF (metalware) and Gral (glass). From autumn 1956 onwards the travelling exhibition form–farbe–fertigung (form-colour-production) toured numerous cities.[30] Later, the circle expanded to form Gruppe 21. This allowed for important contacts to renowned companies in the interiors sector to be made.[31]

Marketing at Braun defined its fundamentals in the 1950s and maintained them for several decades. Alongside factual advertising, PR and exhibition activities this included, above all, fostering intense dealer loyalty, which was highly prioritized. The nationwide points of sale were the small- and medium-sized radio and photo retailers, who were protected against price wars with mail order and department stores. This signalled to end consumers that their Braun appliances were in good hands and that they could expect to receive any service from their local dealership. The fact that there was an on-site repair service was an important selling point. Today, consumer behaviour is again reorientated towards ecological responsibility. We are now seeing that the ability to repair appliances has again become an important reason to buy.

↖
Oversized models of Braun's most important products are transported on the roofs of VW Beetles during Braun's famous 1954 'Ten-City Tour' of West Germany, here through Stuttgart.

→|
In 1955, the Frankfurt Aviation Association bought a Focke-Wulff-44-Stieglitz double decker plane and in 1956 participated in the first competition for German pilots (Deutschlandflug) after the war. Braun sponsored the event which accounts for the logo on the old timer.

E2
BRAUN

combi DL 5

Electric shaver, 1957,
Dieter Rams / Gerd A. Müller

04
Braun Design

Artur and Erwin Braun, the second generation of Braun entrepreneurs, had a vision to establish a company focused on the creation of improved products that catered to everyone. They aimed to produce 'appliances for today'. To achieve this, the company collaborated with Bauhaus students and the HfG Ulm, a design college founded in 1953.

By the late 1950s, Braun had formed its own design team, making a significant impact on the landscape of modern design as we recognize it today.

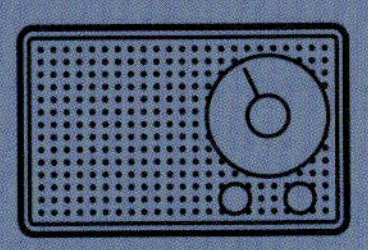

SK 1 1955

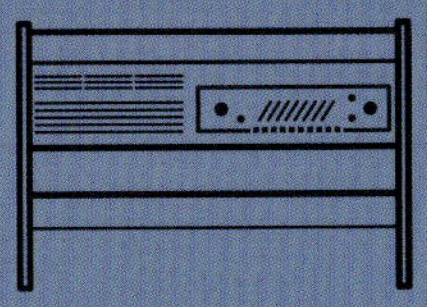

PK-G 1955

G 11 1955

exporter 2 1956

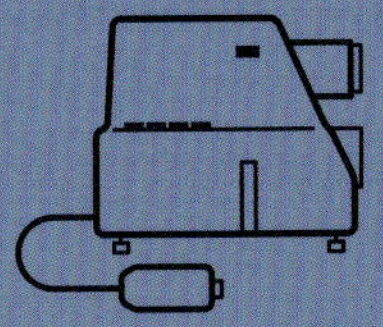

PA 1 1956

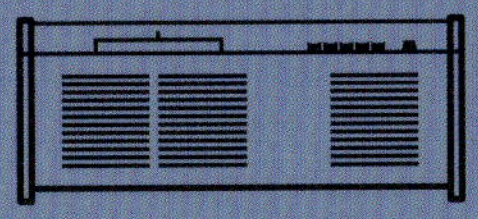

SK 4 1956

KM 3 1957

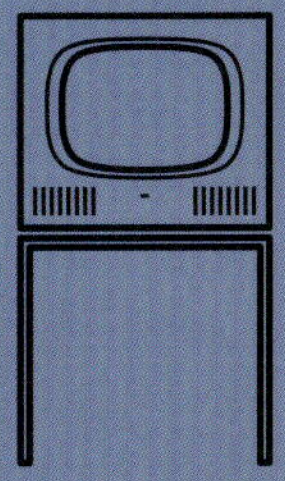

HF 1 1958

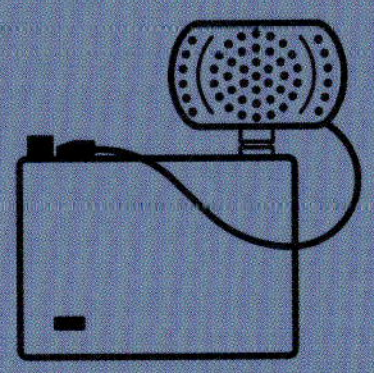

EF 1 1958

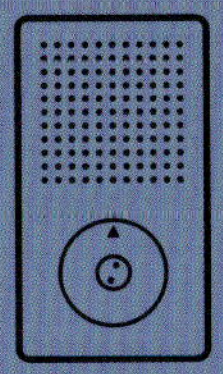

T 3 1958

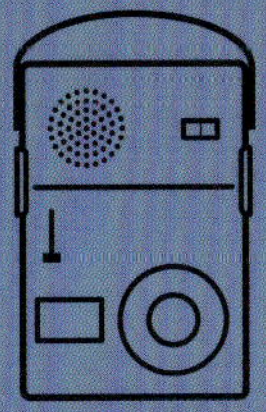

TP 1 1959

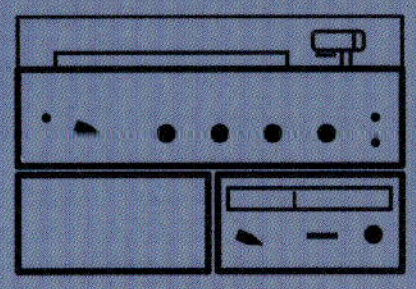

studio 2 1959

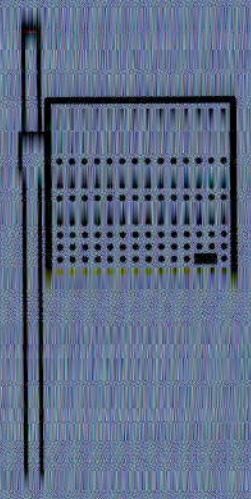

L 01 1959

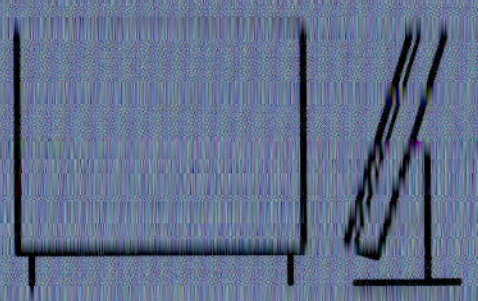

LE 1 1959

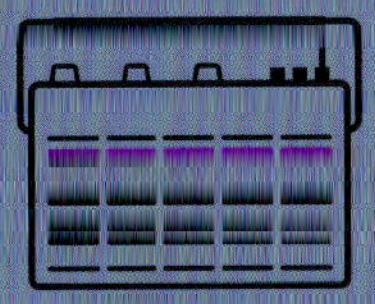

T 52 1961

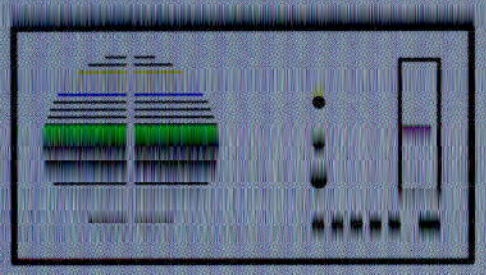

RT 20 1961

Pursuing the New

Erwin Braun pursued the new. As Artur Braun remembered in 2010: 'My brother was undoubtedly the true father of Braun Design. However, mostly in conceptual terms. My brother was a brilliant thinker, but no practitioner when it came to applied crafts.' He went on:

> Because of this, he was always looking for people who could work with us. And so Fritz Eichler, among others, fell into his clutches. An art historian in a medium-sized family business! ... I once wrote 'it's where culture meets commerce'. But this really was a stroke of luck, as in the end it was Fritz Eichler who joined up all the dots. He had the thoughts my brother couldn't convey yet – at least certainly not to me at the time – and he lent them shape. He formulated these thoughts and got them underway ... After things had been tricky with Wilhelm Wagenfeld, I had started to have my doubts when it came to modern design. As an engineer I had hoped that such a collaboration would make my job easier ... But it didn't get easier. On the contrary, things actually got more difficult ... Strictly speaking, it was clear to me afterwards that Wagenfeld's approach was not practicable ... At the time, design and technology were two topics that somehow seemed incompatible ... The contact to Ulm was then established through my brother's constant search.[1]

Erich Untied, managing director of the Jettingen-based Thun company, which produced wooden casings for the Braun radio sets, hooked Braun up to the new Hochschule für Gestaltung (HfG) Ulm university. Roderich, Count of Thun, the company's owner, was also a member of the administrative board of the HfG Ulm.[2] In early December 1954, together with his friend, physicist Gerhard Lander, Erwin Braun travelled to the university, which had been founded just a year before and was still operating in provisional premises while its new building was under construction.

Collaboration with the Hochschule für Gestaltung (HfG) Ulm

During his first visit to Ulm in 1954, Erwin Braun, then just thirty-two years old, must have had considerable sympathies for the people he met, all of whom sought to keep alive the democratic spirit of Hans and Sophie Scholl, resistance fighters executed by the Nazis. Inge Aicher-Scholl certainly saw a possible strategic partner in the industrialist from Frankfurt – after all, the private university was constantly under pressure to generate revenue in the private sector. And their outlook concerning the past, present and future was the same. On 10 December 1954, Aicher-Scholl thanked Erwin Braun in a letter: 'Once again, I approvingly noted just how easy it is to communicate when both sides have the same mindset and certainly not solely because we ultimately share a joint task.' As was also Erwin Braun's wish, she proposed a radio model:

> to which the university can be proud to lend its name. Naturally, such a development project requires a far greater effort than simply aligning to previous standards

I

Hans Gugelot and Erwin Braun on the way to the lecturers' houses at the HfG Ulm.

III

Part of the main building of the HfG Ulm on the Oberer Kuhberg with the large terrace in the foreground and the lecturers' houses in the background. The VW Bus in the parking lot was part of the 'Braun Tent City' promotional tour.

↑
Braun product range, 1956, presented largely with furniture by Knoll International in a furniture store in Wiesbaden. The Bar Trolley by Herbert Hirche from that same year is also on display. This is one of Marlene Schneyder's photographs for the company.

and making minor changes here and there. Nevertheless, from a financial point of view, we would like to make you an offer that is below our usual terms and conditions, as we are interested in continuing our cooperation with you in the future.[3]

A moderate fee of 3,500 Deutschmarks (a figure that also covered all the rights) was offered in the event of production, and only a 'flat rate' of 500 Deutschmarks in the event of rejection. The letter demonstrates the school's self-confidence in the early days, or at the very least the purported self-confidence, as well as the interest in a long-term collaboration. In subsequent years, Braun would become the HfG's single most important client.

The New Radios

In Otl Aicher, Inge Aicher-Scholl and Hans Gugelot, Erwin Braun had thus found partners at HfG who were not only of the same age, but like-minded. He commissioned them to design a new radio set in early 1955. The result was an unostentatious clear, almost white, acorn-wood case: the super G 11 radio by Hans Gugelot. Subsequently, further designs were commissioned and submitted. The first G-series (with the 'G' standing for Gugelot) of 1955 was supplemented by a record player (G 12) and a television cabinet (FS-G). All the devices had the same basic dimensions, which meant they could be stacked. They were joined by a large radio-phonograph with different base units (PK-G). The idea to exclusively present the new designs at the upcoming radio exhibition in Düsseldorf in August 1955 soon took shape. However, it was impossible to create entirely new devices in seven months, which meant that the company inevitably ended up showing the designs for the cases – containing older, pre-existing technology. Nevertheless, the cases were radically different to the old, glossy-brown versions that had come before.

Additionally, Fritz Eichler and Artur Braun designed the small but technically high-quality SK 1 radio set with a varnished thermoset case, a white front made of perforated sheet plastic and a striking, asymmetrically placed rotating dial. In 1998, Hans Wichmann suggested that Artur Braun and Eichler had been inspired by American Fada radio sets.

Collaboration with Hans Gugelot

HfG's early products for Braun were created primarily by Hans Gugelot, sometimes in collaboration with his students. He was not an applied philosopher like Aicher-Scholl but a seasoned architect and a pragmatist with an interest in technology. The development work for the first Braun devices did not take very long. Gugelot's assistant, Helmut Müller-Kühn, later remembered the first collaborations with Braun: 'Gugelot made a few sketches and then said to Paul Hildinger, the HfG's master carpenter ... "we have to knock off a few radio cases." Everything happened so quickly that we students only got wind of it when Gugelot came back with these models from Frankfurt.'[4]

↗
Maike and Hans Gugelot with Fritz Eichler in Ulm.

G 11

Table radio, 1955,
Hans Gugelot

19 m Band
Luxembourg
Italia
Intern. Gleichw
Mte Carlo
Bayer. Rundfunk
Berlin-W
Bruxelles
Praha
SW-Funk
Vaticano
Paris Intern
BBC
Radio-Sweden
NWDR
Lille
Hannover
Italia
Graz
Lisboa
Bayr. Rundf
Sudfunk
Saarbr
Lautstärke
Höhen
Aus

PK-G

Audio cabinet, 1955, Hans Gugelot

Also known as 'Langer Heinrich' (Long Henry) on account of its breadth.

In his obituary for Gugelot, Eichler wrote in 1965:

> I remember our first encounter in Frankfurt. You brought a mysterious package with you that contained the first model for a future radio set. It was an exciting and festive moment – an unveiling of a monument of sorts: there it stood, the thing – a rectangular wooden box – on the front it had a black circular area and a rectangle – that was it.

Eichler also pointed to the difficulties that a new concept entailed:

> You weren't able to start creating something entirely new from scratch but were instead forced to do something that you essentially rejected: to put a new shape around existing technology ... Then one day the radio-phonograph 'PK-G' – which we nicknamed 'Langer Heinrich' – was finished. A short while later there was a new Tischsuper in 2 versions – the one a record player, the other a television set. Proceeding from the wrong starting point, as it were, you had managed to develop devices that were coordinated in terms of dimensions and shape and that could be arranged on top of or next to each other.[5]

Braun on Tour

In 1956, Erwin Braun, Fritz Eichler, Hans Gugelot, Artur Braun and his wife and Richard Rohlf, the architect and engineer who had since 1954 overseen the construction of the Walldürn plant, went on an informational trip to the United States. In New York, the group split into two: Artur, his wife and Rohlf visited factories for shaver production and suppliers, where they were particularly impressed by the American production methods and long assembly lines. This without question provided inspiration for the company at home.

The 'design group' of Erwin, Eichler and Gugelot was less successful. They had wanted to explore the country that had given rise to Knoll International, Ray and Charles Eames and the heroes of 'white modernism', Ludwig Mies van der Rohe and Walter Gropius, to hunt for possible new ideas.[6] The group travelled around California, visited the Eameses, then moved on to New York and Chicago. However, the American design that they had much admired, and which in Europe was represented by a few outstanding companies, proved to be a lot more banal in America. On the return voyage by ship via Naples, the Braun group visited the companies Necchi in Pavia and Olivetti in Ivrea at the behest of Inge Aicher-Scholl. She and Malke Gugelot met up with the group in Genoa.[7] This is once again evidence of Aicher-Scholl's efforts to forge a permanent link between Braun and the private HfG as a business partner.

Almost as a matter of course, the company had also joined the Society of the Friends of the Geschwister-Scholl Foundation in 1955.

↖
Erwin Braun and Fritz Eichler touring the US in spring 1956, accompanied by a Hobby flash unit.

→|
Braun and Eichler's visit to Charles and Ray Eames in Pacific Palisades, Los Angeles, 1956. The present from Germany was a loaf of wholemeal bread.

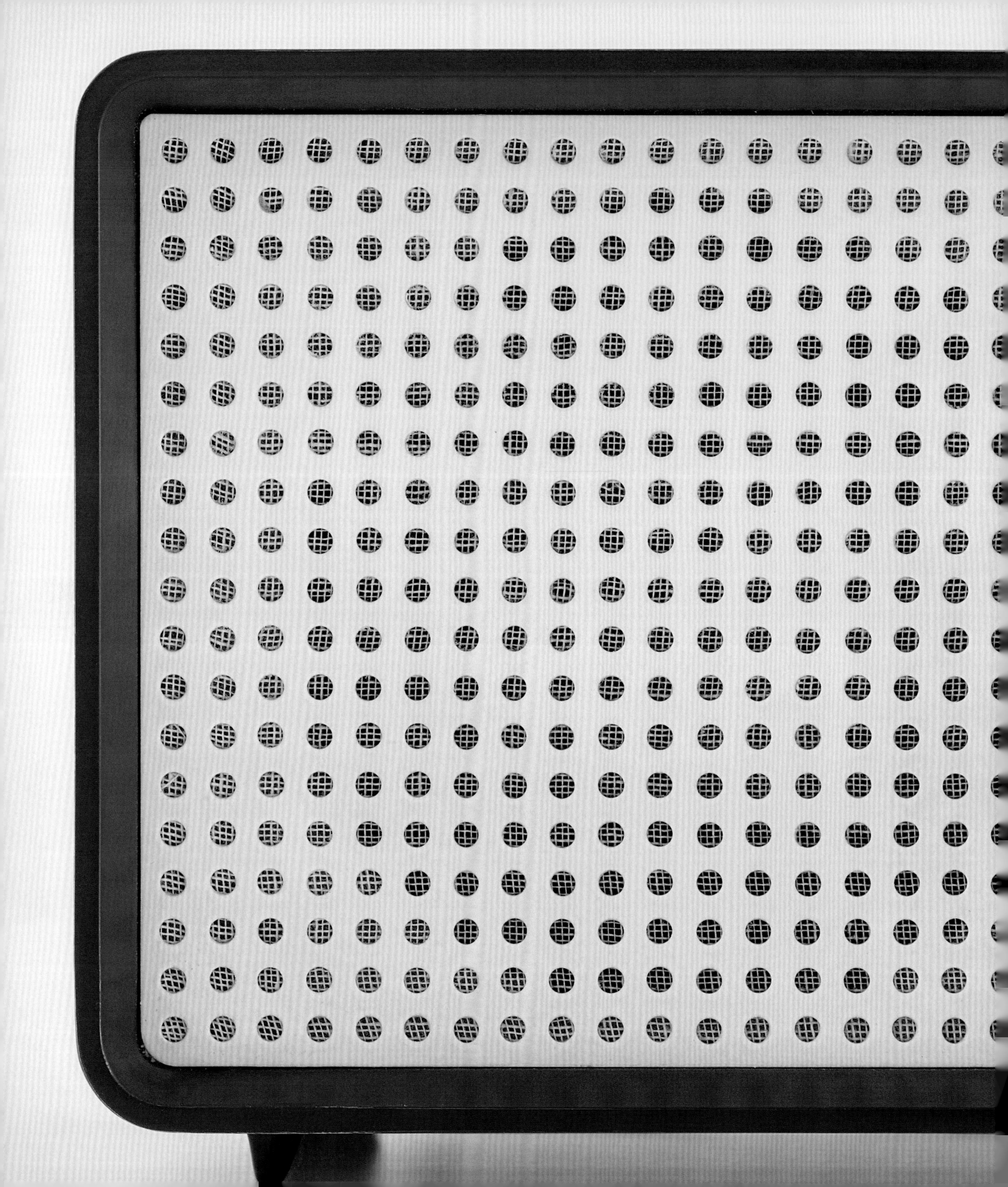

SK 1

Table radio, 1955,
Artur Braun / Fritz Eichler

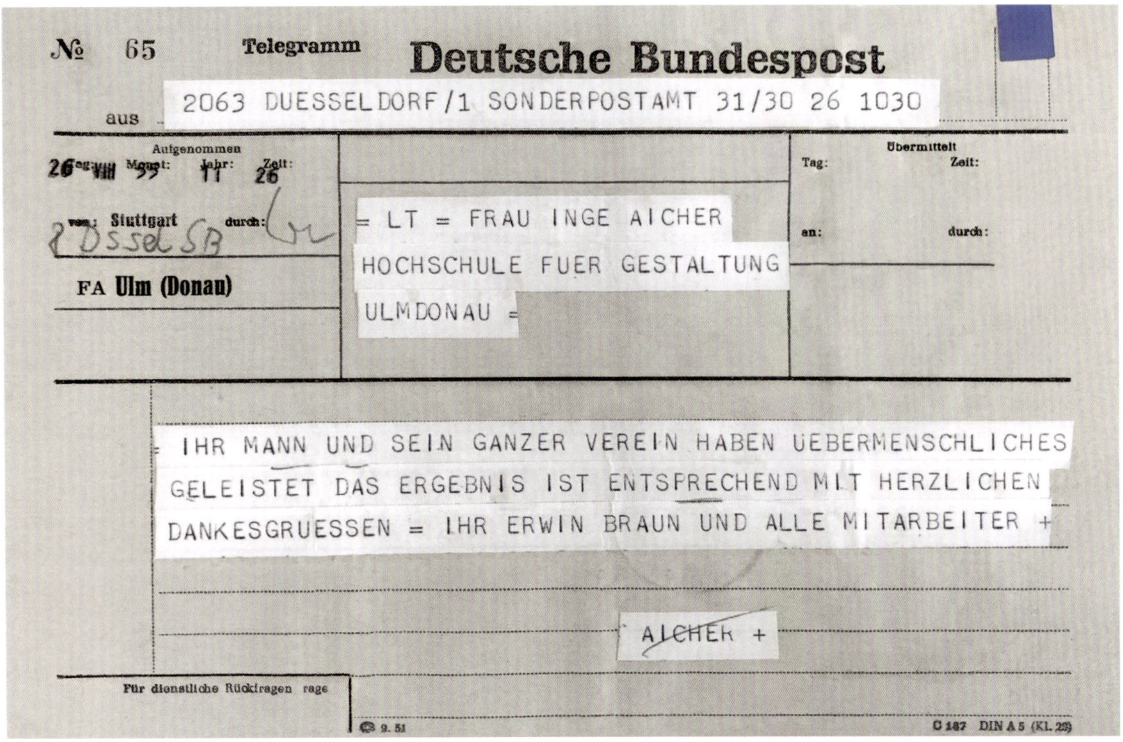

№ 65 Telegramm Deutsche Bundespost

aus 2063 DUESSELDORF/1 SONDERPOSTAMT 31/30 26 1030

Aufgenommen Tag: Monat: Jahr: Zeit: 26 VIII 55 11 26

von: Stuttgart durch:

FA Ulm (Donau)

Übermittelt Tag: Zeit: an: durch:

= LT = FRAU INGE AICHER
HOCHSCHULE FUER GESTALTUNG
ULMDONAU =

= IHR MANN UND SEIN GANZER VEREIN HABEN UEBERMENSCHLICHES GELEISTET DAS ERGEBNIS IST ENTSPRECHEND MIT HERZLICHEN DANKESGRUESSEN = IHR ERWIN BRAUN UND ALLE MITARBEITER +

AICHER +

Für dienstliche Rückfragen

C 187 DIN A 5 (KL. 25)

All New: Graphic Design in the Modernist Tradition

Graphic design at Braun also radically changed in the year 1955, when the Ulm-based designers transformed not only the shapes of the devices but also the company's visual communications. Otl Aicher had introduced his system with its strict grid orientation, which was now also adopted by Braun's advertising department. Aicher designed a four-column grid that was to be used for the brochure layouts and the proportions of photographs. Alongside graphic design, from 1955 onwards there was also a specialized photography department that included Marlene Schneyder from June 1955 to December 1958, as well as Ursula Sehring and Ingeborg Kracht, who would later marry Dieter Rams. They developed an approach to photographing objects so that individual items did not overlap and, through large apertures, devised creative compositions that focused on the products and featured a blur that obscured the backgrounds.

Together with young interior designer Dieter Rams, the department also took pictures in a design studio in Wiesbaden that sold furniture by Knoll. The photographs did away with dramatic shadows, showing off the pure interior design.[8]

New Corporate Communications: The 1955 Radio Exhibition in Düsseldorf

It was not just Braun's photography and advertising that were changing but corporate communications altogether. Primarily, the most radical step taken was Braun's trade-fair booth at the Düsseldorf Radio Show in 1955. To this end, Otl Aicher designed an innovative modular fair stand that stood out against the competitors' bombastic booths, resembling a transparent modernist structure more reminiscent of Mies van der Rohe's Farnsworth House than anything else. The detailed planning was done by Hans G. Conrad, an associate of Max Bill's in Zürich and one of the first students to attend the HfG. His graduation piece would later be an exhibition bus for Braun that was never actually realized. From 1958 through to 1962, he headed the exhibition design section at Braun.

Braun deployed the new trade-fair booth for many years. It was a variable system of steel profiles made up of columns and beams as well as plywood panels with snap-lock joints that could be assembled in a wide range of formats. It likewise heralded the creation of other modular exhibition systems such as the outdoor Braun pavilion for the 1959 Frankfurt Fair or a never-produced, more delicate system of thin supports and beams.

Only the latest radio models by Fritz Eichler, Artur Braun, Wilhelm Wagenfeld and Hans Gugelot went on display in Düsseldorf in-between house plants and furnishings by Knoll International and Wohnbedarf Zürich. Although the Braun sales team seemed to not quite trust the company's design revolution and brought older radio models for display on the stands, the new product and communications design was here to stay.

↖
Telegram sent by Erwin Braun on 26 August 1955 to Inge Aicher from the 19th Great German Radio, Television and Phone Exhibition in Düsseldorf thanking the HfG Ulm for the exhibition stand's success.

→|
Exhibition stand 'd 55', the modular Braun exhibition stand for the 1955 Radio Exhibition in Düsseldorf, was designed by Otl Aicher and Hans G. Conrad. The stand featured furniture by Knoll and was used with only slight modifications for a further twenty years.

BRAUN

↖
A design concept from December 1956 by an unknown designer for an unrealized Braun audio cabinet, labelled 'Sketch for design No. 1 coloured matt varnish'. Whether Braun had commissioned additional designers around 1955 or 1956 or the designs were unsolicited cannot always be determined from the drawings in the Braun Archive.

↑
A design for a Braun portable radio around 1956 by Count Abrecht von Goertz. It is one of several drawings of portable radios. It is highly unlikely that the successful German designer, who had his own studio in New York and whose sensational design for the BMW 507 sportscar was presented one year earlier, sent an unsolicited design. The initiative would have come from Braun.

One, Two, Three Designs

The HfG models for the Düsseldorf Radio Show were a runaway success in the media, but this did not translate into sales figures. Erwin Braun pursued a three-pronged product strategy. First there was the HfG, which stood for a new departure in design. Then there was erstwhile Bauhaus student and successful post-war designer Herbert Hirche, who was hired to create modern yet more or less conventional 'sound furniture' that dovetailed with items offered by furniture sellers WK-Verband.[9] Thirdly there was the small in-house design group, which well into the year 1956 was also in charge of furnishing the company's interiors, and was entrusted with new projects to be conceived in collaboration with the technology department. This was the case, for example, with the PA 1 and PA 2 slide projectors or the exporter 2 pocket radio receiver, both developed in 1956, and with the order for the SK 4 radio-phonograph. The Braun brothers and Fritz Eichler no doubt realized early on that a collaboration on entirely new products would not really work in a 'long-distance relationship' between technology and design. Marlene Schnelle-Schneyder remembers that the HfG designers were in Frankfurt every two months at best, which is also how Dieter Rams and Artur Braun describe the situation in retrospect.[10]

The cooperation with Wagenfeld (until 1956) and the Thun workshops (until 1961) gradually wound down. The relationship with the HfG was upheld through Hans Gugelot until his death in 1965, even if the focus shifted in late 1957 from design work to a consulting agreement. The company also stayed in contact with Otl Aicher, who, as late as 1980, designed the logo for a/d/s/, the distribution company for Braun's atelier sets. Most importantly, by the close of the 1950s, the 'long-distance relationship' had ended and Ulm alumni Reinhold Weiss and Richard Fischer were integrated into the in-house design department.

Alongside this, there were of course links to other external designers. In the Braun archive, there are plans and drawings of never-realized devices. The best-known Braun collaborator is probably French designer Louis Lucien Lepoix, who had come to Germany in 1945 and had set up a large design studio, FTI, in Baden-Baden. The studio worked primarily for the German car industry, in particular designing truck cabs. Lepoix also designed the 1972 Bic lighter J1, to date one of the company's best-known and most successful products. The Braun Archive contains several designs by Lepoix for a Braun music cabinet, dated November 1955, meaning the company drew on the skills of one of the most productive and well-known designers of the period. Numerous other music cabinet designs are signed 'Kurt Gross, Werkkunstschule Offenbach' and dated 17 January 1956. No data on Kurt Gross is available from the school's teachers or alumni of that period. Yet the designs are intricate, elaborated structures with detailed dimensional data, which does not suggest they were initial sketches. Finally, there are numerous sketches and construction drawings under the name M. Rosenhauer, dated 22 January 1955.

These designs, which have never been shown to the public before, were presented to the company

↑
The exporter 2, a portable radio designed by the HfG Ulm and Dieter Rams in 1956.

between January 1955 and January 1956 and must have taken some time to create. They were either commissioned by Thun as a supplier, or by Braun itself. We could interpret the fact that designs were brought in from Werkkunstschule Offenbach as an indicator that they were commissioned from nearby Frankfurt. They show that the orientation towards a future design approach was not in any way fixed at the time and that, after the initial contact with the HfG, several other designers were also consulted.

In 1961, Erwin Braun described his company after its longer phase of exploration as follows:

> The term 'design' is a broad one ... designing the devices ... designing the distribution ... The current Braun device program can be described as follows: state of the art, well-functioning, understated, electrically powered 'helpers' for home and hobby at reasonable prices ... In condensing the flow of products (from production via the retail trade to the client and consumer), all reasonable design options must be kept open. Any abuse of this freedom must be avoided.[11]

Dieter Rams and Gerd A. Müller

On 15 July 1955, Dieter Rams joined the Braun company as an interior designer and, until 1956, primarily focused on office design. At the same time, he was also consulted about the product designs for the upcoming Düsseldorf Radio Show. Alongside producing colour designs for the SK 1/2, his first test was to redesign the exporter 2. The surprising new design, which had been transformed by Rams into a successful model merely by changing the colour and the use of the dial, may have prompted his being employed in a capacity not originally intended. The PA 1 slide projector constructed by Gerhard Lander was then Rams's first real design job in 1956. A small, dedicated product design department was established in early 1956. Since December 1955, Gerd A. Müller had been working as an engineer under Artur Braun in the Idsteiner Strasse factory on the plaster model of the KM 3 food processor. He joined Rams at Rüsselsheimer Strasse, where Rams had been working since July 1955 in the graphic design department, though Rams was soon given his own room one floor up. In March 1956, Roland Weigend, another Wiesbaden Werkkunstschule alumnus, also joined. The architects-cum-designers, all of whom had attended Dr Hans Soeder's Wiesbadener Werkkunstschule with its new orientation, formed the product design sub-section, with Rams as its spokesperson from the very start. Müller was a designer on equal terms with Rams, while Weigend oversaw model-making.

Dieter Rams was tasked with creating a new device in early 1956 about which Artur Braun wrote: 'However, especially in the lower price range, a small receiver with a built-in record player still did not exist in 1956 .... Dieter Rams, hired as an interior architect in 1955, was commissioned to design the "SK 4" radio-phonograph.'[12]

→|
Advertising photograph demonstrating how to make baby food with the KM 3, 1957 or 1958. Photograph by Braun photographers Marlene Schneyder and Ingeborg Kracht-Rams.

BRAUN

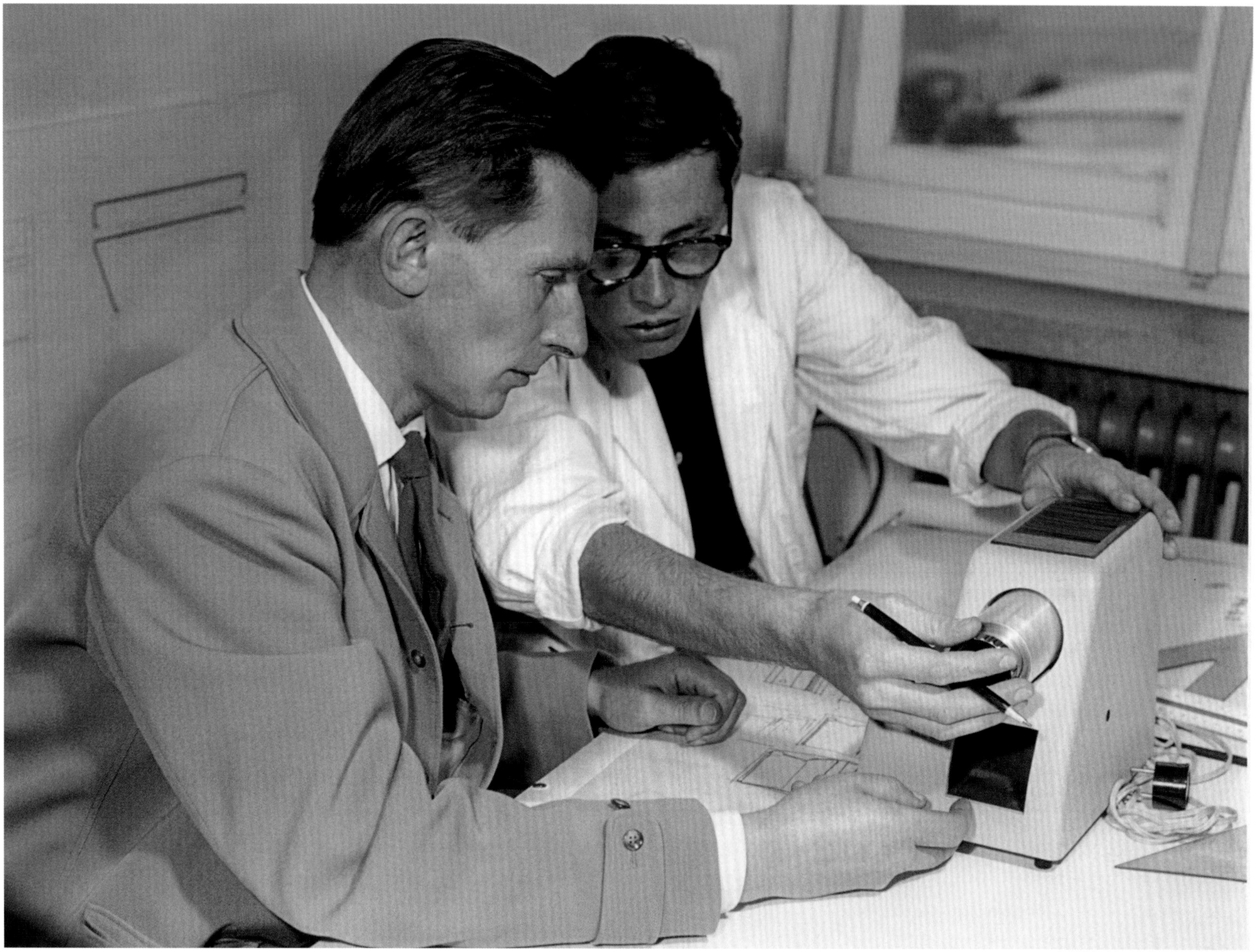

↑
Fritz Eichler and Dieter Rams with the working model for the PA 1 slide projector, 1956.

↑
Gerd A. Müller with the
KM 3 Kitchen Machine,
[illegible]

BRAUN

PA 1

Automatic slide projector, 1956, Dieter Rams

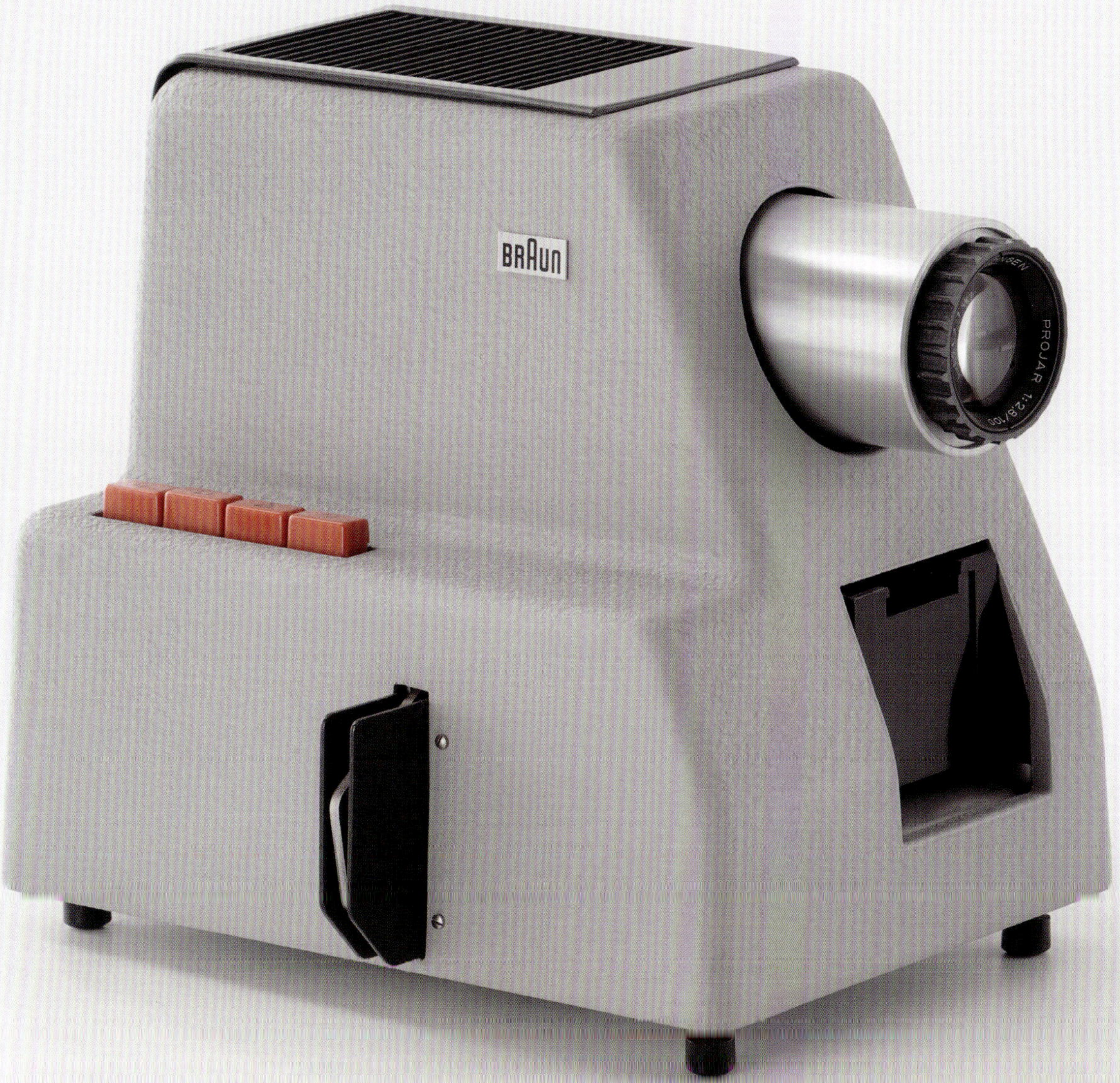

KM 3

Kitchen machine, 1957,
Gerd A. Müller

haus 62

In August 1958, Otl Aicher presented a 'discussion sketch' for a 'developmental program "haus 62"' which must have already been initially considered by both his HfG colleagues and Braun, with the company now planning a structural expansion beyond the existing factory buildings. Aicher was now no longer concerned with individual devices or a corporate design, but with making one big splash. The text reads: 'various producers and the Hochschule für Gestaltung set themselves the task of fully evaluating today's technical options for the creation and furnishing of an apartment by the year 1962 and to realize these in models.'

For this purpose, industrially manufactured components, container systems ('cabinets, boxes, shelves, compartments, baskets, trays, drawers') were to be developed 'for all utensils occurring in the house ... with high combinability'. Furthermore, a 'comprehensive electrical service of coordinated devices' for kitchens, communication and households was to be created, from cooking appliances to radios, televisions, telephones and film projectors to vacuum cleaners, dryers and heating and cooling products. Ultimately, Aicher's call to action was about producing a modular system 'from which the dimensions for each object could then be derived'.[13]

The project was, essentially, a total design for a new experimental office park and housing estate. This space was to 'include an assembly plant of the Braun company, administrative and sales offices as well as demonstration rooms, a housing estate for factory employees, a shopping centre, a primary school and playgrounds, a department store for exemplary consumer goods, a hotel', as well as further 'green factories'.

The concept, according to Aicher, required 'a mindset among the designers not geared towards establishing their reputation through the design of beautiful, individual products. In place of individual artistic ambitions, the initiative advocates working with technologists and engineers to extract the best possible results from given conditions and make them available to the public.'[14]

Was the project a sign of the HfG designer's megalomania or the result of the optimistic-utopian conversations he had been having with Erwin Braun and Fritz Eichler? Presumably the latter. Erwin Braun certainly did think on a large scale. And Dieter Rams remembers that he often spoke of the 'department store of good taste'.

In the late 1950s, the expansion plans became more concrete, and not just within Frankfurt. The company's 1959 annual report announced the first construction phase at the 'Kronberg reserve site' for the end of 1961. In 1958, a 13,000-sq-m area had been acquired in the south of Kronberg, as well as the 'Roter Hang' in the northeast, on which bungalows for senior staff members and guest apartments were planned. These were Erwin Braun's construction plans and Otl Aicher expanded on them in his discussion paper and saw in them great prospects for the HfG. The project sponsor for the factory and the sales department was to be Braun, while a housing cooperative was to be the lead partner for the housing estate and shopping centre. A sales organization, possibly the 'Verbundkreis', would provide financial support for the department store; a production consortium for the steel, aluminium and plastics industries was to be project sponsor for the structural elements; a to-be-determined manufacturer would fund the 'container programme'; and the City of Frankfurt would sponsor the school, playgrounds and public green spaces. Aicher had presumably considered a location in Frankfurt at first. The HfG was to be responsible for the 'overall planning and the development programme', supported by federal research grants and grants from the State of Hesse. Almost all the HfG teachers were earmarked for tasks, as was Jean Prouvé and the Ulm-based landscape architect Günther Grzimek, who later collaborated with Aicher for the Olympic Games in Munich. Founding the initiative and its management was to be the job of Erwin and Artur Braun. Finally, the Ulm students were also to be involved 'in the conception of additional smaller objects such as fixtures, plugs, switches, fittings and so on. Packaging, signets, business papers and the like would have to be designed. This could be done to a large extent within the framework of the lessons.'[15] Otl Aicher produced numerous site plans and module sketches for the project.[16]

'haus 62' was certainly also an attempt to bag a large assignment for the HfG Institute for Product Design established in 1958, in which lecturers from the departments of visual communication, product design and industrialized construction were to work together. What a different turn the history of the HfG would have taken if this project had been realized! But in the era of forced 'construction industry functionalism',[17] this project must have been too ambitious and went over the head of the others involved. It nevertheless shows how the collaboration between the HfG and Braun, the discussions and conversations between equals, led to different, bigger project ideas that were then fleshed out firmly and have lost little of their relevance even today. The driving forces behind the 'haus 62' concept – combining home living and work, building a 'green factory', initiating cooperative thinking and creating a department store with exemplary products – continue to be relevant topics for corporate and civic institutions today.

↗|
Designs by Otl Aicher for the 'haus 62' project from 1957.

→|
Pavilion on the Frankfurt Fair exhibition grounds in 1959, designed by Otl Aicher, Hans G. Conrad and Hans Gugelot.

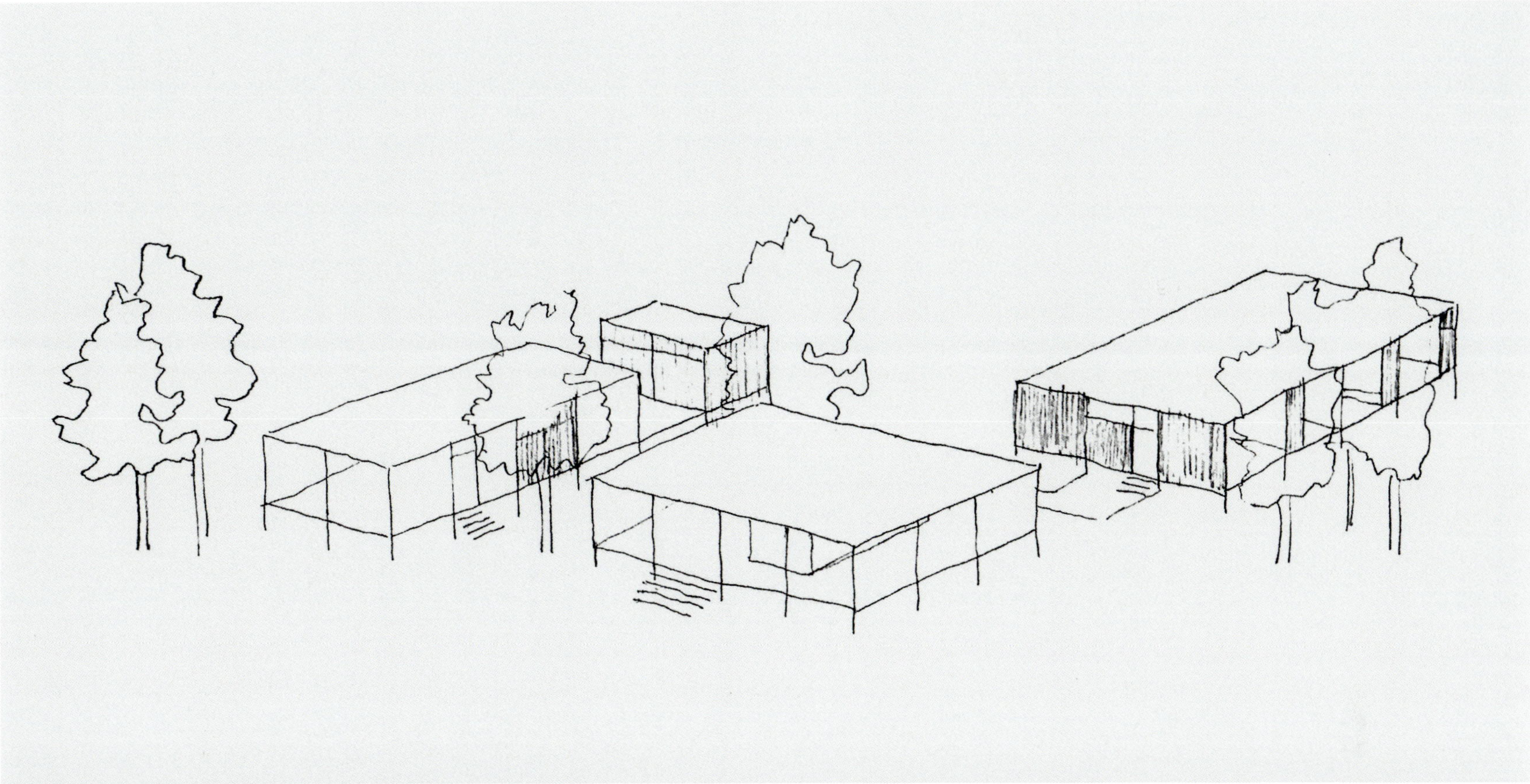

BRAUN

The foundation of the Hochschule für Gestaltung (HfG) Ulm can be attributed to Inge Aicher-Scholl and Otl Aicher, who, after 1945, passionately set about re-democratizing Germany in Ulm. The couple focused not only on design but on political education, which they believed should be founded on Christian-philosophical ethics.

Inge was the older sister of Sophie and Hans Scholl, two members of the 'White Rose' resistance group that fought the Nazi regime. The two siblings were sentenced to death by the People's Court on 22 February 1942 for their part in a leafleting campaign. Inge came from a Protestant family: her wealthy father was a tax consultant and auditor, a liberal politician and a strict opponent of National Socialism. Otl grew up in a small artisan family and was more interested in theology and philosophy than in commerce. He began reading Nietzsche, Kierkegaard and Thomas More as a rebellious teenager and from 1939 onwards was a permanent guest at the Scholls'.

A School for Democracy

After the war's end in May 1945, Inge and Otl organized 'Religious addresses on the Christian worldview' in Ulm. Thanks to the support of the American occupying forces, the project soon developed into an adult education centre with world-class speakers.

The couple soon began thinking about a private university for education in democracy, a 'school of tomorrow' for future journalists, designers and politicians that would go beyond the scope of an adult education centre. The subjects they planned to teach included politics, press, radio and film, as well as advertising, photography, industrial design and urban planning.

The HfG under Max Bill

During their deliberations, the duo came across Swiss architect, artist, designer and later politician, Max Bill, who had launched a travelling exhibition entitled *Die gute Form* and was to become the school's first director. Bill wanted to establish a first-class university of art and design that was to take up where the Bauhaus (where he himself had studied) had left off. He felt art needed to be an elementary part of the curriculum, just as it had been at the original Bauhaus, but Otl and other teachers did not share his view. On 3 August 1953, the new school started operating in temporary premises and, a year later, construction of the school building began on the 'Oberer Kuhberg', on the outskirts of Ulm. Just one year after the new building had been inaugurated, a conflict with Max Bill erupted over the thrust of the curriculum that led to his resignation and departure from the school the following year.

A New Orientation

The HfG thus became 'free of art', and this led to a new turn towards design, architecture and film: the idea was for these subjects to now be taught in an analytical-technological vein. The period between 1956 and 1963 was the most productive phase at the HfG, and this was also the time when the school had Braun as its most important commercial partner, with the company providing support not only through contracts but considerable donations.

A Philosophy of Design

Otl and architect Hans Gugelot, who taught product design at the HfG, without a doubt greatly influenced the design mindset at Braun in the mid-1950s, not just through his introduction of layout grids and initial modular product designs, but also primarily through his philosophy of design based on rationality. As Otl wrote, 'We don't live in a world of being, but in a world of projection. We are not content with what is, but address what should be. It is not nature that is our problem, nor technology, or even culture, but what we can do with nature, with technology, with culture.'[1]

In the mid-1960s, the HfG increasingly headed for crisis. In 1967, Otl opened a separate office dedicated to designs for the forthcoming 1972 Olympic Games in Munich, and Lufthansa was also already a major commercial client of his at the time. Gugelot founded his institut für produktentwicklung und design as early as 1962. There were also financial problems: Even Erwin Braun's donations were no longer sufficient to cover the ever-increasing deficit. On 31 December 1968, the HfG officially ceased operations.

‘We’ll Do the Design Ourselves’

The commission for the SK 4 radio-phonograph in early 1956 was the first big project for the group of designers made up of Dieter Rams, Gerd A. Müller and Roland Weigend. Though the initial design and conceptual phase were still implemented with the help of the Hochschule für Gestaltung (HfG) Ulm, the small design team nevertheless soon became more and more established within the company. With the innovative LE 1 speaker as well as the pocket and transistor radios by Rams, products were launched that had been created in close collaboration with engineers. This laid an important foundation for Braun Design, namely the ongoing alliance of the technology, sales and design departments. In all of this, Müller was responsible for the household appliances, Rams for the radios and record players and Weigend for producing the technical drawings and model-making.

Dieter Rams wrote about this in 1995: ‘I believe audiovisual devices must be understood as technical apparatuses – their formal design must correspond with this view. Making this fundamental view a reality was the true merit of our design department – of course with the support of our colleagues in Ulm.’[18]

Gradually, the three young graduates from the Wiesbaden design college evolved into the company’s new design HQ. The HfG became an expert advisory body, but the new products were now almost exclusively developed in-house through close cooperation between all the departments. This collaborative way of working and the unique atmosphere within the company were ultimately what gave rise to an approach to design that combined design, technical, commercial and, not least, ethical elements. Fritz Eichler evidently contributed to this to a great degree. In Frankfurt there was no ‘monastic’ community, as there was in the HfG, and instead collegial, yet active, competition between the individual protagonists. Added to this was the atmosphere of a big city desperately trying to reinvent itself after the war. While at the HfG lengthy discussions were possible, in Frankfurt things were all about production, which today’s design theorists describe as the intrinsic production of knowledge, meaning that knowledge creation occurs within the design process.

In 1961, when the company was restructured and transformed into a joint stock corporation, the designers (now including Reinhold Weiss and Richard Fischer from the HfG) had become a constant in the company, so a separate design department, headed by Rams with Weiss as his deputy, was established.

In 1977, the internal functional description of the ‘PG’ (product design) department stated that, among other things: ‘PG has the goal of developing new products or product groups in cooperation with the engineering department, with product programme management and business management. Once the decision to manufacture these products has been made, they are supervised by the section until they are ready for production.’ The design approach was also set out in this statement: ‘The design: insightful, consistent, constructed and designed based on

→|
Reinhold Weiss, Dieter Rams and Roland Weigend in the Braun design department, around 1961.

the intrinsic properties of the product and with as many 'unique benefits' as possible (and thus difficult to copy).'

In 1977, Rams's department was divided into three sub-sections: one for product design 'PG-P' (headed by Robert Oberheim), a second for product graphics 'PG-G' (overseen by Dietrich Lubs) and, third, the model workshop 'PG-M' (under Roland Weigend, and later Klaus Zimmermann). These were also incorporated into the company's overall process: 'For all areas of responsibility, but especially for PG-P and PG-G, it is necessary to provide impactful participation in the coordination of interdisciplinary cooperation in accordance with the set goals.'[19]

Yet the design department was anything but a mighty fortress. In an article about the Braun team that was published in the company's magazine in 1980, Rudolf Schönwandt very openly pointed out the danger of external influences on the team. According to him,

> it is the designers in particular who are constantly forced to defend the scope of duties for which they as specialists are responsible. After all, many other teams have a vital interest in poaching on the designers' territory. Every marketing man for example understands the market success of a product is also dependent on its design. And so, it stands to reason that he will try to move the designers to realize precisely that design he as a marketing specialist deems especially promising.[20]

In this respect Dieter Rams must be seen not just as a designer but also a manager for continuous design. In the Braun Archive there are numerous external designs, such as revised models of toasters and portable radios by Rolf Garnich, as well as products by several other designers, that were not ultimately realized.

Fending off external influences on design matters was a key strategic task for those overseeing the design department. Yet to achieve this, the individuals involved always had to be able to table better suggestions, so competition certainly existed. The team did not have a monopoly but was under constant pressure to prove itself.

In such a situation of both collective and individual design outputs, the head of the department's role was inevitably complex. In the 1960s it was no longer possible to reprimand or dictate to individual designers.

Rams granted the strong designers in his team the freedom to express themselves and yet always managed to ensure that a common-sense attitude towards design prevailed. Good design is not something that can be mutually agreed upon or compromised – it is the consistent implementation of a sound idea in cooperation with all parties involved. And, in the end, this was what made the Braun Design of that era so convincing: it did not prescribe fixed rules or repeat itself. Instead, it encompassed a broad spectrum of design ideas within a framework that was jointly developed by its workers.

↖
Prototypes for the Braun T 3 Pocket Radio, designed by Dieter Rams and the HfG Ulm, and the T 41, designed by Rams, c. 1957 or 1961.

→|
Model by Dieter Rams for a portable radio, around 1960. The design is similar to that of Rams's first wall-mounted device.

BRAUN

This spectrum can be described as follows:

1. Common-sense design: small radii for the cases; a reduced use of colour with an emphasis on the achromatic colours white, black and grey, combined with small contrapuntal colour accents; a focus on functionality; an interest in culture; and the willingness time and again to take up, process and implement external influences.

2. Use of a permanently advanced canon of details: green power button; initially concave, then convex pressure switches; black dials; yellow sweep hands; precise product graphics using an Akzidenz Grotesk font; and the additive (for example, interlocking cylinders on the KF 20 coffee maker) or subtractive (such as indentation on the MPZ 2 citrus juicer) design or further development of design details. In other words, an emphasis on the adoption of individual elements related to the shape (for example, in the case of the convex pressure switches, which were used in several devices).

3. Inclusion of different characters, ideas and design approaches (up to a certain degree of deviation) in a discursive process; respecting individual 'styles' within the fundamental attitude.[21]

From 'Classical Modernism' to the Present Day

The head designer and his team managed to bridge the gap between the so-called 'classical modernism' period (of, for example, the Bauhaus and New Frankfurt movements) and the present day. And this bridging did not entail, as was the case at the HfG, the simple swapping of pre-war modernism for design-engineering technicism, but rather the creation of a formal idiom that was as precise as it was poetic. At Braun, good conduct was defined in relation to the users of the products. As Rams said in 1975:

> In effect, I believe indifference towards people and their everyday reality is the only sin a designer can commit. Functional design grows out of an intense, comprehensive, patient, and thoughtful examination of people's lives, needs, desires, and feelings. The design of a product reflects a designer's image of humanity ... Just as, conversely, you can understand a great deal about a person when you see how his living space has been furnished.[22]

In the 1960s, all of this turned Braun into a well-received and often copied design pioneer. In the year 1997, Panasonic brought its SL ph7 mobile radio complete with the CD player by designer Shin Nishibori to market, which directly quoted Rams's pocket receiver. The predecessor was essentially doubled and mirrored, and now had two round, perforated speaker surfaces, while the window showing the transmission

↖
Members of the Braun design department, Robert Oberheim and Dietrich Lubs, with an unknown visitor in the early 1960s.

↑
World Expo in Brussels, 1958, visitors examine the studio 1 radio and record player on the shelf. Photograph by Braun photographer Ingeborg Kracht-Rams.

↑
World Expo in Brussels, 1958. On the sideboard, the SK 2 table radio can be seen. Photograph by Braun photographer Marlene Schneyder.

→|
World Expo in Brussels, 1958. Braun's transistor 1 portable radio can be seen in the window. Photograph by Braun photographer Marlene Schneyder.

setting had become a narrow longitudinal rectangle. Alongside this singular product citation, there is also a younger generation of designers that seem fundamentally committed to Braun's design ethic. Achim Heine has for many years supervised design at the German camera manufacturer Leica. His Digilux 2[23] both stands in the tradition of the company's legendary M series and shows the same attitude to design referenced by Rams, Jasper Morrison and Naoto Fukasawa.

At Apple, Jonathan Ive, with his iMacs, iPods and iPhones, also exhibited a special relationship with the design ethics of Rams and his team. The products Ive created are based on an approach to design that is certainly comparable to that of the devices created by Dieter Rams and his respective teams at Braun.

Awards

With Gugelot, Hirche, Rams and Müller at the helm, Braun's innovative design approach was soon held in high esteem by the professional community. International awards and exhibitions recognized the step forward that the company symbolized – among them the Grand Prix at the 11th Triennale Milano in 1957 or when the company participated in the major exhibition 20th Century Design from the Museum Collection at New York's Museum of Modern Art in 1958. Braun's new devices were a great success at the Interbau Berlin in 1957. 'When selecting series models, all interior designers independently opted for Braun devices. Even though not all the apartments had been furnished at the opening reception of the Interbau, there were already more than 60 Braun radios and televisions in the showrooms,' noted Braun's in-house magazine Betriebsspiegel. This prompted the idea of inviting 120 radio dealers from across Germany to a generously hosted two-day visit to Berlin.[24] While previously many dealers had refused to showcase the SK 4 radio-phonograph combination in their window displays due to its unusual design, acceptance grew markedly after their visit to Berlin.

Braun was also represented by numerous products at the World Expo 58 in Brussels. The German contribution was highly praised for its pavilion buildings designed by architects Sepp Ruf and Egon Eiermann, and the restrained presentations, although not by the German press. A special edition of the Werkbund-magazine werk und zeit published excerpts from articles in the international press. Most importantly, however, the special edition, of which 300,000 copies were made, was enclosed as a supplement with news magazine Der Spiegel. This was made possible 'by a donation from industrialist and Werkbund member Erwin Braun'.[25]

→|
Promotional photo for the portable TP 1, a combined record player and radio, with the T 4 portable radio, 1959.

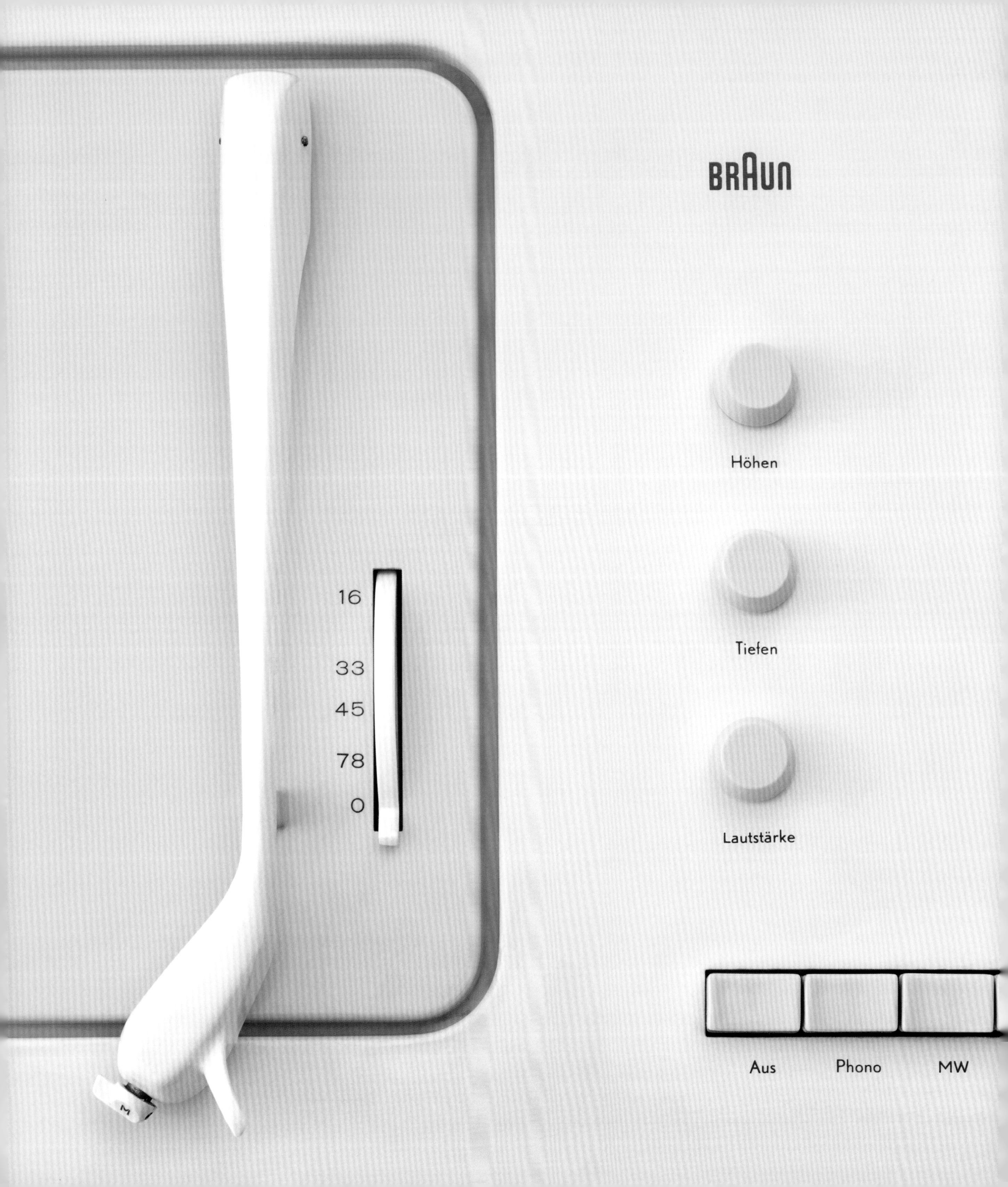
BRAUN
Höhen
Tiefen
Lautstärke
16
33
45
78
0
Aus
Phono
MW

SK 4

Combined radio and record player, 1956, Dieter Rams / Hans Gugelot / Gerd A. Müller / Werkstatt Wagenfeld

On account of its transparent Perspex lid, this famous Braun audio device was nicknamed 'Snow White's Coffin'.

HF 1

Television set, 1958, Herbert Hirche

The HF 1 has just one visible switch, which can be used to turn it on. All the other less frequently used switches are concealed under a flap on the top. This became the prototype for a major Braun design principle, namely intuitive operation. There is also a clear semantics to the switch: 'Press me and watch TV.' Apple would take up this one-button principle years later for its iPhone.

Kanalwähler
Feinabstimmung

EF 1

Flash unit, 1958, Dieter Rams

Departing from the Hobby devices' conservative black boxes and round reflectors, Braun launched an elegant pale grey flash unit.

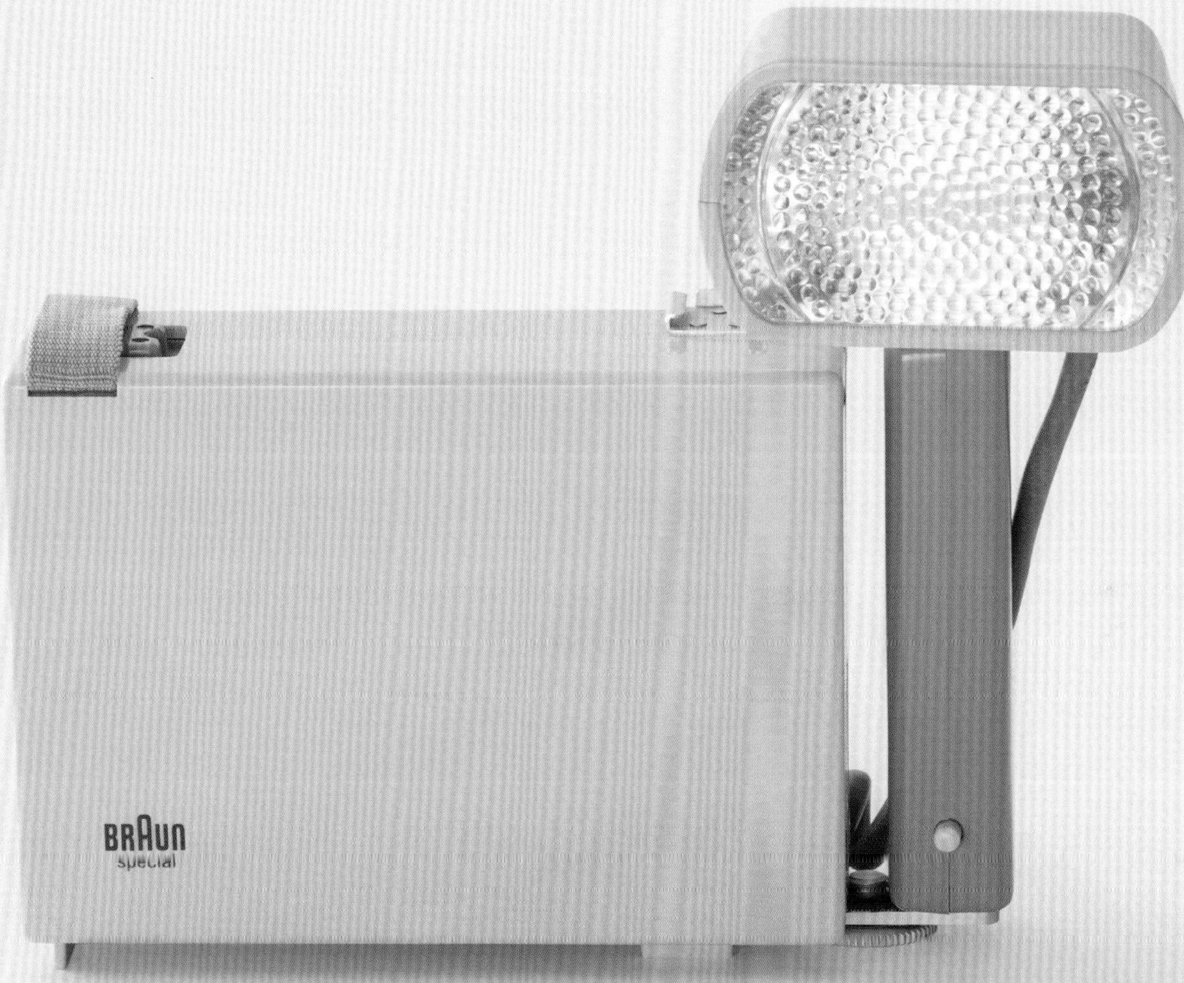

TP 1 (P 1 and T 4)

Portable transistor radio and phonograph, 1959, Dieter Rams

With the TP 1, Braun created a device for music on the go. The T 4 pocket radio could be slotted into the record player for single records.

The idea of listening to music on the road existed, but the right technology was missing as the small records could only play two to three minutes. It took twenty years for someone to come up with a good solution in the form of the compact cassette – it was Sony, who had the right idea with the Walkman.

T 3

Pocket Radio, 1958,
Dieter Rams

studio 2

Hi-Fi, 1959, Dieter Rams

This device marked Braun's entry into the Hi-Fi sector. The CE 11/12 receiver, the CV 11 power amplifier and the CS 11 mixer plus record player made for an excellent Hi-Fi unit, especially when combined with the innovative LE 1 electrostatic loudspeaker.

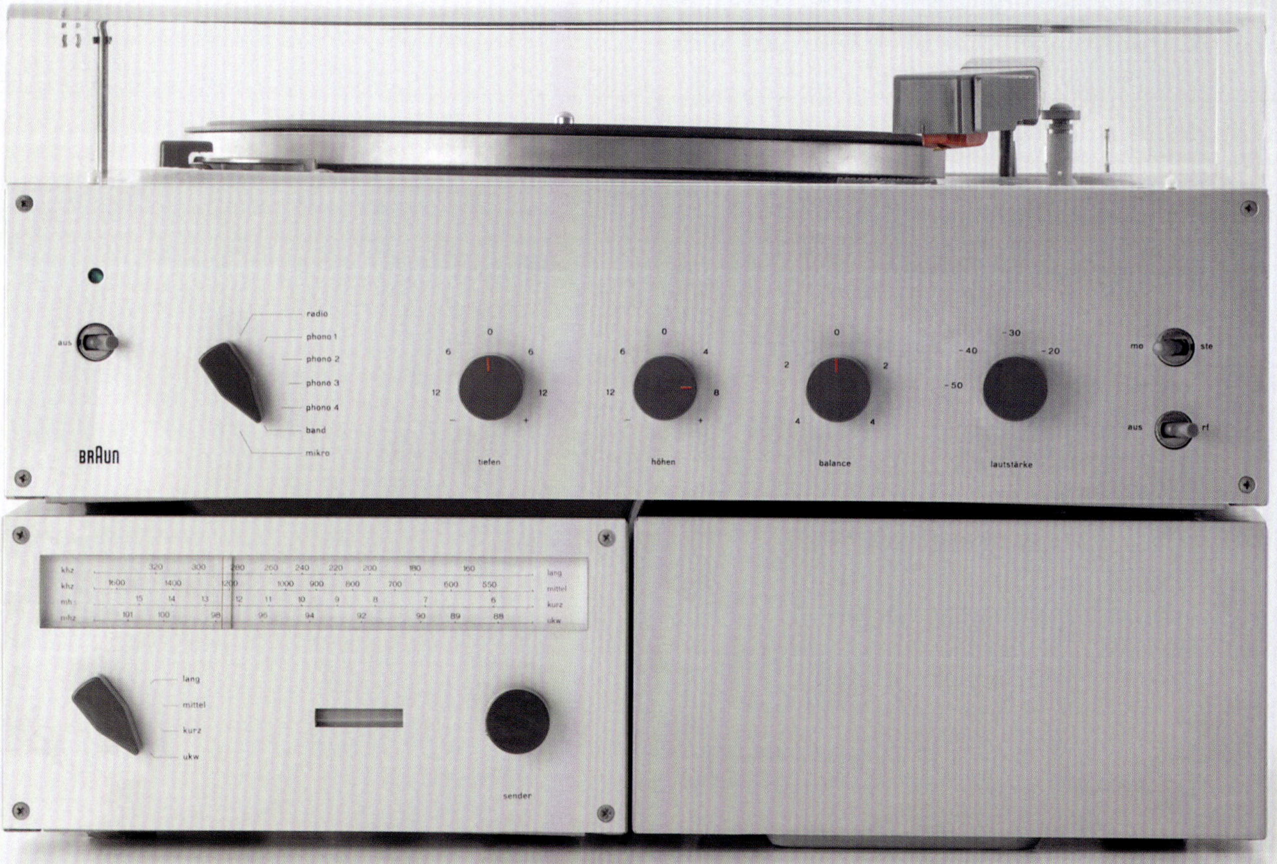

aus
BRAUN
radio
phono 1
phono 2
phono 3
phono 4
band
mikro
tiefen
höhen
balance
lautstärke
mittel
kurz

LE 1 and L 01

Electrostatic loudspeaker and auxiliary speaker, 1959, Dieter Rams

Braun had a licence from the British Acoustical Mfg Co Ltd Huntingdon to use its Quad Technology in this electrostatic loudspeaker, which features a diaphragm surface and can work with 1650 and 6500 volts. The design is inspired by metal industrial products, and the perforated front plate is slightly convex, producing progressive light reflections. Such small but crucial details bring many Braun products to life.

T 52

Portable radio, 1961, Dieter Rams

This portable radio can also be fitted into a car. Its somewhat longer knobs can be more easily operated by the driver without averting their gaze from the road.

RT 20

Table radio, 1961,
Dieter Rams

The radio takes up the design elements of the SK 4 but transforms them to create an extremely elegant device made of wood and metal with excellent sound. The RT 20 set new standards for tube radio technology, form and function.

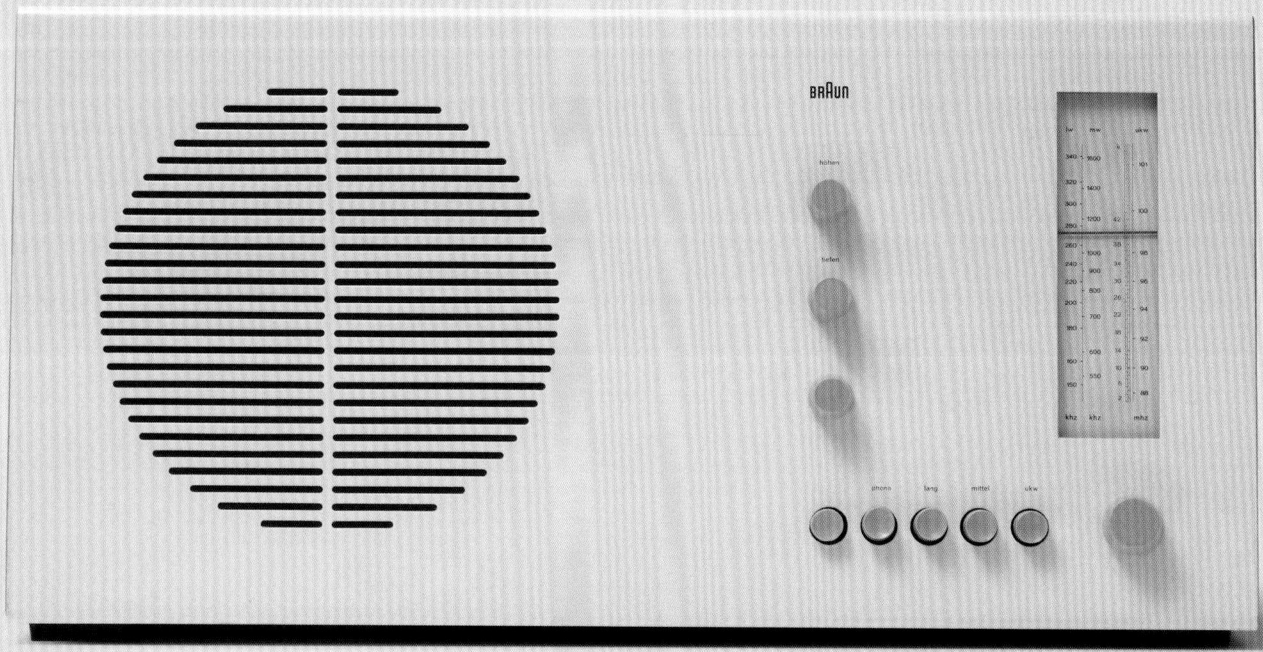

The 1950s: A New Focus

By the mid-1950s, the end of the Occupation Statute heralded an end to post-war reconstruction, the establishment of West Germany's sovereignty and a return to full employment as of 1956. The food situation also had returned to normal, and consumption now once more went beyond the essentials of survival.

'America, You Have It Better'

In the United States, creatives such as Ray and Charles Eames and Raymond Loewy shaped design together with numerous designers who had fled the Nazis in Europe. With them, the United States had achieved a 'push' in terms of design, one that had been embraced by industry there and greeted by a public with deep pockets and whose home country had scarcely been affected domestically by the war. In the 1950s and up until the early 1960s, the United States was a model nation in terms of politics, business, culture and morals, admired in western Europe and Asia. As the ultimate vanquisher of the Nazi regime and Japanese imperialism, the Americans established themselves as leaders and heroes of the 'free world'. This 'Americanization', which was pronounced across the globe, was most strongly felt in West Germany, Japan and South Korea, and was received favourably in many countries.

From the mid-1950s onwards, US cultural influences increased in West Germany. Coca-Cola became a cult drink, the first cars with tail fins were produced, young people listened to jazz and later rock'n'roll and the long-legged kidney table ousted oak furniture from many an apartment.

Those who were liberal-minded now perused Reader's Digest and believed in a better future. Girls swapped their pigtails for short hairstyles and wore petticoats, while boys slicked their hair back with ample grease and dreamt of owning a moped.

Alongside America, Italy was the primary new guiding light. After 1945, the Italians had returned to 'la dolce vita' without much ado – a way of living that turned out to be a hit export to Germany, where everything from the Vespa scooter and Rimini beaches to ice cream parlours came into fashion. Germans brushed their Nazi past under the carpet and delighted in the 'economic miracle' that soon took place. 'Wir sind wieder wer' (We are who again), the name of a show in a Munich cabaret in the mid-1950s, quickly became a popular catch phrase in Germany.

Uninventive Design in Germany

Cone lamps and kidney tables, 'artistic' curtains and the streamlined baby stroller – these stereotypical products of the decade took their cue from the American designs of the 1940s, which emphasized streamlined and organic shapes. In Germany, however, these trends were short-lived and furthermore tended only to be second-rate adaptations of the American versions. Even real design innovations from abroad, such as Eames or Knoll furniture from the United States, Danish designer Arne Jacobsen's 1962 Ant chair or the 1955 French Citroën DS, remained the exception rather than the rule in Germany. The average German living room of the war generation remained dominated by the furniture of a past era – and this consisted mostly of second-hand ornamental cabinets and chests of drawers with precious wooden, polished veneers.

Iron Curtain

More and more people in Germany fled from East to West. In the summer of 1961, the exodus to the West peaked. On 13 August 1961, East Germany reacted by building the Berlin Wall and closing all borders to the West. The 'iron curtain' was now firmly in place.

The year 1957 saw the first Russian 'Sputnik' satellite orbit the earth, which whipped the United States into a frenzy. The Space Race had begun. Braun's boom years coincided with the Cold War, a tense international situation and simultaneous economic recovery in western European countries. The Soviet Union also experienced a prosperous reconstruction phase, which, after Stalin's death at the end of the decade, even led to a degree of liberalization under Nikita Khrushchev. By and large, the world was deeply divided but economically optimistic as Braun entered its reform phase.

05
Kronberg and the World

With Dieter Rams heading the Design Team, an exemplary form of product design arose at Braun that became famous around the world. The phrase ‘designed in Germany’ became synonymous with outstanding quality and is still held in high regard in many countries to this day. With its clear design approach, Braun successfully weathered changing fashions and created products that were simple, useful and built to last.

HL 1 1961

HE 1 1961

HT 1 1961

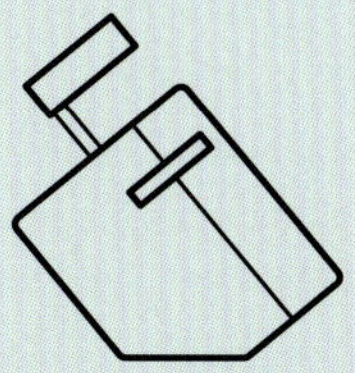

D 5 1962

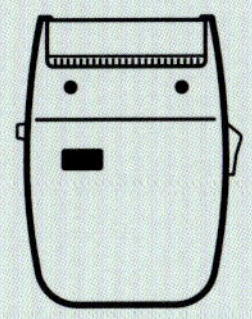

sixtant 1962

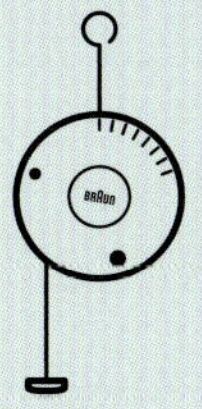

Tonearm Scale 1962

audio 1 1962

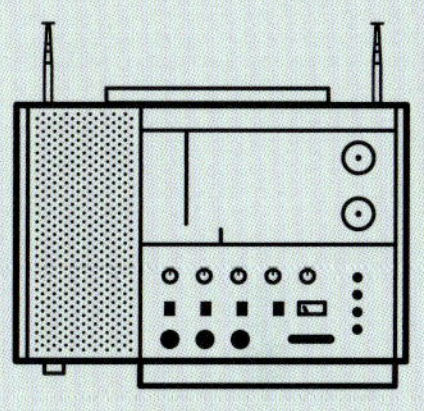

T 1000 1963

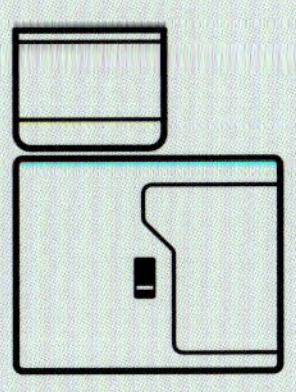

KMM 1 1965

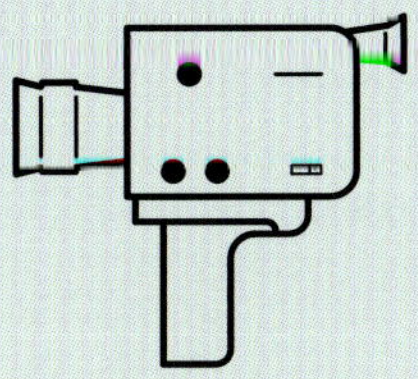

Nizo S 80 1965

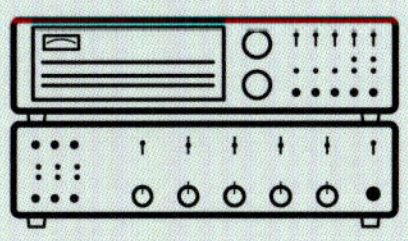

studio 1000 1965

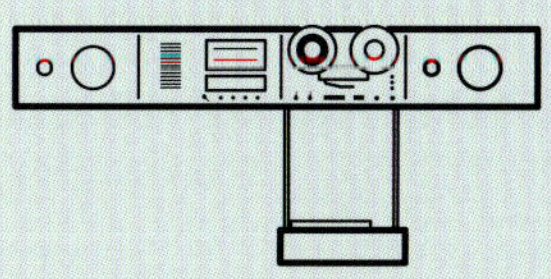

L 450, TS 45, TG 60, PCS 52 E 1965

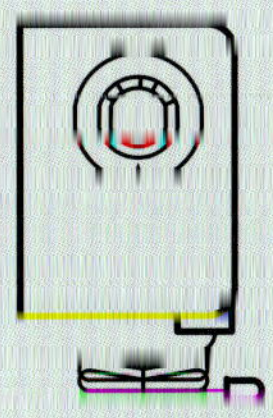

F 100 1966

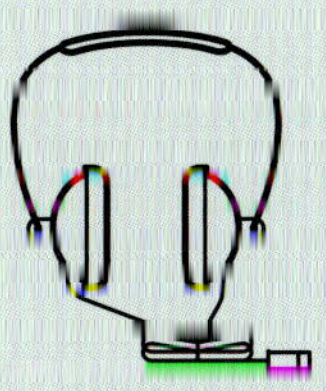

KH 1000 1967

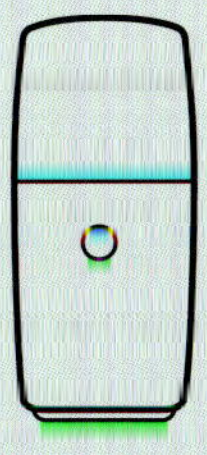

KSM 1 1967

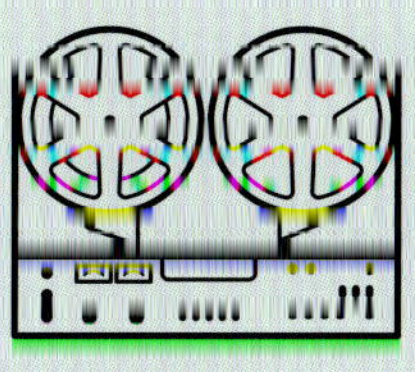

TG 1000 1970

Kronberg and the World

In 1959, Braun's payroll comprised 2,213 factory workers and 951 office workers, a twelve per cent increase on the previous year. At this point, sixty per cent of the workforce were women, however, they worked primarily on the assembly lines and not in management. Exports accounted for 34.4 per cent of goods. Life was getting crowded in the two Frankfurt plants.

As early as 1953, the company established a plant in Walldürn, some 75 km southeast of Frankfurt. It specialized in the production of shavers, which, as a Braun subsidiary under the management of Richard Rohlf, consisted of about 500 employees in 1959. A factory for kitchen appliances was planned in Marktheidenfeld, about 85 km east of Frankfurt, and commenced operations in March 1961. And, in 1958, a so-called 'reserve site' was acquired in Kronberg, where construction also began in 1961. Nevertheless, as early as 1959 the production capacities in Frankfurt and Walldürn could no longer keep up with increased demand, such that, for the first time, subcontractors were also employed to produce entire devices. Distribution companies in the Netherlands, Finland and Canada began operations. Customer service was also expanded: there were 15 customer service centres in West Germany and around 200 customer service points worldwide. Braun appliances were at the time, and continue to be, repairable and designed for long use.

In 1959, sales rose by eleven per cent to reach 101.9 million Deutschmarks, and the very next year 1960 turnover rose again by the same margin. This growth was mostly attributable to household appliances and photo-technical devices as well as shavers. A total of 1.3 million Deutschmarks was paid out as staff bonuses and under the profit-sharing scheme. The return was five per cent. Braun Electric International S.A. was founded in Baden, Switzerland, to manage all the company's foreign interests. Here, again, the Braun brothers remained the owners.

Braun Becomes a Joint Stock Corporation

On 1 October 1961, the Max Braun oHG was turned into a joint stock corporation.[1] Erwin Braun stated that the reason behind the change in company structure was the possibility of raising capital. He now switched from his position on the executive board to the supervisory board. Artur initially acted as chairman of the executive board and, on 19 December 1962, he joined his brother on the supervisory board. Dr Rudolf Gros replaced him as chairman. In the newly incorporated company, too, the return on investment ran at five per cent. Together with the five subsidiaries, the company had almost 4,000 members of staff in business during the 1961–62 fiscal year. New acquisitions included the Munich-based camera manufacturer Niezoldi & Krämer and Braun Electronic GmbH for the production and sale of electronic measuring and control equipment in Waldkirch in South Germany. The company's annual report listed the expansion of the Hi-Fi range, in particular the new audio 1 radio-phonograph system and F 80 electronic flash unit for professional use, slide projectors,

→|
Blimp emblazoned with the Braun logo above the headquarters in Kronberg, 1970s.

Braun sixtant
BRAUN

|←
Shaver assembly line in Walldürn, 1955. This first factory for shavers was opened in Baden-Württemberg, not far from the Rhine-Main area.

↗
Braun plant in Kronberg, Am Schanzenfeld development with shed halls, 1964.

especially the new D 5 combination unit, a 'new type of cooking grill for sophisticated requirements' and, finally, the sixtant shaver as the most important innovations.[2]

The annual report from 1963–64 shows the new executive board structure introduced two years earlier, with its four (vertical) product areas and three (horizontal) business units. In addition to the chairman, Dr Rudolf Gros, two other executive board members were now also responsible for the business units: Gros for administration, Karl Buresch for the technology and electronics teams, Hagen Gross for the household department, Ernst Krull for the photography department and Albrecht Schultz for sales and the shaver teams.

The company went public on the Frankfurt Stock Exchange in October 1964, floating six million preferred shares. At the end of 1964, the Braun Group payroll for the first time topped 5,000 people. Another first: the 1963–64 annual report was lavishly illustrated with product and exhibition photos.[3]

Sales Figures Soar

In the mid-1960s, the company was in fantastic form. In the 1964–65 fiscal year, annual gross earnings had risen by 28.4 per cent, with the company's net income now amounting to 7.6 million Deutschmarks, 4.1 million of which was distributed to shareholders, representing a return of investment of seventeen per cent. Braun AG's sales increased by twenty per cent from 145 to 175 million Deutschmarks.[4] With its exemplary product design, fully developed and innovative technologies, efficient and modern advertising, intact dealer loyalty, high growth figures and an exemplary operational structure, Braun had become an exceptionally attractive enterprise, not least on the stock market.

After 1 April 1966, Hagen Gross assumed responsibility for the newly created executive board position for new products. The supervisory and executive boards wrote to the company's management: 'It is becoming increasingly crucial what development capacities are available to a company ... We consider the new area of responsibility to be so important that it should be handled by a dedicated department. And we consider Hagen Gross to be particularly suitable for the management of this division.'[5] Gross went on to travel the world to discover new product ideas.

New Headquarters in Kronberg

Though Otl Aicher's large-scale project, 'haus 62', an innovative production and residential hub in Schwalbach and Kronberg, had come to nothing, Braun continued to build in Kronberg, albeit in a 'conventional modern style'. In August 1961, construction started on a multipurpose hall with a sawtooth roof. A second major construction phase took place in 1965–67 based on designs by Frankfurt architectural practice ABB (Apel, Beckert & Becker), with office partner H. D. Scheid

↑
Information Centre, Braun and Knoll International, Frankfurt/Main, c. 1962.

↗
Braun sixtant advertising on a Porsche 914 (the so-called 'People's Porsche') at a sportscar race, early 1970s. Erwin Braun, who was a nature lover, found the connection between Braun products and sportscar races unfitting.

in charge. Gradually, the various departments of the company moved to the new premises spanning 13,000 sq. m in Kronberg. By so doing, Braun left its native Frankfurt and relocated its main hub to the smaller town of Kronberg in the Taunus mountains, which had already become the preferred residence of wealthy industrialists, merchants and bankers in the nineteenth century. The widow of Emperor Frederick III and mother of Wilhelm II had resided at the newly built Friedrichshof Palace in Kronberg since 1893. The Braun management lived in this town or the neighbouring areas of Königstein and Bad Homburg. In 1970, Dieter Rams moved into a bungalow in the 'Roter Hang' settlement. The small town of Kronberg provided a sharp contrast to the metropolis of Frankfurt, but Braun had meanwhile become a sphere of its own, more orientated toward the world than Frankfurt. Management flew from nearby Frankfurt Airport to the many branch offices or to business meetings worldwide.

'Grand Design'

In his first concept for the 'haus 62' of 1958, Otl Aicher had mentioned a 'department store for exemplary consumer goods'.[6] What exactly he meant by this cannot be gleaned from the records in the Ulm archive, but it is possible that the idea was influenced by Rinascente, the Milan department store renowned for its commitment to good design. In 1962, Braun was awarded the Premio Compasso d'Oro prize funded by the store.[7] There was undoubtedly a more intensive exchange of ideas between Otl Aicher and Erwin Braun about future product design in the late 1950s.

As early as the start of the 1960s, Erwin Braun had initiated a 'Grand Design' at Braun, which was now becoming increasingly visible. After he had moved to the supervisory board in 1961, freeing him from day-to-day operations, he started thinking about products that were particularly outstanding in terms of technology and design, especially in the Hi-Fi sector. The first of these was the T 1000 shortwave receiver of 1963, developed with immense technical efforts by head engineer Joachim Fahrendholz and design engineer Harald Hauppenberger. The device marked a high point in German engineering. Even the most distant broadcasting station could be received in optimum quality.

The T 1000 was more like a two-way radio than a portable radio set, although it could theoretically be operated on the FM wave in the kitchen. At a cost of 1,400 Deutschmarks, however, it was luxurious for a kitchen radio. There had been no budget cap on Dieter Rams's design work on this project. All possible variations were investigated and turned into models. Rams achieved the highest possible degree of elegance and a visually stunning product that also boasted fantastic functionality.

Without a doubt, the T 1000 is one of his most iconic pieces and established his position as a guiding figure in design at Braun. More and more Hi-Fi components such as receivers,

Braun sixtant
BRAUN

sixtant SM 31

Electric shaver, 1962,
Gerd A Müller / Hans Gugelot

The sixtant was Braun's breakthrough in the shaver field. Basing the shaver's form on Gerd A. Müller's SM 3 from 1960, Hans Gugelot lent the sixtant a housing of grained black plastic and a shearing head of brushed stainless steel. There was also a new, efficient shaving foil with hexagonal perforations that gave the shaver its name. The sleek device was also available with a practical leather case. Numerous variations of this successful model were produced up until 1993.

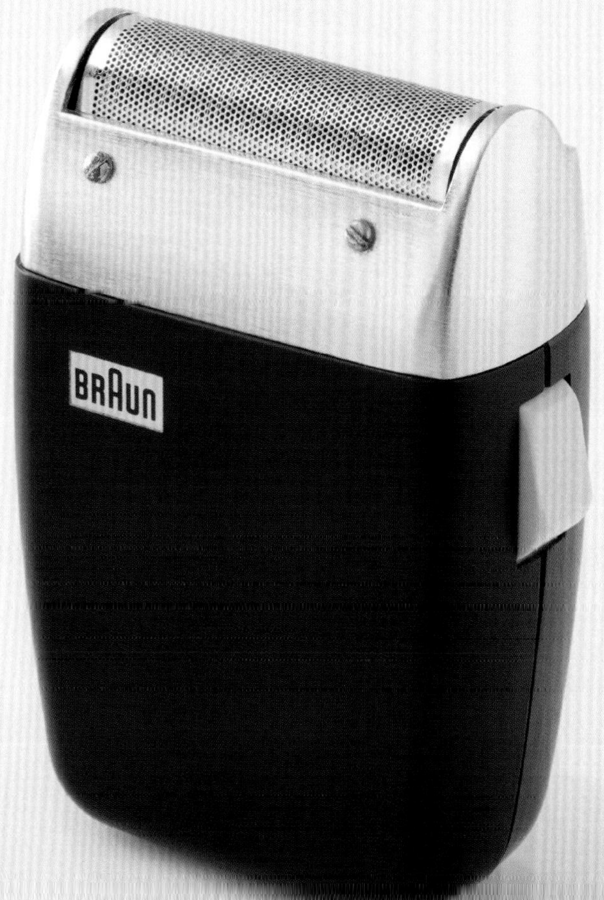

amplifiers, turntables and loudspeakers were created around the same time, the highlight of which, in terms of both technology and design, was the studio 1000 system of 1965.

The sixtant

In the 1960s, Braun stood for German design and quality products, and became increasingly internationally renowned. New products considerably expanded the company's portfolio and shored up its reputation as an exemplary and design-orientated enterprise. The sixtant shaver was the success story of the decade, not only in technical and design terms, but also economically, and was especially well received in Europe and Japan. The platinum-coated shaving foil was only 58/1,000 mm thick.[8] Engineer Bodo Fütterer, who had been working for Braun since 1950, was the brains behind this revolutionary shaver. The body of the previous SM 3 model designed by Gerd A. Müller had remained almost unchanged, with the exception of its brushed black casing and brushed stainless-steel unit head, features introduced by Hans Gugelot. Such alterations transformed the clinically white bathroom accessory into a highly aesthetic product. Shavers were also interesting in the company portfolio because customers needed to purchase new shaving foil roughly once a year, depending on the thickness of their beards, which generated regular follow-up revenue. It was thus not just the device itself but also the repurchases that brought in sales.

Braun for the Few and for All

Around 1965 Braun's Hi-Fi division ranked among the best in the world, despite not being particularly lucrative. A stereo for 13,000 Deutschmarks was a luxury product and something only a few people could afford. Nevertheless, the high-end Hi-Fi system was the brand generator for Braun. A sixtant cost 94 Deutschmarks, came with an equally great design and almost everybody could afford one, allowing them to claim a product by the high-end Braun brand their own. Erwin Braun had realized early on that he could transfer the upmarket image onto less expensive products, and this remained the Braun strategy until well into the 1980s.

The German kitchen also opened its doors to Braun Design thanks to a key product. The KM 3 food processor, designed by Gerd A. Müller with technical insides devised under Artur Braun's guidance, replaced the whisk and stirring spoon and, with additional equipment, could also be used to shred and mix food. Together with shavers, fruit juicers, hair dryers, alarm clocks and coffee makers, it was mainly these devices by Braun that were distributed widely at home and abroad that made the company's products 'design for all'. Braun design became part of upscale everyday culture and most midcentury middle-class families would have boasted at least one Braun product in their homes.

Expansion

In the mid-1960s, Europe's economies were headed for a crisis. Germany, too, had gone through a recession from 1965 to 1967, with corresponding decreases in GDP. Before Braun could float shares on the stock exchange to raise additional capital, it had to have published two consecutive annual reports, so this was not an option until 1963 or 1964 at the earliest. However, before that could happen, a further increase in the company's appeal was also required. After the takeover of Niezoldi & Krämer, Spanish household appliances manufacturer Pimer, based near Barcelona, became Braun Española. The design would come from Frankfurt, but the company brought on constructor and designer Gabriel Lluelles to act as the local representative. In Spain, he also co-facilitated Braun's most successful food processors: the citromatic citrus juicer which was initially developed specially for the Spanish market.

Subsidiaries were created under 'Braun electric' in Baden in Switzerland, Barcelona, Copenhagen, Gothenburg, Helsinki, Paris, The Hague, Toledo (USA), Toronto, Vienna and Yokohama.[9] In September 1962, Braun AG's capital stock increased from 12 million Deutschmarks to 22 million Deutschmarks. The Braun family held ninety-five per cent of this. Within three years, Braun AG had escalated its sales from 103.7 million Deutschmarks to 145 million Deutschmarks. Exports now accounted for thirty-five per cent of the total.[10]

Finally, at the end of 1967, when Braun was sold to Gillette, sales amounted to 276 million Deutschmarks with an export share of fifty-three per cent. By the end of the family business era, Braun had already become a global player, albeit still a small one.

Marketing in the 1960s

In the 1960s, the company's marketing strategy relied on the promise of quality technology and design. In 1963, a 'Braun studio as a centre for discussion and demonstration' was established

→|
Braun products from 1960 to 1974 – consistent design in a range of appliances.

BRAUN
Braun Multiwerk
Hochleistungs-Küchenmaschine
Braun Multiquirl M 121

↑
Jim Morrison with a Nizo super-8 camera. Morrison studied film at the University of California, Los Angeles, under director Josef von Sternberg, amongst others. With his fellow student Ray Manzarek, he founded The Doors in 1965 and became their lead singer. In both film and music, Morrison was equally experimental and excessive.

→|
The F 21 electric flash from 1962 is so compact that it can be mounted on a camera.

BRAUN F 21

9679573
Schneider - Kreuznach
Variogon 1:1,8/8-40
Nizo S8

Nizo S 8

Super-8 camera, 1965, Robert Oberheim

This camera from the large model series established the design of future Braun cameras for years to come.

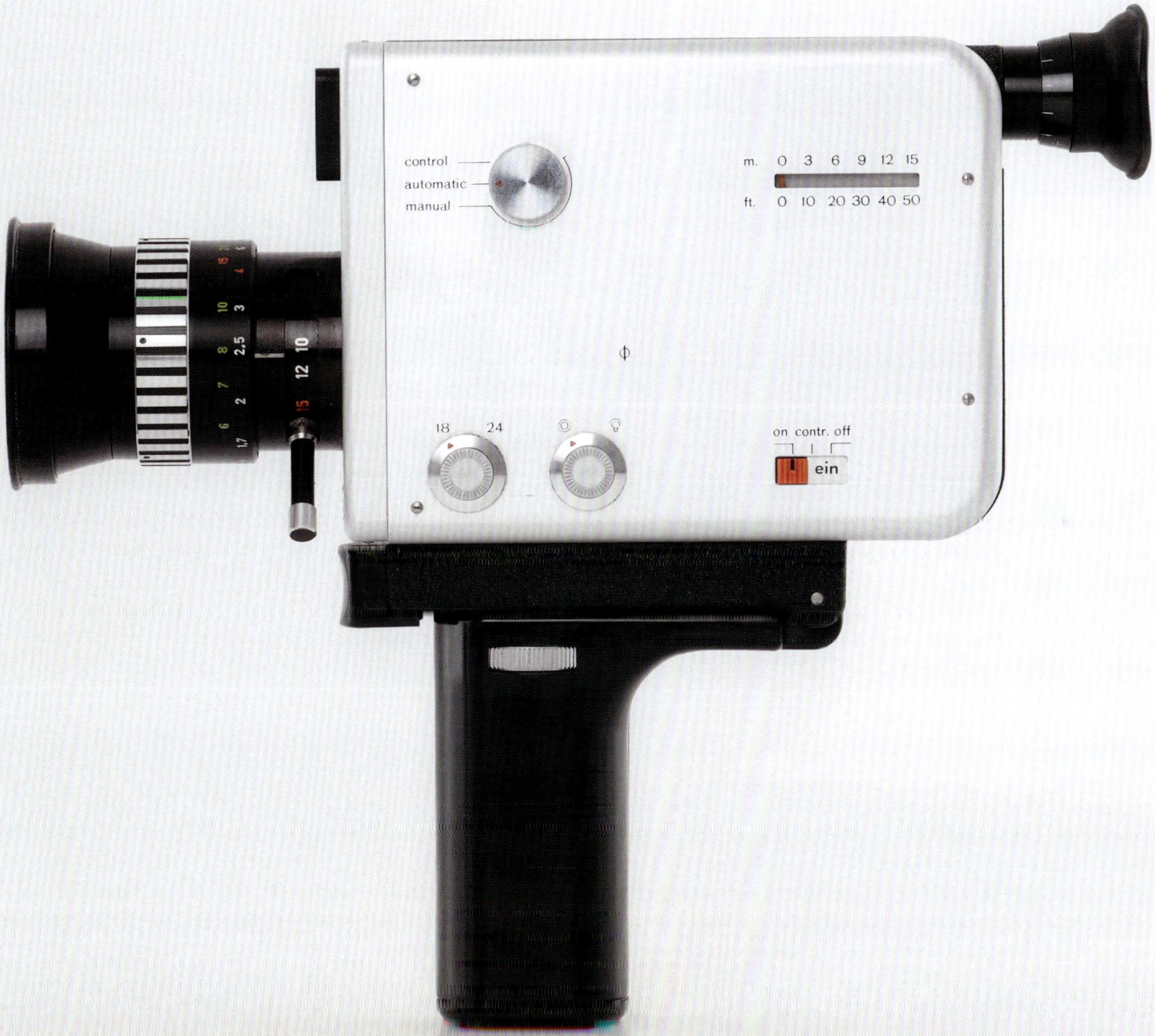

in the rooms of Knoll International in Frankfurt. It was inaugurated on 26 April with an exhibition of sound sculptures and prints by American artist Harry Bertoia,[11] who had also designed steel wire chairs for Knoll. In this case, the network of friendly companies and the combination of design, spatial experience, furniture and contemporary art proved its worth and left a lasting impression. 'Record concerts'[12] with Braun's Hi-Fi equipment were also held at the Braun studios. Four studios in cooperation with Knoll International existed until 1965, in Frankfurt, Berlin, Hamburg and Düsseldorf. A further one was established in 1966 in the Milan department store Rinascente, which was co-initiated by the Triennale Milano. Together with the furniture firm Dansk, Braun also ran a showroom on London's Sloane Street.[13]

New products were, for the most part, presented at trade fairs in Frankfurt, Cologne, Hanover, Lyon, Milan and New York. New Hi-Fi equipment was also shown at the audio fair in London. For many years, the company continued to use the trade fair booth that Otl Aicher and Hans G. Conrad had designed for the radio exhibition in Düsseldorf back in 1955.

Braun also used targeted product placement. In one famous example, Norbert Sakowski, the head of Braun's press office, wrote to Joachim Peter, the head of Frankfurt's municipal press and information office, in May 1963 on the occasion of President John F. Kennedy's visit to Germany with the suggestion of inviting the president to the Braun information centre.[14] The modern company would be a good fit for the youthful president and already had a growing presence in America both as a purveyor of objects to buy and objects to study (due to the presence of some products in New York's Museum of Modern Art, or MoMA). This was smart political thinking and corporate political strategy. While Kennedy could not squeeze in a visit to Braun on his short tour through town, the suggestion led to the 'compromise' of equipping the speaker's lectern at Frankfurt's Paulskirche with a Braun HL 1 fan for the president's speech on 25 June 1963 – the corresponding press photo promptly went around the world.

There were also unorthodox ideas, for example, setting up a Braun-sponsored 'vitamin and shaver bar' during a car rally or giving out awards at tennis tournaments.[15] An expedition to Lofoten was also provided with Braun shavers.[16] In 1968, Braun gifted the National Olympic Committee a promising young horse for equestrian sports, on the condition it be named 'sixtant'. The committee accepted.[17] At the time, Germany was already feverishly anticipating the 1972 Olympics in Munich.

Braun in the Museum

Braun devices were increasingly also exhibited in museums. After the early success of its 1958 MoMA exhibition, the company was represented at an ICSID[18] conference in the Louvre in Paris in 1963, again at MoMA in 1964 and at documenta III in Kassel and the Triennale XIII in Milan.

↖
In 1961, Braun received the renowned Italian industrial design prize the Compasso d'Oro. The prize was awarded in March 1962 at the Italian Embassy in Bonn, and the ceremony was attended by the German Minister of Economic Affairs and Vice Chancellor Ludwig Erhard.

→|
American President John F. Kennedy giving a speech in Frankfurt's Paulskirche on 25 June 1963 during his visit to Germany. The fan mounted on the podium and used to cool the President was the Braun HL 1.

HL 1

Table fan, 1961,
Reinhold Weiss

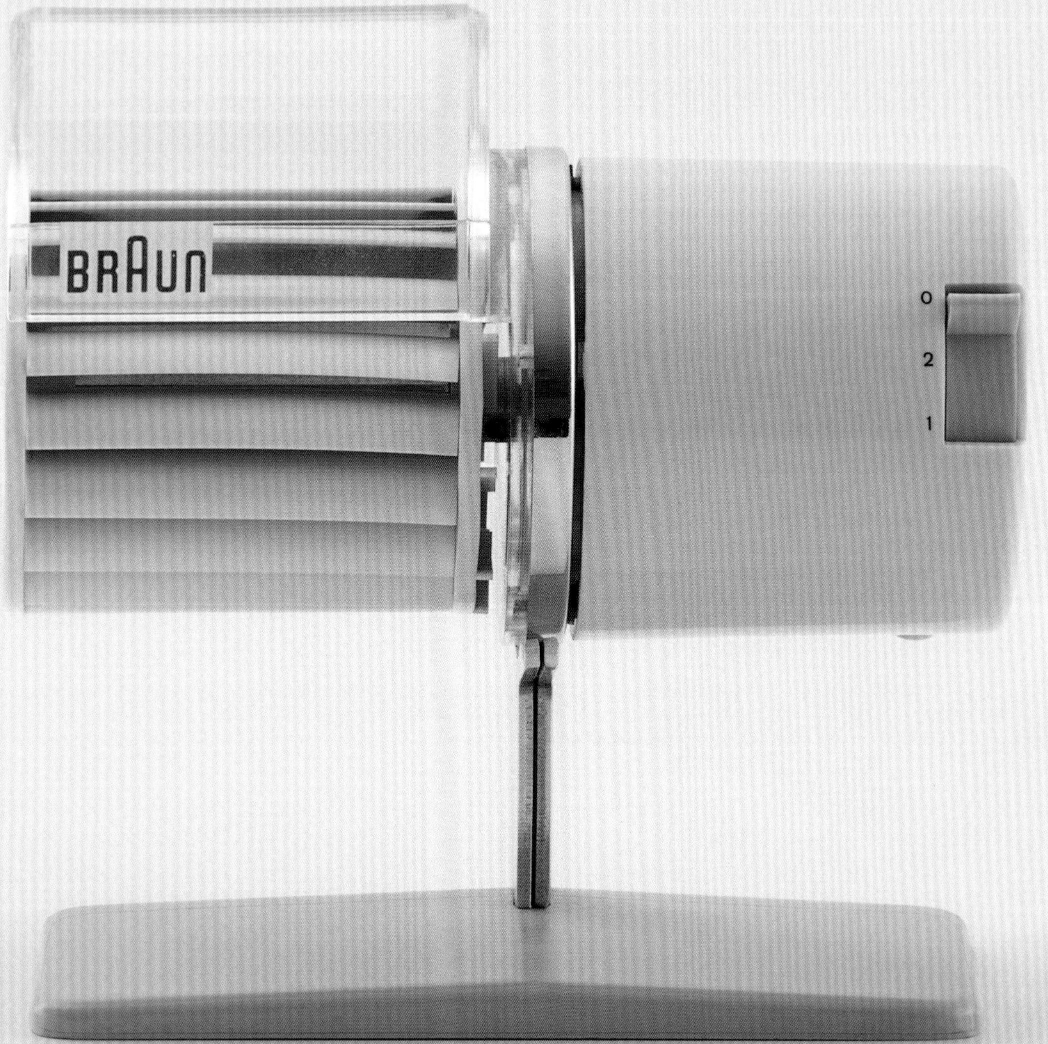

BRAUN

There was hardly an international exhibition of contemporary design in the 1960s at which Braun was not on display. The company became a synonym for German design par excellence. This was what gave rise to the myth of objective-functional 'German' design. And yet Braun wanted to be anything but 'German' – after all, the end of the Third Reich's bombastic chauvinism was only twenty years in the past. In this way, Braun had much in common with Otl Aicher and his work for Lufthansa. Aicher also avoided designing anything 'typically German', but tried to represent an international approach. In the end, it was precisely this which was then became identified with 'typically German' design.

A touring show of Braun Design was also conceived, Braun – The Face of a Company, relying on the special exhibition system designed by Herbert Hirche. The exhibition travelled not only to German and European museums, but also several Japanese cities.

The Product Design Department

Towards the late 1960s, the Design Team was firmly established at the company and over the coming decade would further expand its position. In a 1980 catalogue contribution, Rudolf Schönwandt described the product design department as follows:

> Many Braun employees seem to be slightly disappointed when they first get to see the design department. They evidently expect to walk into some sort of art studio atmosphere. Instead, they encounter a large, quiet, sober seeming space. It looks much like a construction department would: There are desks along the window front. In the room next door there is a row of large drafting machines. Along the wall you'll find the flat file cabinets and a shelf for models and products.[19]

The place where Braun Design was created was as clear and uncluttered as the products themselves.

↑
When held for the third time in 1964, the documenta art exhibition in Kassel had an Industrial Design section. On large low tables with teak veneer there were products by Olivetti, IBM, HfG Ulm, Wilhelm Wagenfeld, Arne Jacobsen and Gerrit Rietveld, as well as Ray and Charles Eames. There were also a great many Braun products featured.

The 1960s and 1970s

The 1960s and 1970s were a period of immense social and psychological upheaval. Major events included the Vietnam War (and the massive media coverage of it), the moon landing, the warnings of the Club of Rome and the Woodstock music festival. In Germany, a post-war generation unsparingly questioned the country's more recent history. Parents were confronted with critical questions about German's Nazi past. Outwardly, shaped by Willy Brandt's Ostpolitik, a new relationship with neighbouring countries was evolving, while domestically the nation's motto had become 'daring to be more democratic'. Longer hair was now fashionable, while the mini skirt was accompanied by its opposite, the maxi. Anglo-American pop won the day and even among the older generation critical thought about the past and an optimistic view of the future came together.

A political student body developed at universities in Europe and the United States, especially in protest of the Vietnam War, which students saw as being particularly unjust and brutal. This trend developed into a civil rights and peace movement with large demonstrations across the world.

The open-air music festival in Woodstock formed the climax of the hippie movement. In the 1970s, it gave way to rebellious and provocative punk, and musical dominance shifted from The Beatles to the Sex Pistols. The Braun studios, which in several cities had also been fitted out by Knoll international, especially for the presentation of Hi-Fi systems, hosted concerts where jazz and classical music albums were played.

Image Flood

The 1960s and 1970s were also decades that were as eminently political as they were visual. All political, social and cultural events were accompanied by a flood of images, and for the first time in human history almost all of those who wanted to witness them could. Magazines, periodicals and TV programmes reached even the most remote places in the world. In 1968, in Germany alone, amateurs and professionals took 1.5 billion photographs. In 1964 the Syncom 3 news satellite ushered in an era of intercontinental television broadcasting, starting with coverage of the Summer Olympics in Tokyo. What's more, this was the first time that a convincing and modern graphic design was presented.

Design in Frankfurt

Industrial design was exhibited for the first time at documenta III in Kassel in 1964, including items by Olivetti, IBM and Braun, among others. However, the national and international perception of design in the Frankfurt of the 1960s and 1970s was quite ambivalent. The city had opposing positions emerging side by side: on the one hand, Braun championed progress through product design, becoming a leader in design and business on the world stage and, on the other, Frankfurt's students took an anti-consumerist stance precisely against that world enamored with progress. Frankfurt defined itself by way of these contradictions.

Prosperity and New Crises

While the 1960s were characterized by a largely unchecked belief in progress, which climaxed in the United States' successful moon landing in 1969, the following decade saw an awareness of the world's emerging and continuing problems. Notably, the Club of Rome, an association of intellectual and business leaders, started to point out the limits of growth in 1972.

In the mid-1970s, West Germany was as much characterized by prosperity and a booming society as it was by new crises. In 1973 and 1979, oil prices caused a severe recession in industrial countries. The drastic hike in the price of crude oil in 1973 and the concomitant flattening of the plastics boom were as much a part of the 1970s as was the 'German Autumn', when, among other events, the extremist far-left terrorist group RAF murdered Jürgen Ponto, the Chairman of the Board of Dresdner Bank, and Hans Martin Schleyer, President of the German Employers' Association.

In terms of aesthetics, modernism had now finally regained a foothold in West Germany. All of Braun's products went well with flokati carpets, mini and maxi skirts, bungalows and the new BMW 3 Series. By the mid-1970s, the popularity of Braun was at its peak.

The era of the personal computer also began at this time. The Apple II was brought to market in 1977 and decentralized PCs would subsequently revolutionize processes in business and science.

Design in Montreal, in Munich and on the Moon

Swiss design magazine <u>Graphis</u> reported of Expo 67 in Montreal that:

> Everything was designed: Luminaires, trash cans, telephone booths, fountains and sculptures, the architecture, the landscape, canals and bridges, the model Habitat housing estate, the uniforms of the hostesses, the canopies of the Expo Express stops, even the myriads of miniskirts in the new 'psychedelic' colours of the subconscious mind; whether good or bad or neither: they absolutely captivate the senses. Millions of visitors here face an entirely novel environment.[1]

Design was now a central topic. And, in the German Pavilion, Braun and its products were featured extensively.

Design on the Moon

On 20 July 1969, US astronaut Neil Armstrong was the first human to set foot on the moon. On board with him was a small plate intended to explain the human species to possible extraterrestrials. The signatures of the three astronauts were engraved in it alongside that of US president Richard Nixon, while the remaining text about the peaceful intent of this mission was inscribed in factual sans-serif upper case Futura type. The plaque was left behind on the moon. While this was clearly a somewhat inept attempt to communicate with aliens, it was perhaps also a testament to the inherent human need for communicative and creative expression.

The 'moon shot' was more than just an arbitrary technological-scientific achievement and the triumph of the United States over Russia. It was primarily an expression of progressive thinking, which in the late 1960s still consisted of a faithful hope in a better world borne out of the spirit of technological innovation. Within the western world – and, with certain limitations, also the eastern bloc – the 1960s were above all a time when technology became usable by all. Telecommunications, automobiles, stereo systems, colour TV, inexpensive cameras and all kinds of technical appliances for the household, right down to the Teflon frying pan (a spin-off deploying material used spacecraft for the moon shot), became widespread: technology for all.

The 1960s' faith in progress was one side of the coin, while a ruthless depletion of resources and production of emissions as well as consumption of products with short lifecycles were on the other. In this regard, the technical and design longevity and repairability of its products made Braun back then an exemplary company from today's perspective.

Olympic Games

Much like Expo 67, the Olympic Games of the 1960s and early 1970s were mega-projects for an unprecedented form of international design. When Germany won the bid for the 1972 Games, the level of expectation for the Olympics' graphic design was already high due to the success of previous campaigns. The debate about the identity of the Olympic Games in Munich led to Otl Aicher and his office winning the commission to come up with a cosmopolitan and cheerful corporate design.

Around 1970, corporate design seemed to be the ultimate design task. Braun was certainly not alone in this but far ahead of most competitors. Yet the corporate design boom also motivated Braun. Under Wolfgang Schmittel, an exemplary corporate identity was created that exactly corresponded with the design attitude with which the products had been conceived.

HT 1

Toaster, 1961, Reinhold Weiss

The toaster's long slit
accommodated the larger
slices of German rye bread.

HE 1

Rapid-boil kettle, 1961, Reinhold Weiss

D 5

Combiscope slide viewer and projector in one device, 1962, Dieter Rams

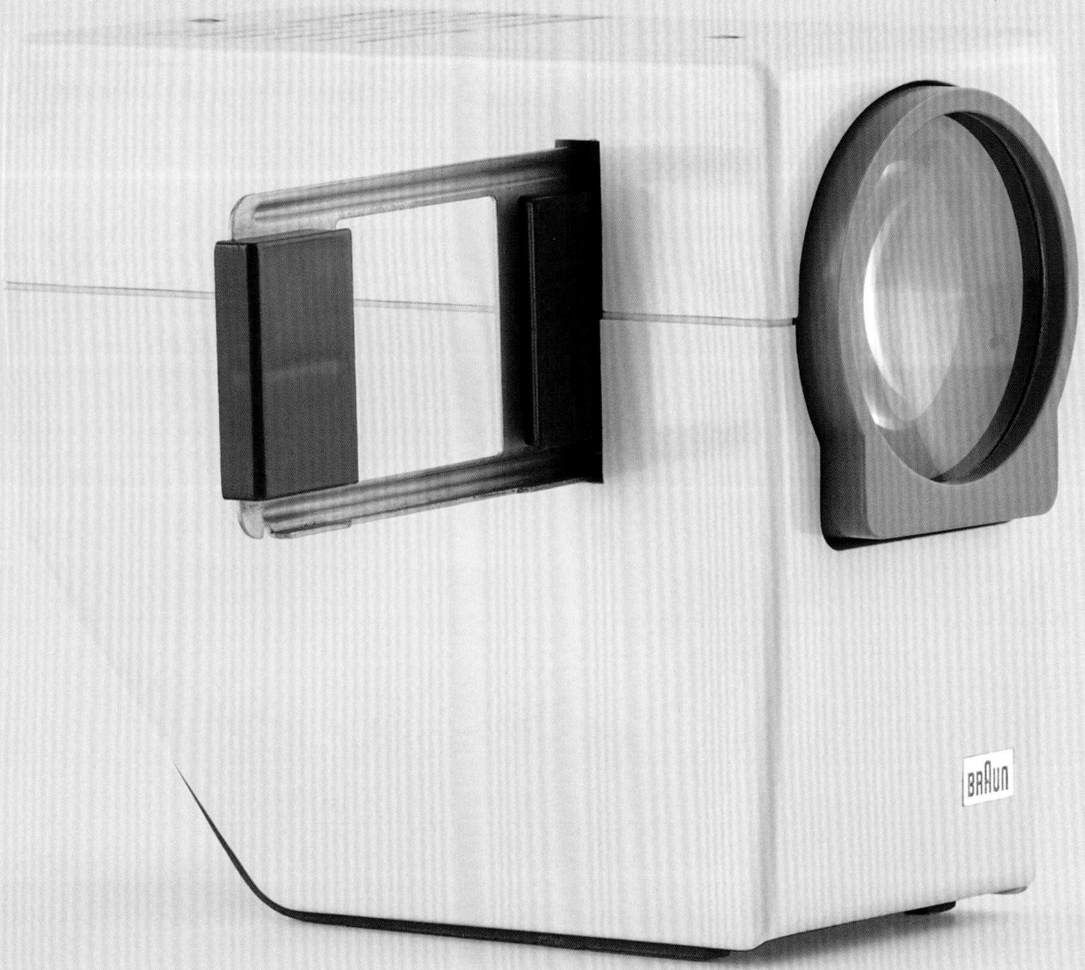

T 1000

World receiver, 1963, Dieter Rams

The first product to be designed according to Erwin Braun's 'Grand Design' concept whereby the technology and design departments could work without a cost limit.

tuning
volume
tone
low
high
voice
bfo
mgc
el. bandspread
afc
tape
am
fm
off
sharp
dial light
on
battery
broad
on / batt. c.

mayadent

**Electric toothbrush,
1962/63, Willi Zimmermann**

Braun's first electric toothbrush was designed by the head of advertising in the company's Swiss location and only sold in Switzerland. The mayadent laid the foundations for the later development of electric toothbrushes at Braun.

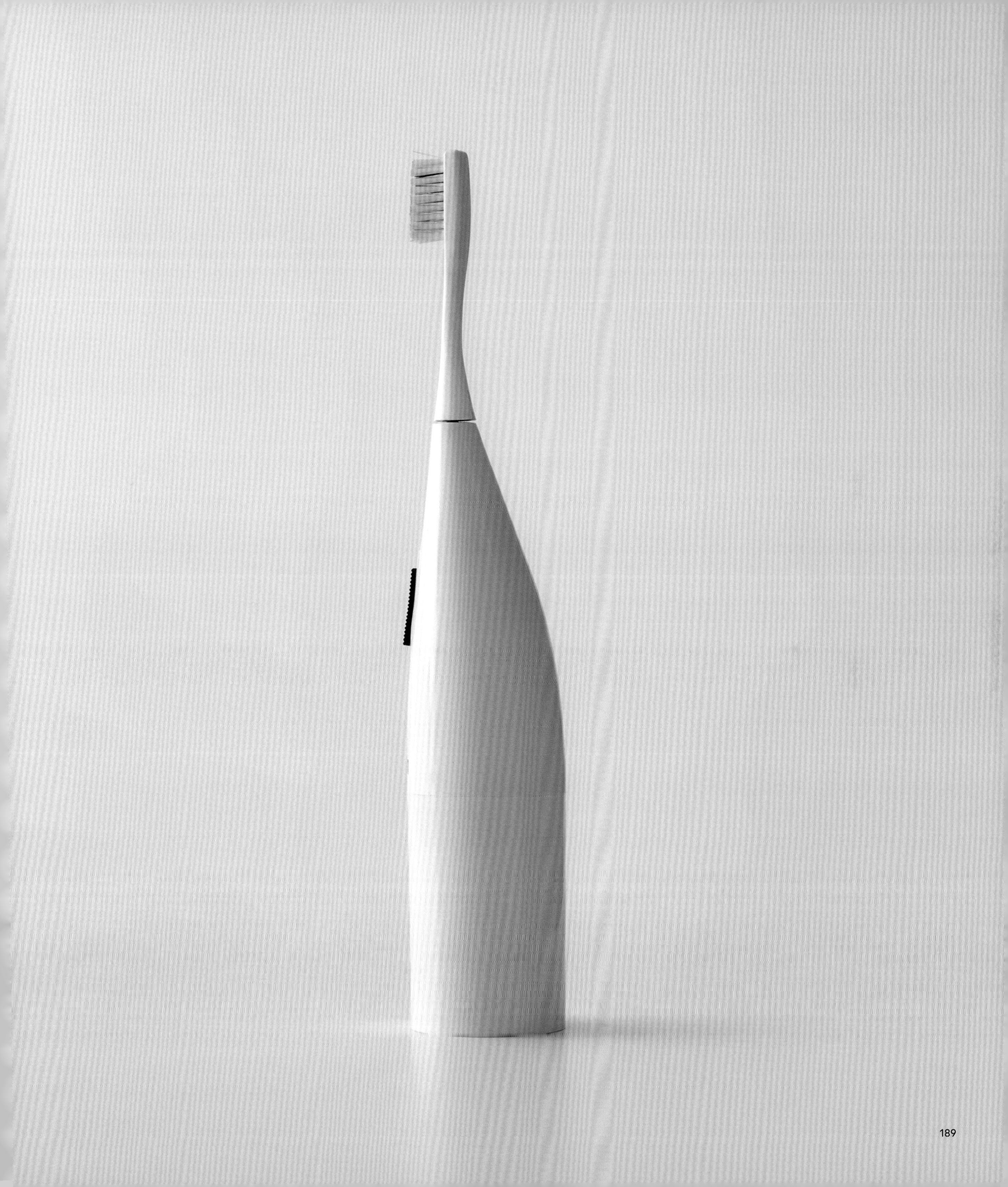

Communications Design at Braun

In 1973, a study carried out by the IFAK Institut for market and social research found that ninety-seven per cent of the West German population was familiar with the Braun brand.[20]

Communication about a product is thus integral to the product itself, and this includes the way in which the product is presented to and packaged for consumers. Though packaging waste is today a critical topic, it was relatively ignored during the 1950s and 1960s.

Packaging Design at Braun

The early Braun devices were mostly presented in brown or grey packaging that simply bore the product name. The thin brown cardboard box that held the tonearm scale is legendary. It did not even bear the Braun name.

Braun's early radio sets were installed on-site by radio retailers, so there was no need for sales packaging. All that was required was perhaps a blanket to protect the items during transit. For the first shaver, the S 50, Max Braun conceived a black or white moulded box with a transparent polystyrene lid: one could say that this was a miniature progenitor for the SK 1 case. The packaging features a product illustration, the Braun logo and the slogan 'Your electrical dry shaver',[21] anticipating later Braun packaging. In the 1960s, too, the advertising department delivered product packaging that showed only the product, product name and logo. Such realistic depictions of the content on its packaging became very successful.

'The product is the message' was the slogan at Braun. Up until the 1980s, the company adhered to this formula, and it has been reinstituted, albeit in an updated form, today. Contemporary companies follow Braun's packaging tradition – adhering to the simple idea of focusing on a product without any assisting images goes back to the object posters by Lucian Bernhard, Julius Klinger and Ludwig Hohlwein before and after World War I. Even then the product – be it a car spark plug by Bosch or the Adler typewriter – was accompanied by nothing more than the company name.

Pure Product

After Otl Aicher had established design grids for Braun advertising in the 1950s, these were further refined by Wolfgang Schmittel, who had already reworked the Braun logo in 1952 and was made head of the advertising design department in 1961. In 1968, he outlined every design rule in a manual that became the standard for all visual design at Braun. These 'guidelines for the design of information and advertising items' stipulated the proper use of the logo, colours and design grids as well as the exclusive use of Grotesque fonts. They also gave precise information about the type of photography to be employed as the preferred visual means of representation. A 1975 manual for packaging design reads: 'Product name at the top. Font: semi-bold Helvetica; product photo over entire area. Not overlapping on side areas. Company logo "Braun" set off from Braun in the product name by position or distinguished by colour.'[22]

→|
Braun packaging design from the 1960s. Braun packaging in this early stage advertises the product without any additional embellishment, putting special emphasis on the object 'held in the hand' to demonstrate its easy operation and haptic qualities.

BRAUN
KSM 11
BRAUN
BRAUN
sixtant
System
BRAUN
BRAUN

↑
In 1967, the KH 1000 headphones designed by Reinhold Weiss were the only element featured on the top of their packaging. The aim was to persuade consumers through the product's design quality, not witty words or colourful decoration.

→|
The yellow cassett travel shaver by Florian Seiffert in matching yellow packaging is presented both with its case and as pleasant to hold.

braun
cassett
battery
BRAUN
Made in
W. Germany
BRAUN
Braun cassett
5537-701

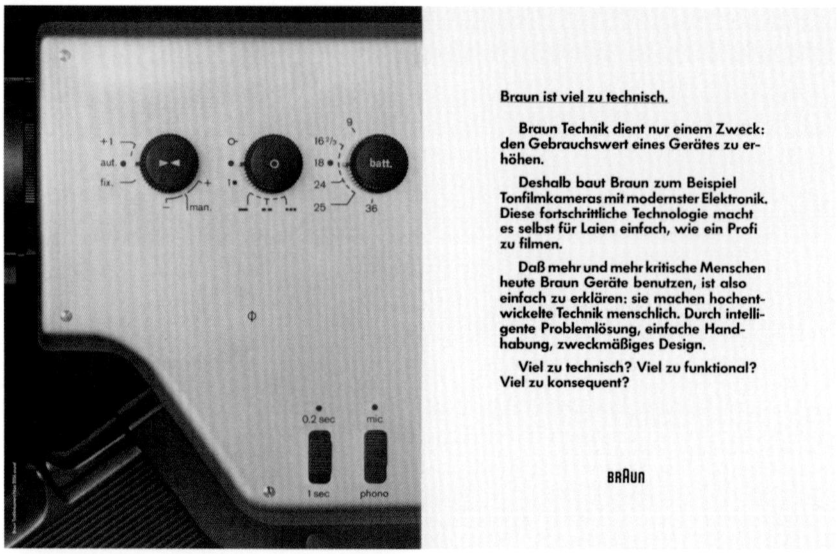

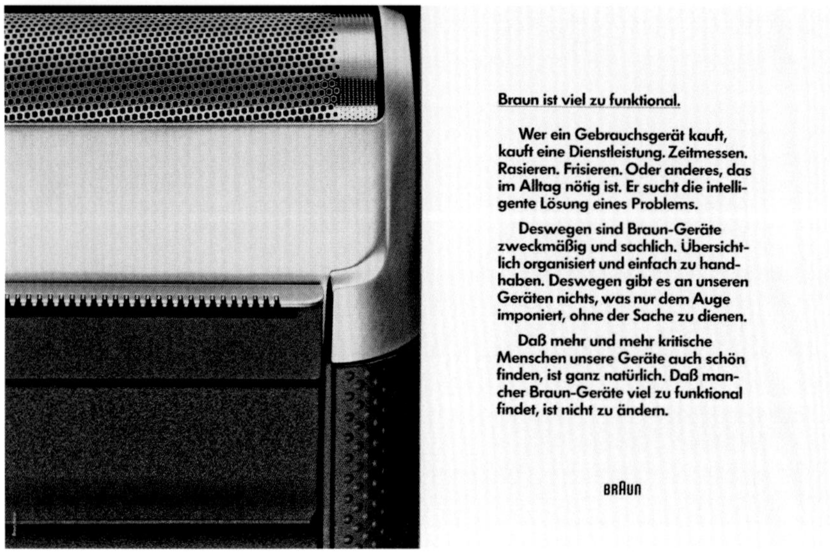

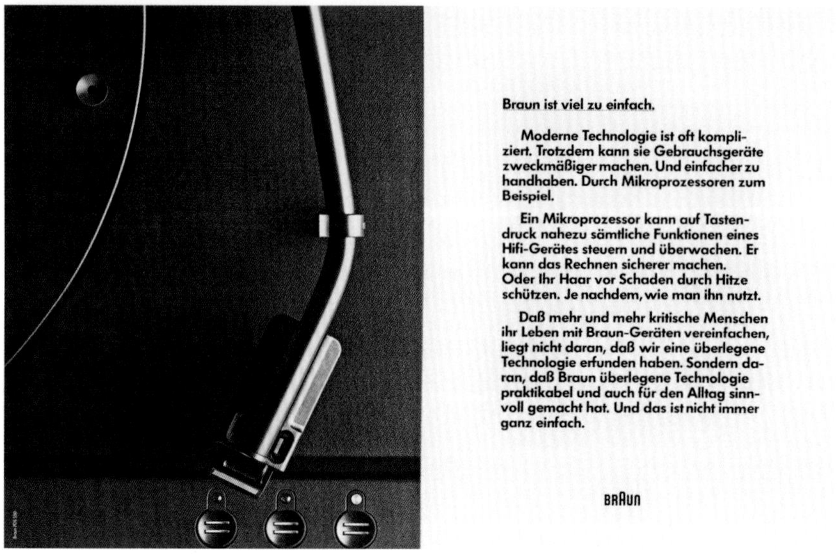

Braun is much too masculine.
Why should appliances for women be less useful than appliances for men? Let alone less intelligent?

The latest example: the Braun electronic Sensor hair styler. Its microprocessor controls the flow of heat. And protects your hair from being damaged.

There's a simple reason why an increasing number of women use Braun appliances: they are simply practical, down-to-earth and useful. And now just a bit more intelligent thanks to technology.

Braun is much too functional.
Anyone buying a household appliance is buying a service. Checking the time. Shaving. Hairdressing. Or one of the other things that are part of everyday life. They're looking for an intelligent solution to a problem.

This is why Braun appliances are useful and down-to-earth. Well laid out and easy to use. And why our appliances feature nothing that merely impresses the eye without serving a purpose.

There's a simple reason why an increasing number of discerning people also find our appliances beautiful. The fact that the one finds Braun appliances much too functional can't be helped.

Braun is much too technical.
Braun technology serves one purpose only: to enhance the utility of an appliance.

This is why Braun makes film cameras featuring the latest electronics, for example. This advanced technology makes it easy even for amateurs to film like professionals.

There's a simple reason why an increasing number of discerning people use our appliances: they bring high-tech back to the human level. By offering intelligent solutions to problems, simplicity of use and fit-for-purpose design. Far too technical? Far too functional? Far too consistent?

Braun is much too simple.
Modern technology is often complicated. Nevertheless, it can make household appliances better fit their purpose. And easier to use. By means of microprocessors, for example.

At the touch of a button, a microprocessor can control and monitor almost all the functions of a Hi-Fi system. It can ensure greater accuracy in calculations. Or protect your hair against damage from heat. All depending on how it's used.

There's a simple reason why an increasing number of discerning people simplify their lives by using Braun appliances: it's not because we have invented a superior technology. It's because Braun makes superior technology practical and meaningful for everyday use. And that's not always that simple.

↑
This series of advertisements responded to criticism of Braun objectively and humorously, thereby demonstrating the company's mindset. These ads were designed in-house in 1978 by the Braun advertising department.

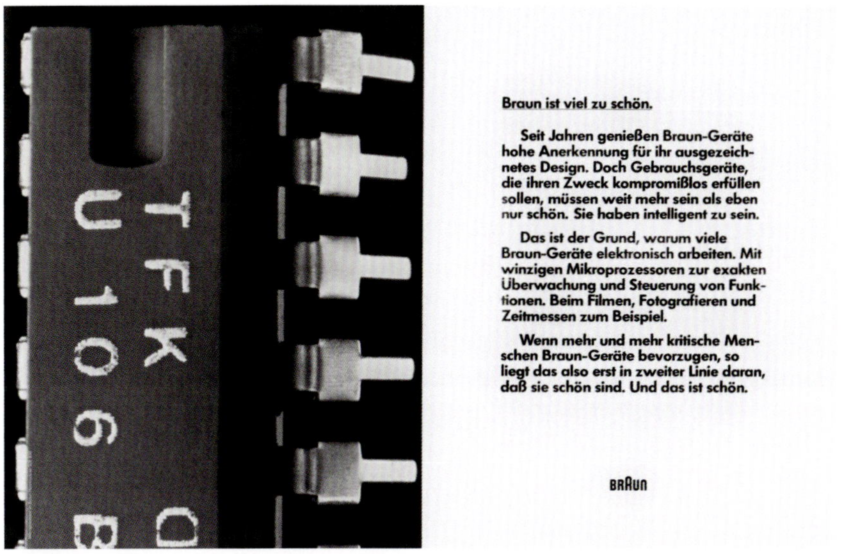

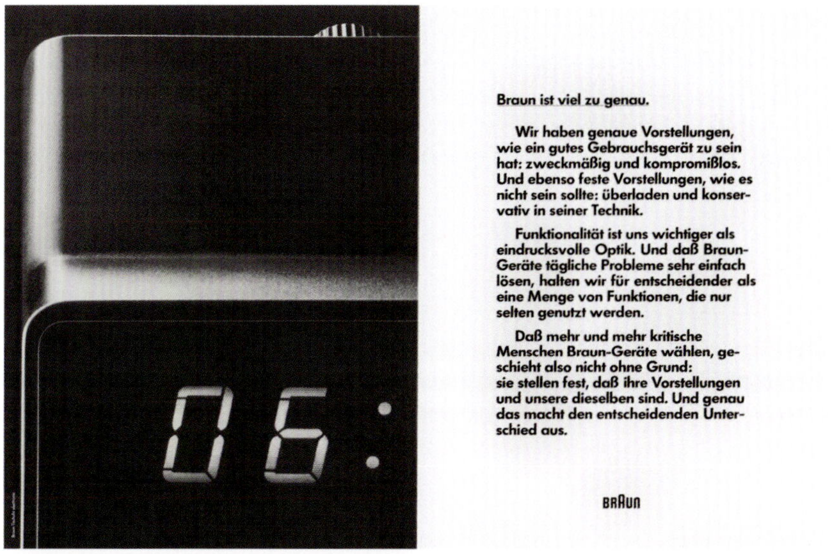

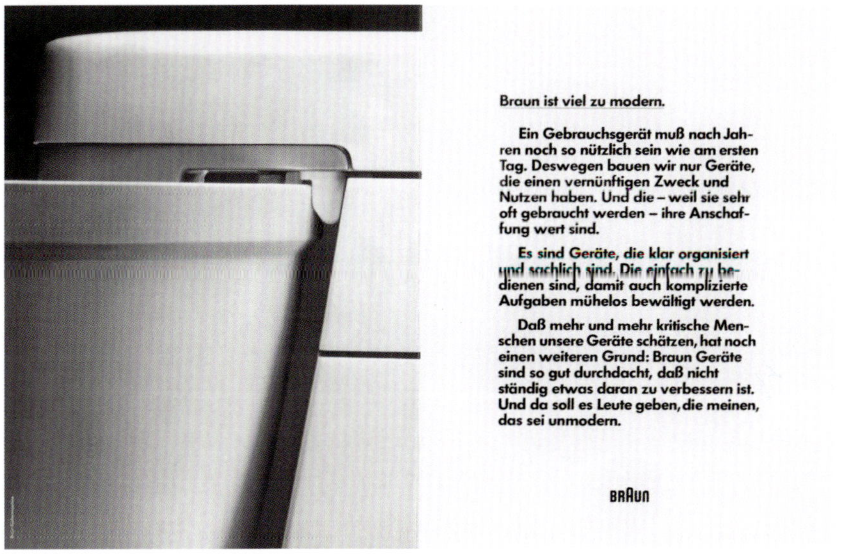

Braun is much too beautiful.
For years, Braun appliances have been highly regarded thanks to their outstanding design. That said, household appliances that are destined to fit their purpose need to be far more than just beautiful. They need to be intelligent. That's the reason why many Braun appliances function electronically. Using tiny microprocessors to perfectly control and monitor the functions. For example, when you're filming, taking a photo or checking the time.

There is a simple reason why an increasing number of discerning people prefer Braun appliances. And the second reason is because they're beautiful. Which is just the way it should be.

Braun is much too modern.
Years after you buy it, a household appliance needs to be as useful as it was on the day you bought it. Which is why we only make appliances that fulfil a rational purpose and use. And are well worth their price simply because they are used so often.

They are appliances that are clearly laid out and down-to-earth. And are simple to use so that even complicated tasks are easy to perform.

There's another reason why an increasing number of discerning people appreciate our products: Braun appliances are so thought through that there's no need to constantly improve them. Although some people insist on claiming that isn't modern.

Braun is much too exact
We have a very clear idea of what goes into making a good household appliance: useful and uncompromising. And just as clear an idea of what it should not be: too much or too little technology.

For us, functionality is more important than an impressive appearance. And for us it is far more important that Braun appliances solve everyday problems very simply instead of having a whole series of functions that are only rarely needed.

There's a simple reason why an increasing number of discerning people choose Braun appliances: they realize that their ideas are the same as ours. And that is the decisive difference.

Braun is much too consistent.
We make household appliances that solve very simple everyday problems. We also make ones that solve highly complex tasks. For everything in between, Braun appliances essentially share a single principle: they expand users' options for solving problems better and more intelligently. In an inconceivably simple way.

There's a simple reason why an increasing number of discerning people understand and honour this principle: we are consistent enough never to abandon it. Even if some still can't seem to get this.

The following year, all communications, from target group definition, typography and the design of in-house printed matter and TV commercials to criteria for media decisions, were brought together – all into one group.[23]

The adverts from that era appear tidy and factual. Their headlines are tightly set, with letters often touching. This style of typography takes its cue from American role models and in 1960s Germany was used by Willy Fleckhaus for the magazine Twen. As in Twen, the proportion of white in the Braun adverts was relatively high, and therefore immediately attention-grabbing.

A Distinct Visual and Textual Language

Much like Dieter Rams's product design team, Wolfgang Schmittel's advertising design department emancipated itself from the initial specifications laid down by the HfG and developed a textual and visual idiom all its own. After all, the department was also under constant pressure to produce, and this demand led to a constant stream of new ideas.

The department developed a communications design canon grounded in matter-of-fact graphics and typography and light and subtly humorous textual communication. Headlines read, for example: 'It tells you who it is even before the shave. A Braun.' Or 'Your beard grows without a system. Get to grips with it systematically.'[24]

A 1975 internal design manual reads:

> Many Braun products are among the market leaders. They are in no way specialties for a small, select circle. Yet there is such a thing as a typical Braun customer. And this is who we have in mind as we work. We imagine he is critical, attentive, quality conscious. Suspicious of any kind of manipulation. Reasonable, sensitive to the human dimension and natural. We envisage the image of a user who is 'learned', enlightened and experienced, and not just in relation to one country – Germany – but internationally.[25]

The advertising design department also intensively monitored the market and carried out surveys. For example, ads were tested and found to have high approval ratings. As in product design, general criteria were developed:

> Our communications should be informative.
>
> ... We do not use elements designed solely for garnering attention and interest that have nothing to do with the product.
>
> ... Our communications should be credible and express quality.
>
> ... We are honest. We strive to show the products and their benefits the way they are.
>
> ... Our communication should be intelligent and progressive.
>
> ... We do not underestimate the intelligence and experience of our target group. We encourage people to think for themselves.
>
> ... Our communications should be personal, easygoing, and likeable.
>
> ... We avoid anything that would also seem inappropriate in a direct personal informational conversation: excessive volume, declamatory gestures, pompous solemnity, boastfulness, formality.[26]

Communications Team

By the year 1980, the advertising department, now called 'Central Communications', encompassed thirty-one employees and was twice the size of Dieter Rams's design department. After the company relocated from Frankfurt to Kronberg, the communications team had spacious facilities at its disposal, including a two-storey film and photo studio. Everything, from the product photos to the advertising poster, brochure design and 35-mm advertising films for cinemas, was done in-house here. This was communications 'from a single source', which made the company one of the first in Germany to achieve a consistent and coherent corporate brand identity made up of products, communications and corporate culture.

However, in 1980, Wolfgang Schmittel was dismissed. The technically complex and expensive equipment he had acquired to fit out the communications team had earned it the nickname 'Klein-Geiselgasteig'[27] after the Bavarian film studios near Munich, and it seems that in-house production was becoming too costly for management. Advertising was now delegated primarily to agencies such as J. Walter Thomson or McCann, the approach of which corresponded more with American marketing. Yet this meant that the uniformity of corporate communications was forfeited along the way.

→|
Redesign of the Braun logo – originally designed by Will Munch in 1933 and updated in 1952 by Wolfgang Schmittel. The precise shape has only changed slightly up until today.

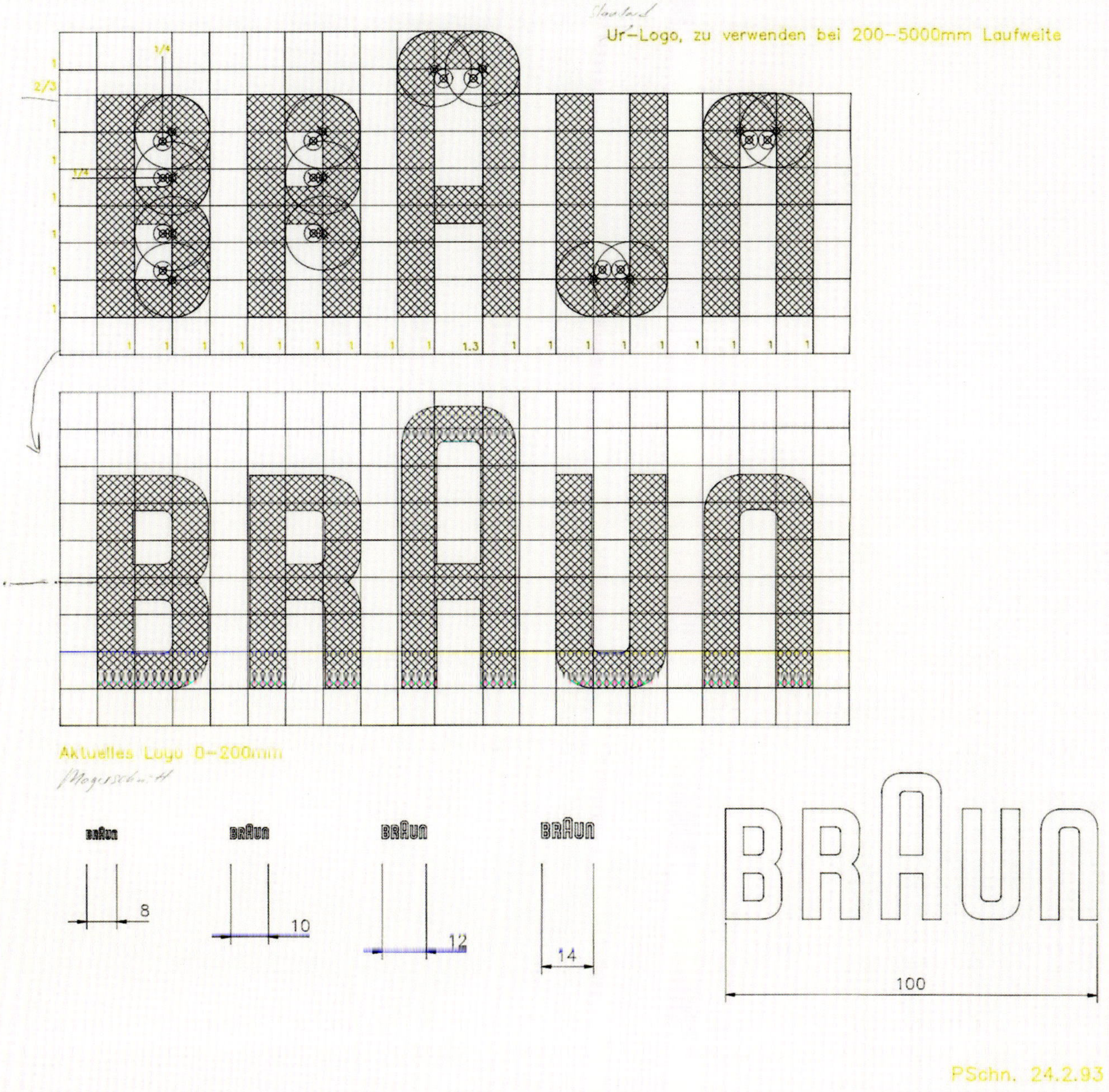

Ur-Logo, zu verwenden bei 200–5000mm Laufweite
Aktuelles Logo 0–200mm
8
10
12
14
100
PSchn. 24.2.93

Photography at Braun

Braun had a photography department as early as the mid-1950s, where Ursula Sehring, Marlene Schneyder and Ingeborg Kracht, later the wife of Dieter Rams, worked. The department developed an approach to black-and-white photography of interiors and products that was as matter of fact as possible. Braun's new photographers employed a non-overlapping representation of the objects and used larger apertures for a creative combination of focus and blur that obscured the background.

Wolfgang Schmittel initially had more of a personal interest in photography and became a specialist in shooting jazz concerts. His highly emotional jazz photographs were also used in Braun advertisements and on trade fair stands as well as in exhibitions. These elements were used to juxtapose the consistently matter-of-fact product illustrations.

New Concepts

The 1980s and 1990s were a time of upheaval and experimentation, even in communications design. A filtered version of what was first perceived by a cultural elite as a post-functionalist view of the world – colourful, trippy and vague, a gaze that focused on the enigmas of art – now reached marketing departments and corporate boards. Braun was likewise ready to try out all things organic, decorative and emotional. A new approach to advertising was the first step towards a change in the design mindset, and once Dieter Rams left the company, product design also shifted in favour of greater emotionality. Ads and packaging were now illustrated with all sorts of subjects, ranging from fresh vegetables for household appliances to women's legs for epilators – compared to the previous packaging design, this at times constituted a complete about-face. Certain elements, such as the matter-of-fact typography, remained, but this was no longer a Braun brand identity.

Today we are witnessing Braun product packaging return to its roots. It is once more designed by the in-house Design Team and places the products and their primary innovations firmly at the centre.

↖
A member of the advertising department has to 'lend a hand' when Wolfgang Schmittel wants to photograph a product 'hands on'.

→|
The film and photo studio of the advertising department in Kronberg grew to a considerable size, earning the nickname 'Klein-Geiselgasteig' after the giant film studio near Munich. The new American finance managers at Braun were sceptical as to whether it was all really necessary.

Tonarmwaage

Tonearm scale, 1962, Dieter Rams

If you are very particular about the weight of your tonearm and did not trust the dials on your turntable, you would need this little device. Even if used on other competing turntables, the bright yellow logo still advertises Braun.

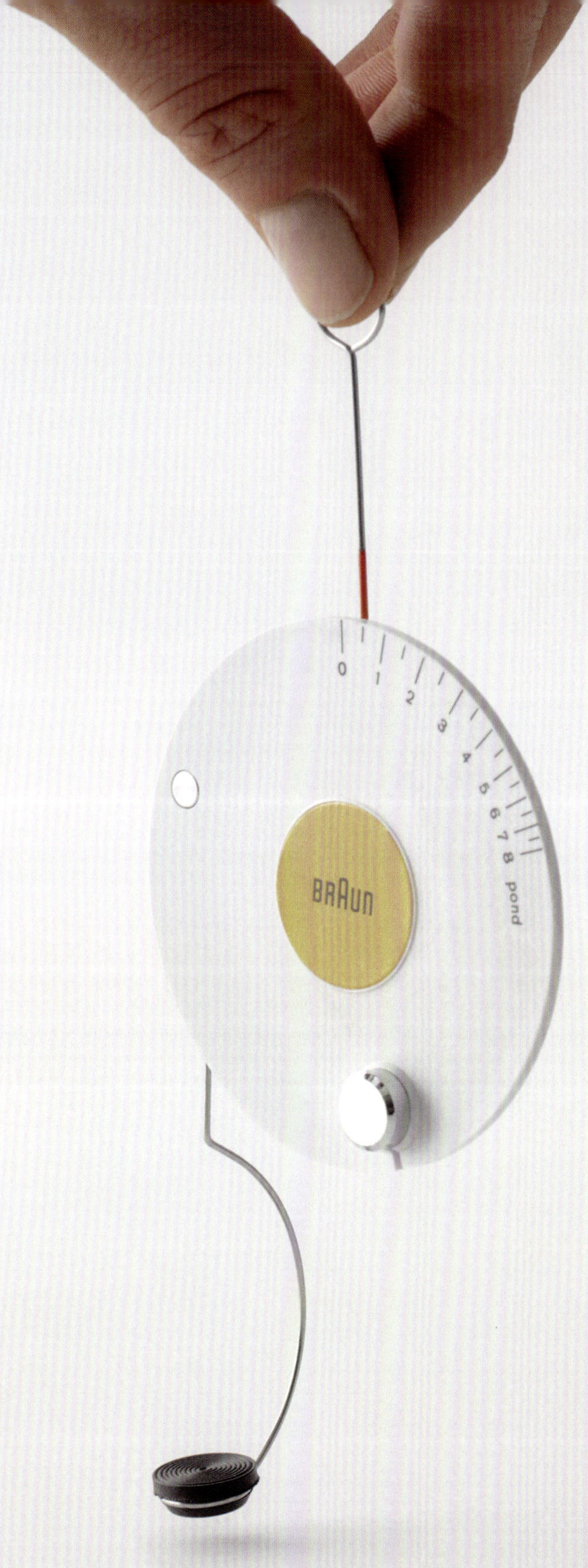

audio 1

Compact unit, 1962, Dieter Rams

The new transistor technology meant that the SK 4 could be advanced further to create this flat appliance. As the loudspeakers were now separate, it could be made entirely in metal and later in Perspex. This produced a new form which continued to be made at Braun until 1977.

studio 1000

Hi-Fi unit (CE 1000 receiver, CSV 1000 amplifier and PS 1000 record player), 1965, Dieter Rams

This audio unit is a 'Grand Design' in the Hi-Fi field highly ranked for both its technology and design. It came at a price, however: if you added the L 1000 loudspeakers, you would have paid the equivalent of a Mercedes in 1965.

Wall-mounted Hi-Fi unit: TS 45, TG 60 and L 450

Controller, tape recorder and flat loudspeaker, 1964 & 1965, Dieter Rams

Hanging Hi-Fi on the wall like a picture was wildly innovative in 1965. It referenced the visual and artistic qualities of the devices. Later, other manufacturers would gladly and successfully copy this idea.

sender
lang
mittel
kurz
khz
mhz
ukw
autom.
band mono
band stereo
lautstärke
balance
tiefen
hohen
BRAUN
netz

KH 1000

Stereo headphones, 1967, Reinhold Weiss

Boasting an excellent design and innovative technology, these headphones were created to match the 'Grand Design' of the studio 1000 Hi-Fi.

2 x 400 Ω
BRAUN

KMM 1

Electric coffee grinder, 1965, Reinhold Weiss

This grinder is the perfect translation of a two-dimensional sketch into a three-dimensional object with excellent proportions. This is where the architect merges with the product designer.

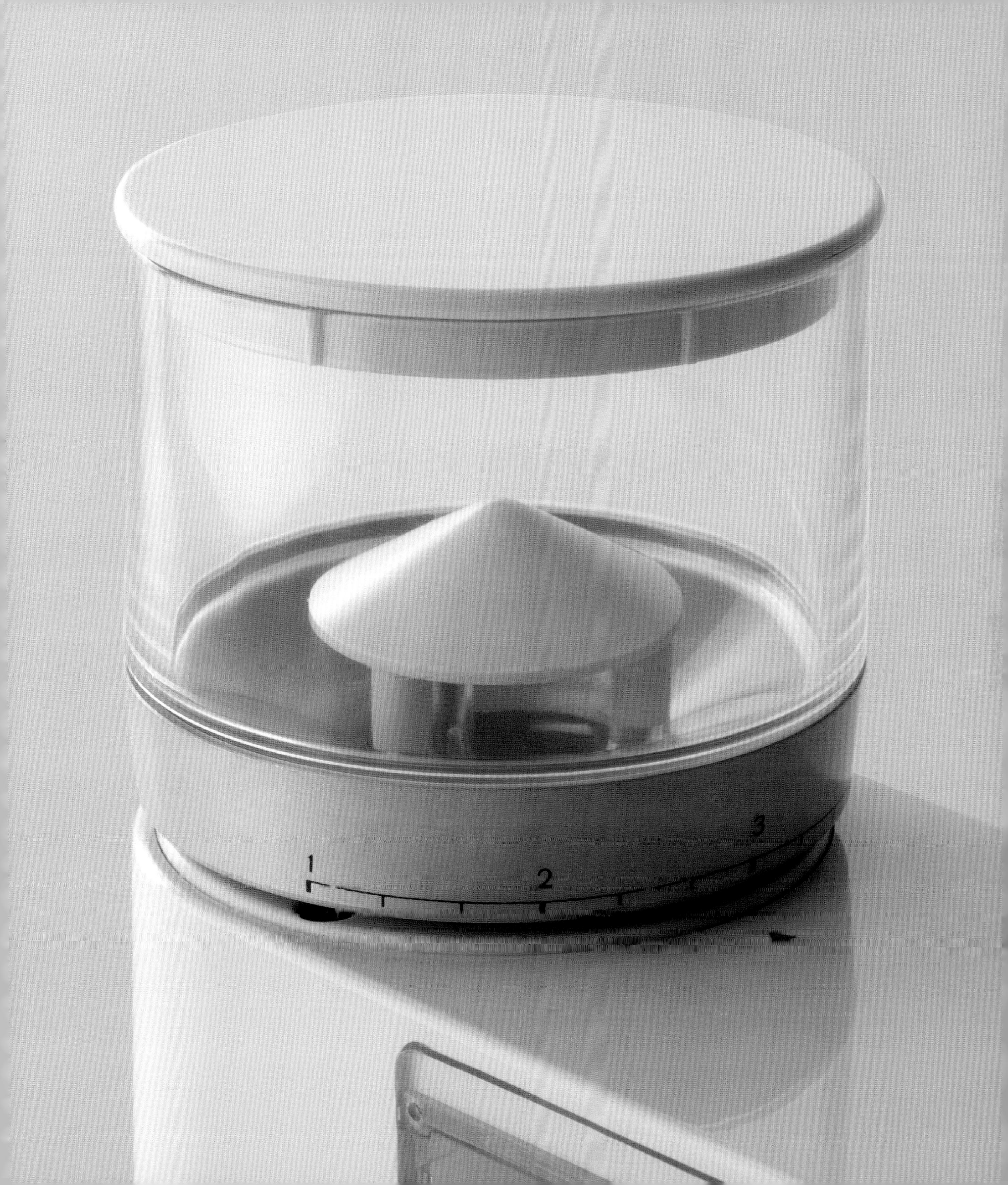
1
2
3

BRAUN
BRAUN

KSM 1

Electric coffee grinder, 1967, Reinhold Weiss

This type of coffee grinder is popular in Germany as it can also be used to grind nuts and spices. And the zeitgeist of the late 1960s called for colourful kitchen appliances.

TG 1000

Tape recorder, 1970, Dieter Rams

This second-generation tape recorder was essentially a semi-pro tape system with outstanding technical features. And the design reflects both the slightly older studio 1000 devices and the regie systems of the 1970s.

0000
pause
netz
computer
4,75
9,5
19
rücklauf
stop
start
vorlauf
aufnahme

06
'The Times They Are a-Changin'

In 1967, Braun became part of the global Gillette group. Initially, the American corporation allowed its Kronberg subsidiary to continue its successful design strategy in the hope that this would have a positive effect on Gillette's own core brand and its other group member companies.

This did not succeed across the board, but Braun was able to remain true to its design approach into the early 1990s. After that, management expected the company's product design to exhibit more emotionality in its form and appearance.

T 2 / TFG 2 1968

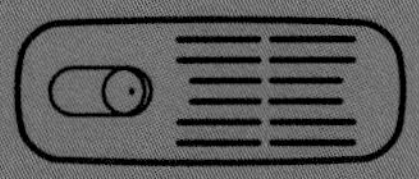

HLD 4 1970

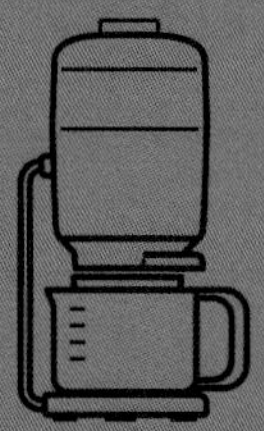

KF 20 1972

audio 308 1973

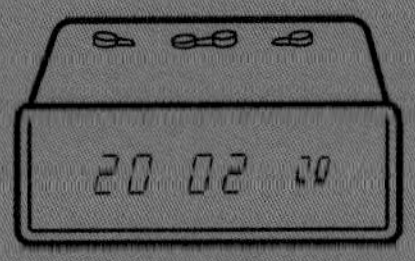

functional 1975

regie 550 1976

DW 30 1978

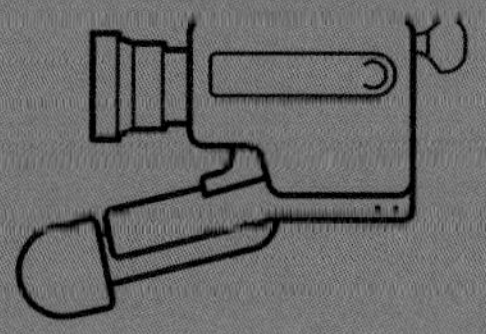

Nizo 7 integral 1979

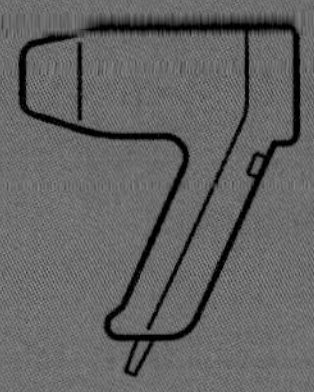

PGS 1200 1982

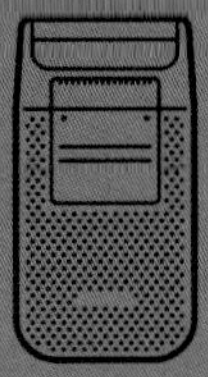

micron plus universal 1982

atelier 1984

KF 40 1984

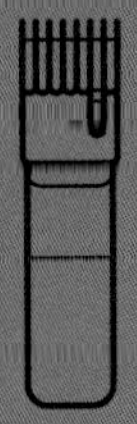

exact universal 1986

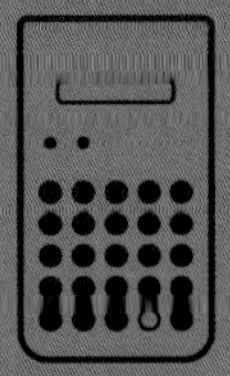

ET 66 1987

AB 1 1987

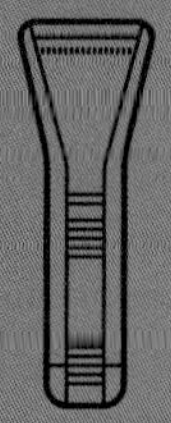

Lady Braun style 1988

New Big Mother: Braun and Gillette

The 1960s and 1970s saw a considerable surge in growth for the Braun business. With the takeover by Gillette in 1967 it became part of a US company with global operations. In the mid-1960s, Braun was at its peak in terms of business, technical and design success: the company reached growth rates of around twenty per cent, obtained market leadership in shavers that extended beyond Germany, produced a Hi-Fi range that was superior to all those offered by German rivals technically and design-wise, included a photo-film division that was held in high esteem and made kitchen appliances that could be found in almost every household, and not just in Germany.

Braun had established a new, contemporary design standard that soon found imitators. Nevertheless, there were both economical and highly personal reasons to sell. Erwin Braun's biographer Hans Wichmann stated in 1998: 'Ultimately, Erwin Braun wasn't happy in the world of management, the pomp of big money, the restless, noisy bustle with its constant demands and emphasis on the wrong things.'[1]

For some years, Erwin had spent periods on his farm near Lucerne, practicing organic agriculture. He had already retired from the management board and day-to-day business when the stock corporation was founded in 1961, and on 1 November 1965 he also withdrew from the supervisory board, handing over the chairmanship to Artur. Erwin did not become a full-time alternative organic farmer, but remained an entrepreneur with many ideas, operating for some time a 'Braun 2.0', a new company of sorts in Switzerland.

Braun at Big Mother Gillette

Braun expanded considerably from 1961 onwards. It is unclear, however, whether this was initially in preparation for the sale or whether the expansion, as was officially stated, made the sale necessary. There were certainly business reasons that made a sale in this phase appear sensible. The radio and Hi-Fi markets were now characterized by increasing competition with a growing number of international players. Above all, Japanese manufacturers were busy entering the European market with high-quality devices that were significantly cheaper than those produced locally. Sony acquired German radio makers WEGA and continued the latter's cooperation with Studio Esslinger, which was later renamed frog design. The competition from Japan also meant that European industry needed to make ever greater technical development efforts to keep up. Braun management was thus faced with the decision to either eliminate high-cost product divisions (which would, without question, have been the electronics sector) or to team up with companies with deep pockets.

In only sixteen years – from 6 November 1951, the date of Max Braun's death, to 19 December 1967, the day the company was sold – sales had risen from 14 to 276 million Deutschmarks, and the payroll had increased over the same period from 800 to 5,700 employees. Though Braun was successful, the major international competitors booked sales of 1–2 billion Deutschmarks and could thus lock into entirely different development resources and opportunities. Takeover negotiations with Bosch and Siemens in Germany and Canon in Japan ultimately failed due to lack of interest on the part of the potential buyers. In the end, in early 1967, a takeover offer was made by the much larger American company Gillette.

A Braun AG press release of 18 October 1967 reported Gillette's offer to acquire a majority stake in Braun for $50 million to be paid in Gillette stock and cash. The press release stated that Gillette could 'provide valuable assistance in further developing the world market' through its distribution network in 145 countries. At the time, Gillette had 14,000 employees and sales of $400 million, or about 1.6 billion Deutschmarks.[2]

Both Braun brothers remained down-to-earth after the sale. Neither of them used their considerable fortune to buy an ocean-going yacht in Nice or even a thatched-roof villa on Sylt, where German big business flocked for the summer season every year. Artur invested his money wisely in real estate and founded a medium-sized engineering firm, which was located on the Braun factory premises. The office, which had up to ten employees at any given time, was primarily involved in shaver development until the mid-1970s. And Erwin could now finally involve himself in his dream field of medical development on a grand scale.

A New Man at the Helm for Braun

Vincent C. Ziegler, Gillette's highly successful CEO from 1965 to 1973, during whose tenure the company's sales rose from $340 million to $1 billion, was certainly the most important advocate

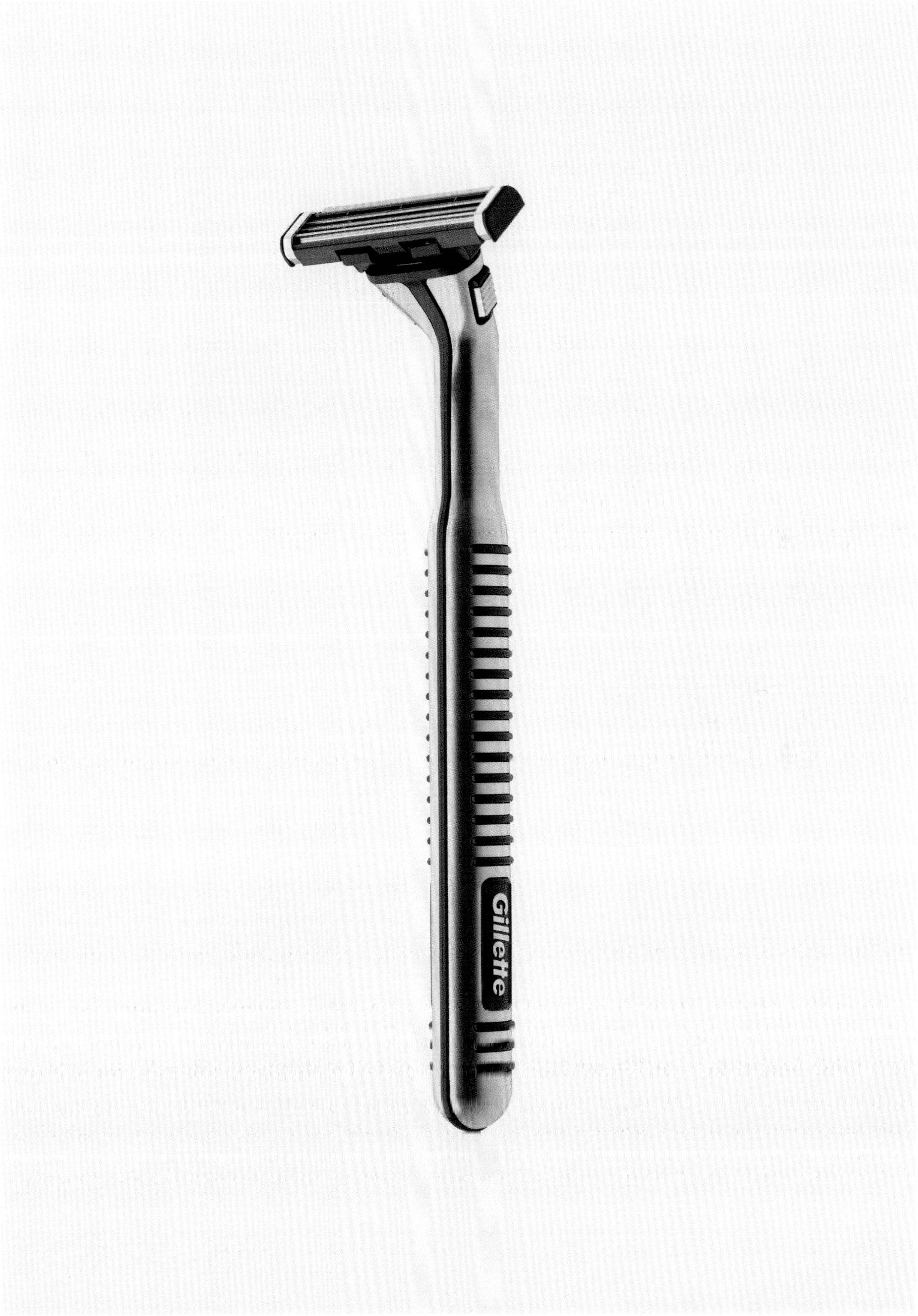

→
The Gillette Sensor was a wet shaver from 1989. Dieter Rams and Jürgen Greubel designed a new kind of grip for the Gillette blade system that used a hard-soft technology. Engineered by the Gillette R&D team in Boston, it became the most successful Gillette shaver at the time, with sales in the hundreds of millions.

for Gillette's decision to acquire the much smaller German company. As the son of a German mother, he spoke fluent German and may have had some affinity with the country of his ancestors. But he had probably also been impressed by the whole corporate culture and not least by Braun Design.

On 19 December 1967, the day the contract was signed, Ziegler wrote to the Braun workforce on paper with a letterhead reading 'The Gillette Company':

> At Gillette, we have long been aware of the good name that Braun AG has created for itself through the high quality and advanced design of its products. In fact, it was these qualities that first attracted Gillette to Braun, because we feel our image and reputation for quality are similar to those of Braun. For this reason, and many others, our two companies are an excellent fit.
>
> We intend to maintain Braun's good name and design policy, through which the company has established a very good place for itself among manufacturers of consumer goods.
>
> We are sure you will be interested to hear that Gillette will maintain all employee welfare facilities and programs and we certainly expect to be offering at least the same range of services as before.[3]

Fritz Eichler defined the goals and distribution of tasks for the design operating division for its new owner as early as February 1968.[4] In so doing, he emphasized that Braun's image required high-quality design, technology and marketing and was thus the 'entrepreneurial responsibility of the company management'. Experience had shown that it would be 'impossible to achieve the focused unified effect that characterizes the Braun image by relying on outside forces'.

From 14 to 20 January 1968, the Braun board and other senior executives, ten people in all, were invited to Boston.[5] Tours of Gillette factories, a main conference at the hotel in Boston, several meetings, factory tours in St Paul and finally an invitation to spend the evening at Mr and Mrs Vincent C. Ziegler's home were on the programme. Then it was off to Montreal for the factory tour.

Traugott Malzan, the head of the communications department, wrote in an obituary for Vincent C. Ziegler: 'Dear Vin ... turbulent years followed, mistakes were made, and the learning process was painful on both sides of the Atlantic. Yet in spite of all these difficulties, you kept your promise. You protected Braun Design from too much interference by the managers, who came and went.'[6]

Focus: Shaver Research

On 18 October 1968, a 'Braun Shaver Study Group' that included internal and external members was set up for the purpose of 'exchanging experiences on all problems of shaver research', the 'utilization of the knowledge gained in research institutes (including universities) for in-house exploitation in the development of new products' and

↖
The first Supervisory Board meeting after the Gillette takeover with CEO Vincent Ziegler, 20 February 1968. Pictured from left to right: Ziegler, Siepenkothen, Gross, Cobarg, Steuernagel, Schultz, Eichler, Steybe, Rholf, Buresch and Malzan (cropped).

↑|
The Dual Aqua shaver was designed around 1990 by Dieter Rams and Roland Ullmann. Braun's first waterproof shaver, the Dual Aqua was developed specifically for the Japanese market. It did not pass the test phase with Japanese men, however, and thus remained a prototype.

↗|
This model for a radio-controlled alarm clock, designed by Dietrich Lubs in 1977, actually needed a large reception antenna and many components. The idea existed, but the technology did not. It was not until 1984 that the first radio-controlled clocks came out in the US and Germany.

'publications in the sense of PR work'.[7] As the new parent company, Gillette knew where the company's future benefit would lie and where special efforts needed to be made. This signaled a gradual shift towards prioritizing sales figures over outstanding design achievements, which previously had primarily taken place within the audio sector.

In retrospect, Dieter Rams declared he believed that Ziegler had hoped to transfer the design ethos at Braun to Gillette, which seems to be confirmed by a 1968 interview with Traugott Malzan. Asked whether Gillette wanted to retain the face of Braun, Ziegler had said: 'We are very impressed by the good design of the Braun products. We not only want to retain it but hope that our current and future programs will profit from it, as well.'[8]

New Product Ideas

In 1970, the product design department made a few suggestions directly to its new owners as to how things might continue. In a memo for a meeting with Ziegler on 7 December 1970,[9] Dieter Rams reported that, as early as 1968, they had been working on 'functional and container units for small kitchens' as elements conceived in a uniform square grid system of 40 × 40 cm, such as a 'refrigerator, freezer, oven' or 'container unit'. Another topic was waste expenditure, 'from disposable dishes as an alternative to dishwashers in small households as well as in offices and cafeterias ... An essential, as it were, central functional unit within the system was therefore the destruction of waste. Thought of in terms of shredding, compressing, and neutralizing.'[10] The system is strongly reminiscent of Otl Aicher's failed 1958 'haus 62' project and would have been an addition to the existing household range.

The following product idea Rams promoted was more visionary and certainly novel at Braun: 'A further suggestion: temperature-controlled clothing that allows the wearer to regulate their body temperature in a completely innovative way (for example through semipermeable films). This would also offer the possibility (at least initially for leisure wear) of entirely redefining the antiquated notions around clothing, and especially men's clothing.' Today, functional clothing is widely used and not only for winter sports. Entering the fashion world, however, would have been a whole new field for Braun.

For the shavers, the proposal was to develop devices that could be 'cleaned under running water' – an idea that would indeed be realized but at a much later date. Rams proposed preparations for aftershave products, which were meant not so much to provide cosmetic fragrance as a sense of 'coolness and freshness', intended to help take away 'the feeling of a hot shaver head and mustiness caused by hair dust' during the shaving process. The problem was then solved in a technically effective way twenty-five years later by adding an automatic cleaning station for the shaver.

For the film camera product line, Rams suggested using films with a soundtrack even for smaller cameras: 'a compact sound/film camera unit with radio microphone would be conceivable as a top range camera.' In terms of slide projectors, 'special products' were suggested, 'for example, miniature projectors for Kodak film strips ... without frames' or 'constant projection of 4–5 slides in homes, exhibition spaces, etc.' A projection instead of wallpaper – this was an idea that half a decade ago essentially anticipated the media walls (large screens) of our day. And finally, modularity was once again invoked: 'all the more I would like to express the hope that in the next project, where the starting point is to replace the single device, namely the world receiver T 1000, a mobile, compact, coordinated program of devices will be delivered with the emphasis on reception performance.'[11]

Changes in Management

In the early 1970s, Braun might have reached its creative peak, but its business performance faced difficulties. Recognizing the need for improvement, Gillette took proactive measures and implemented changes in the executive board to bolster business numbers. The entire board, excluding Albrecht Schultz, was replaced. This strategic decision aimed to bring in fresh perspectives and expertise to drive positive results. Additionally, on 1 May 1973, Fritz Eichler transitioned to the supervisory board, contributing to the overall efforts to enhance business outcomes.

In 1971, former Langnese boss Hans-Wilhelm Hermann became the new chairman of the executive board but he left prematurely in April 1975. Corporate profits had fallen dramatically from 14.3 to 2.2 million Deutschmarks for the domestic group, which also significantly reduced overall revenue for Gillette. The weekly Die Zeit stated in May 1975:

> Ever since American company Gillette bought into Braun in 1967, staff changes in the managerial positions have been constant. Even in the first few years, the Americans made it clear that they were king of their castle. And so the foreign branches were staffed with Gillette managers, board members in Kronberg had to leave or, like Board member responsible for electronics Karl Buresch retired voluntarily ... Of the eleven seats on the Executive Board, four are currently unoccupied.[12]

Rudolf Gros, who had served as the vice chairman since 1971, underwent a change in his position when Gillette appointed Dr Hans-Wilhelm Hermann as the new chairman. As a result, this decision led to Gros's resignation at the conclusion of the 1973–74 financial year.

It's Over and Out for 'Radio Braun'

On 1 January 1975, Gillette was able to end the 1954 Ronson contract, which allowed only Ronson to sell Braun shavers and their licences for North and South America, the United Kingdom and the entire Commonwealth, one year ahead of the original expiration date. It became increasingly clear that Gillette was only interested in the lucrative shaver business and wanted to spin off the other product areas. To this end, according to the editor of Die Zeit, negotiations were being held with Sony about the electronics sector, but they decided to acquire WEGA instead.[13]

In 1975, Braun opened a plant in Carlow, Ireland, to manufacture alarm clocks and lighters there at lower costs. In 1977, the prototype of a radio-controlled clock went on show at the Domotechnica trade fair in Cologne. Simultaneously, loss-making divisions were increasingly sold off at the end of the decade. Braun in Kronberg, Frankfurt and Munich closed its loss-making high-tech divisions: in 1974, the last slide projector was brought to market; in 1976, the last film camera was launched, followed in 1979 by the last film projector; in 1981, the last sound film camera was marketed; and in 1983 the last flash unit hit the market. At the end of 1981, the Hi-Fi line was spun off into Braun Electronic GmbH together with the US company a/d/s/. All devices were already being manufactured in Japan and by other Asian manufacturers. The termination in 1990 of the Hi-Fi segment with the 'atelier last edition' system was spectacular: it was promoted like a limited-edition artwork by an advertising campaign that cost millions, before being sold by more than 250 select traders across Germany.

A Disappointed Erwin Braun

Erwin Braun, who had never lost his interest in the Braun company, was deeply disappointed that the core business had been abandoned. He wrote about this in a 1990 letter to the Max Braun Circle,[14] an association he had founded and financed to regain the interpretive sovereignty over Braun's history and perhaps future:

> In the 1967 negotiations and explicitly in the contract, the contracting party GILLETTE made the continuation and maintenance of all divisions of Braun the basis of the negotiations ... Under this condition, 'Radio Braun' had been entrusted to the wet competition [the wet shaver manufacturer, Gillette] – after all, the fate of the S-Division[15] also had to be considered. 'Radio Braun' was then unfortunately run down – but was still until now the epitome of European industrial culture ... now I must endure that exactly on the 100th birthday of Max Braun, living European industrial culture is being disposed of with 'cunning and trickery' (!).[16]

According to its articles of association, the Circle was founded to 'try to influence the company's reputation and corporate sponsorship'. That was a clear message to Gillette. For what would have been Max Braun's 100th birthday in 1991, Erwin planned an extensive book publication with Otl Aicher as designer, Tom Rautert as photographer and Rudolf Schönwandt as author. The project, however, never happened due to disputes.[17]

Navigating Changing Tides

There were, however, undoubted technical and economic reasons for shuttering the radio division, including growing competition from Asia and significant innovations in the technology itself. Hi-Fi and video technologies from Japan more than matched the quality of European products but were considerably cheaper and manufactured more quickly. The strategic realignment of the company with its focus on a few core areas made sense and proved to be economically successful. In 1984, for example, sales amounted to more than 1 billion Deutschmarks, seventy-five per cent of which came from exports. By 1992, that figure had risen to 2 billion Deutschmarks and in 1997 to $1.7 billion, with an operating profit of $304 million.

Albrecht Schultz reported in a 1984 lecture:

> In the past 15 years, total sales of the Braun Group have increased three-and-a-half-fold from 278 million Deutschmarks to 968 million Deutschmarks. Domestic sales accounted for 250 million of this in the past fiscal year. Today, Braun employs about 7,700 people worldwide, including about 4,000 in Germany. 60% of all Braun equipment is manufactured in Germany.[18]

Braun had eight factories worldwide: Kronberg (parts manufacturing), Marktheidenfeld (household appliances), Walldürn (shavers), Carlow (personal care), Barcelona (household appliances) and specific market-related factories in Mexico, Brazil and Argentina respectively. Sales were now booked in more than 120 countries thanks to Gillette's distribution structure.

New Builds

In Kronberg, structural extensions were already being made in the 1970s. The so-called 'Black Star', a four-wing building around a central structure with a dark facade and large canteen, was a striking feature. This 'Germany Headquarters' (for the German market) was assembled entirely from prefabricated parts and the facade consisted of black fair-faced concrete with granite chips. No external architect was hired: instead, Dieter Rams, who had also been involved in the concept for the high-bay warehouse built in Marktheidenfeld, developed the construction together with a prefabrication company. The initial idea was that if further extensions were needed, new 'stars' would be docked, but this concept ultimately failed because the materials changed and the 'stars' could not be duplicated.

Market Leader

Braun was the global market leader for shavers with foil systems and remains so to this day.
In Germany, it held a sixty-seven per cent share of the market for all men's shavers, a seventy-five per cent share of the market for compact food processors and a seventy-seven per cent share of the market for battery-powered alarm clocks. In 1989, sales of Braun shavers in Japan were greater than those in Germany, and as early as May 1988, the 100 millionth Braun shaver was manufactured in Walldürn. In 1999, the Syncro system for shavers with its cleaning and charging station was launched. Its cleaning fluid cartridge set a new standard of hygiene and both the shaving blade and cleaning fluid needed to be replaced or refilled with use, necessitating their regular repurchase. In 1996, the company took over the American firm ThermoScan, a manufacturer of thermometers, yet did not initially redesign the devices in a way typical for Braun.

↗
Team meeting on the RS 1 and PC 1 Hi-Fi components, designed in 1977 by Dieter Rams and Peter Hartwein. From left to right: Dieter Rams, Robert Oberheim, Roland Ullmann and Ludwig Littmann (Peter Hartwein and Peter Schneider not pictured).

→
Roland Ullmann's workstation, c. 1975.

Dieter Rams: Ten Principles for Good Design

'I expressed in ten principles the basic considerations that determine my work as a designer, and which represent the main features of my design philosophy. However, they are not intended to be, nor can they be, regarded as incontrovertible, because the ideas of what constitutes good design continue to evolve – just as technology and culture also advance.'

Dieter Rams formulated his Zehn Thesen für gutes Design (Ten Principles for Good Design) – note they are principles and not set rules – during his time as a designer at Braun and Vitsœ, not least to define the in-house design approach of the Braun company. From 1975 onwards, he gradually developed, expanded and added to them.

Dieter Rams first mentioned six principles in February 1976 at an international marketing meeting. These were as follows: '1. For us, the function is the starting point and goal of every design, 2. Experience with design is experience with people, 3. A design only makes sense for us if it has organizing lines, 4. Our design seeks to feature all the individual elements in the right proportions, 5. For us, good design means as little design as possible, 6. Our design is always innovative because people's behavior is constantly changing.'

It was in 1985 that Dieter Rams articulated what were ultimately ten principles for the first time in a lecture during the ICSID Congress in Washington. He added the following four principles: 'Good design means a long life, good design means consistency down to the smallest detail, good design means respect for the environment and good design means as little design as possible.' The last one is undoubtedly a response to the overflowing nature of postmodernism in the mid-1980s but it also represents a central aspect of Dieter Rams's work in general.

01
Good design is innovative

The possibilities for innovation are not, by any means, exhausted. Technological development is always offering new opportunities for innovative design. But innovative design always develops in tandem with innovative technology, and can never be an end in itself.

02
Good design makes a product useful

A product is bought to be used. It has to satisfy certain criteria, not only functional, but also psychological and aesthetic. Good design emphasizes the usefulness of a product whilst disregarding anything that could possibly detract from it.

03
Good design is aesthetic

The aesthetic quality of a product is integral to its usefulness because products we use every day affect our person and our well-being. But only well-executed objects can be beautiful.

04
Good design makes a product comprehensible

It clarifies the product's structure. Better still, it can make the product talk. At best, it is self-explanatory.

05
Good design is unobtrusive

Products fulfilling a purpose are like tools. They are neither decorative objects nor works of art. Their design should therefore be both neutral and restrained to leave room for the user's self-expression.

06
Good design is honest

It does not make a product more innovative, powerful or valuable than it really is. It does not attempt to manipulate the consumer with promises that cannot be kept.

07
Good design is enduring

It avoids being fashionable and therefore never appears antiquated. Unlike fashionable design, it lasts many years - even in today's throwaway society.

08
Good design is consistent down to the last detail

Nothing must be arbitrary or left to chance. Care and accuracy in the design process show respect towards the consumer.

09
Good design is environmentally friendly

Design makes an important contribution to the preservation of the environment. It conserves resources and minimises physical and visual pollution throughout the lifecycle of the product.

10
Good design is as little design as possible

Less, but better – because it concentrates on the essential aspects, and the products are not burdened with non-essentials. Back to purity, back to simplicity.

T 2 / TFG 2 cylindric

Table lighter, 1968, Dieter Rams

With its striking push-button and availability in countless different materials, surface finishes and colour versions, the cylindric lighter was one of Braun's most successful products.

BRAUN
20.00
BRAUN
20.00

DW 30

Wristwatch, 1978, Dieter Rams & Dietrich Lubs

After the mid-1970s, Dieter Rams and Dietrich Lubs designed the models for Braun's digital watches together (seen here on the left). This elegant men's watch was their second model to go into production.

HLD 4

Hair dryer, 1970, Dieter Rams

A tangential blower made this dryer's shape possible, with the air intake and out-vent both on one side, allowing the device to be held on the other side. And around 1970, a Braun product was also colourful for once.

sprint

Electric shaver, 1977, Roland Ullmann

This battery-powered travel shaver was marketed as an affordable second shaver in black and with a slot-in case. The coloured versions seen here never made it into production.

Pop Art and Minimalism, Functionalism and Anti-Functionalism

Pop art was a major art trend in the 1960s, especially in the United Kingdom and the United States. It was colourful, opulent and realistic and its themes were lifted from the everyday world or from advertising. It was as affirmative of consumption as it sought to be critical. It was often ironic or erotic – or both at once – and was deliberately ambiguous. This was a turn away from abstract and non-figurative painting. The techniques of Pop art, such as silk-screening, had thus far not been part of an artist's usual repertoire, while the use of consumer goods as art objects (the stacking of boxes of detergent, for instance) took the readymades of previous avant-garde movements one step further. The leading protagonists of Pop were Richard Hamilton and Peter Blake in the United Kingdom and Roy Lichtenstein and Andy Warhol in the United States.

Pop Art and Minimalism

Pop art was popular. It could be profound but also just as easily perceived as superficial. Ultimately, together with new trends in music and fashion, it became part of 1960s youth culture. This had an impact on design, too, which became more colourful and lively. Even Braun offered its cylindric table lighter in red, blue and orange between 1968 and 1970, the KMM 2 coffee grinder came out in bright red in 1969, the HLD 4 hair dryer in red, yellow and blue in 1970 and the KF 20 coffee maker in seven colours in 1972 instead of the tried-and-tested light grey. Nevertheless, apart from these exceptions, during a short period Braun was more or less 'Pop-resistant'.

Braun Design had an affinity with a separate artistic trend at the time, but for different reasons. It was with the Minimalism of such artists as Donald Judd, Carl Andre, Sol LeWitt or Ulrich Rückriem, with their geometric structures geared towards logic and endeavours to de-personalise art, that Braun connected.

Design Without Art

Over the course of the second half of the twentieth century, industrial design for the most part emancipated itself from art and the artists who focused firmly on formal criteria. This was a long process often accompanied by fierce debate. While the Arts and Crafts movement[1] in England and the Bauhaus in Germany sought to fuse art and design, from the late 1950s onwards, the HfG Ulm sought to do away with art altogether, attempting to kill off the floral wallpaper and the kidney table, the porcelain décor and the tail fins on cars. A systematic and objective approach to the design process was thus founded that is now widely used as a matter of course.

Today, art is no longer the mother of industrial design but, at best, its distant cousin.

Braun 2.0

Even after retiring from the Braun GmbH supervisory board, Erwin Braun remained connected to 'his' Braun company, both in business and spirit, and developed a 'Braun 2.0' of sorts, a smaller, parallel world to the large company in Kronberg that had been sold.

Erwin had relocated to Switzerland as early as 1966. Together with Professor Aehnelt from Hanover, he addressed topics such as the soil quality, plantings and fertility of domestic animals on his farm. But being wrapped up in the alternative rural life was not his thing – the forty-five-year-old was still too much of an industrialist and bubbled over with ideas. His entrepreneurial activities, however, were now focused primarily on his old passion: healthcare.

Collaboration with Manfred von Ardenne

In early 1971, Erwin Braun worked together with the private research institute of Manfred von Ardenne in Dresden. Von Ardenne was an ambivalent character in East Germany among other places. His curriculum vitae reads accordingly: he dropped out of university, then became a successful radio and television pioneer in the 1920s and 1930s, during the Third Reich he was involved in atomic research, after the war he worked towards the development of the Russian atomic and hydrogen bomb, was awarded the Stalin Prize and, upon returning to East Germany, founded his own institute in Dresden with 500 employees. In the early 1960s, he turned his attention to medical care, most notably cancer therapy. Braun and von Ardenne's partnership resulted in two former 'radio guys' thus coming together in the field of medical technology.

For many years Erwin provided considerable financial support for von Ardenne's work. The financing of a 'Manfred von Ardenne Foundation for Oxygen Physiology GmbH' in West Berlin was also considered. In retrospect, however, there was some discord, and Erwin Braun gradually disengaged from the project: 'I signed off from the oxygen front in November [1991] and am requesting an extension of my leave.'[19]

Erwin Braun and Braun

On 5 December 1961, Erwin and his brother founded the company Interelectric AG in Sachseln in Switzerland, where, under the direction of former Braun designer Bodo Fütterer, the shaving foils for shavers were manufactured, an in-house development department was set up and long-term contracts with Braun were concluded. Initially Erwin Braun owned the company together with his brother Artur but later took it over as sole proprietor. After Gillette's acquisition of Braun, the patents were soon sold to Gillette, and Interelectric dedicated itself to the production of devices for preventive medicine, life-saving devices and machines that gently grind natural products (such as the Power pack elements system). There was also the project for a self-sufficient autarc system consisting of a crank generator and rechargeable battery with plug-on devices for shavers, lamps, a radio, a Geiger counter or a walkie-talkie. A small shaver was developed, the nineteen patents for which were also sold to Gillette in 1967 due to a lack of sales opportunities. However, neither this nor the radio combination system ST programme of 1964–65, which was developed within the framework of autarc, ever went into production.[20] Later on, the Braun Design Team would work on a similar modular project for kitchen appliances that was also never realized.

Heading for New Shores

After discontinuing the production of shaving foils for Braun, the development of miniaturized DC motors with iron-free rotors became the focus of Interelectric AG's activities, with Bodo Fütterer still leading the charge as head engineer. In 1976, the company achieved a commercial breakthrough with these miniaturized DC motors: even the NASA moon and Mars robots were – and still are! – equipped with these drives. In 1999, the enterprise was renamed MAXON Motor AG, an allusion to 'Max's son'. Erwin's son Karl-Walter Braun is now the owner of the company that has, since 2019, gone by the name of 'maxon'. Also in 2019, maxon achieved a turnover of 557 million Swiss francs.

Back to Medicine

Nevertheless, Erwin Braun's primary focus in Switzerland was on medical devices. At his institute in Engelberg he collaborated with Gregor Gorges, Rido Busse, Karl Dittert, Herbert Hirche and Peter Raacke – almost all of them

old confidants from his time in Frankfurt. Today, his commitment to medicine is continued by his younger son, Werner.

The company's design partner was gugelot design, established by Hans Gugelot in 1965 and continued by his wife Malke as a limited company after his death. Erwin Braun was the capital investor. At times the studio, with its seventy employees, was the largest design outfit in Germany. Among other commissions, there were numerous orders from Erwin, including designs for kitchen appliances for hospitals, that were never implemented. In 1988, the 450 W warming infrared lamp was initially created for the treatment of skin cancer. Designed by Werner Braun, Karl Dittert and Rido Busse, it offered an infrared source with a spectrum like that of the sun, but without harmful UV radiation. The 450 W is still manufactured in an advanced version by the hydrosun company and is sold for therapeutic purposes.

Erwin bought up the Ritterwerk GmbH in Pasing near Munich as early as 1969. The company also produced small electric kitchen appliances and had been working with designer Karl Dittert since 1965. During his studies, Dittert encountered Bauhaus graduates through his teacher Karl Warnecke in Schwäbisch Gmünd, which undoubtedly had an influence on his functional design approach. Since the Pasing plant primarily produced cutting equipment, Erwin would have seen the company as a possible partner for his raw-food-orientated health projects in Switzerland. Throughout his life, he remained connected to design in general, and Braun products in particular, and he was always informed about what was going on at Braun, especially through Claus C. Cobarg.

↗
450 W

Infrared hyperthermia emitter, 1988, Karl-Walter Braun, Karl Dittert & Rido Busse

Originally developed for treating skin cancer, this device is still produced to this day by hydrosun for therapeutic purposes.

A Design Prize That's Really Worth Something: The BraunPrize

In 1967 the Swiss architecture magazine Bauen + Wohnen announced a new design prize:

> The Braun AG, Frankfurt, together with the Design Circle in the BDI Federation of German Industries, has endowed a BraunPrize for technical design. This international prize is intended to support young industrial designers and engineers that have solved a technical design problem in an outstanding way. All industrial designers and engineers who are still in training or who have not practiced their profession for more than two years may apply. The maximum age for applicants is thirty-five, allowing designers to apply who have worked in a trade or other profession before their studies. The BraunPrize is endowed with 25,000 Deutschmarks. This amount can be awarded either in total or in partial amounts to individuals or groups of individuals. It is awarded every two years, for the first time in 1968. A jury will decide on the winners. The jury will be composed as follows: Chair: a representative of Braun AG; Members: a representative of the Design Circle in the BDI, a representative of an educational institution for industrial designers.[21]

The BraunPrize, initiated by Erwin Braun, remains to this day one of the few true design awards. Winners are financially rewarded and do not need to pay a participation fee or for a catalogue entry, as has become customary for many other design prizes with a more commercial approach. The BraunPrize is an explicit advancement award for young designers. A deliberate emphasis was also placed on allowing students and graduates of technical fields of study to apply from the very beginning. It was Erwin Braun and Fritz Eichler's intention not simply to create another prize for already-successful designers but to help promote young designers – and to bring a function-orientated understanding of design to the world. As an additional promotion, travelling exhibitions of the finalists were and are being organized by museums and design centres. The jury consists of a small group of professionals, allowing for intensive discussions.

1968 Launch

The first round of the competition saw 122 designers from 15 different countries apply to compete for two prizes endowed with 15,000 and 10,000 Deutschmarks respectively, a solid prize amount at the time. The first awardee was twenty-five-year-old Florian Seiffert, a student of Essen's Folkwangschule, for his design for a 16-mm film camera. Japanese industrial designer Masanori Umeda, who had already worked for the Castiglioni brothers in Milan, won second prize for a system of mobile living units. The two awardees had certainly chosen the right topics for their work when it came to wowing jury members Fritz Eichler and Otl Aicher. Florian Seiffert went on to join the Braun Design Team that same year.

Initially, the BraunPrize was awarded biennially and, later, every three years. Seiffert would not be the last designer to join Braun through the award. Braun designers Ludwig Littmann and Peter Schneider won the competition in 1972. In 1989, awardee Peter Eckart joined the Braun team for two years, and prize-winners Björn Kling and Sven Wuttig joined Braun in 1992 and 2001, respectively.

Design Is Also Worth Money

In 1980, there were twice as many applicants (274) from twice as many countries (29) as the first year, and the total prize money was increased to 35,000 Deutschmarks. Every tenth applicant was female. By 1992, the number of applicants had once again doubled. The 2018 prize had 3,087 applicants from 107 countries. The prize, or rather industrial design, went global. The prize money was correspondingly adapted. In 2021, $100,000 in total was offered as the prize money, and there are now two categories: students and designers with a maximum of five years of professional experience. The best works worldwide from both categories were each awarded $10,000 and the prize winners also had the opportunity to take part in an internship at Braun in Kronberg.

Juries

The award has always involved renowned jury members, in previous years including Otl Aicher, Kenji Ekuan, Herbert Hirche, Stephan Lengyl, Herbert Lindinger, Friso Kramer, Alessandro

↗
The 1968 BraunPrize, which came with what was at the time the immense sum of 15,000 Deutschmarks, went to Florian Seiffert (centre), who promptly joined the Braun Design Team. Second place went to Japan's Masanori Umeda (left), who is congratulated by Fritz Eichler.

↗
Between 1968 and 1989, Fritz Eichler developed the BraunPrize, originally initiated by Erwin Braun, into Germany's leading design award.

Mendini, George Nelson, Herbert Ohl and Yuri Soloviev. In 2012, new chairman Oliver Grabes brought Naoto Fukasawa and Jane Fulton to the jury team; in 2021, Virgil Abloh, Sukwoo Lee and Gary Hustwit joined. In contrast to the early days of the prize, the jury increasingly consisted of women, including, for example, Heather Martin, Vivian Wai Kwan Cheng, Lidan Liu, Rosario Costa and Alexandra Deschamps-Sonsino, as well as Cecilie Manz, Li Ningning and Johanna Shoemaker.

State of the Art (of Design)

The BraunPrize ceremony is always an event that reflects the state of current design and enjoys considerable media impact. This is due both to the contributions of the award winners and to those speaking at the event. In his opening speech, Mark Breitenberg, Provost at the California College of the Arts in San Francisco and Oakland, President Elect of the ICSID (now the World Design Organization, WDO) and jury member in the year 2009, explained:

> One of the most interesting developments in design training in recent years is a renewed interest on behalf of the current students in the traditional crafts ... We can probably agree that our quality of life, our economic strength and our ability to solve the world's toughest problems in the 21st century will for the most part be shaped by design ... One of the reasons why Apple did so well was that Dieter Rams nurtured our sensibility – he showed the world what good design should look like and how it should work ... The BraunPrize is the best organized, strictest and most renowned design competition for your designers world-wide.[22]

Nowadays, the competition is run by the Braun Design Team and supported by Procter & Gamble Corporate Design. Oliver Grabes, Chairman of the BraunPrize since 2012, describes the conception of the prize as follows:

> Today, good design must satisfy much higher demands than ever before and must successfully integrate sustainability and complex technology into people's lives and social behavior. The quality of our future depends on how well we are able to use knowledge and creativity in finding the solutions for tomorrow's challenges. In this respect the BraunPrize wants to encourage responsible action and innovation in product development and design today, to strengthen the resolve required to generate ideas that have not been thought up yet and to create product concepts that no one can imagine yet. Ideally, this is a kick-start for design talent and new ideas on their way to changing the world. The BraunPrize has evolved while remaining true to its original intent: To promote excellence in design and thinking.[23]

↑
BraunPrize 1992. (Left) Competition entries at Stadthalle in Kronberg. (Right) Jury members deliberate. Pictured left to right: Claus C. Cobarg, Yuri B. Soloviev (president of the Society of Soviet Designers Moscow), Dieter Rams, Peter Eckart, Vittorio Magnago Lampugnani (director of the German Architecture Museum in Frankfurt) and Peter Schneider.

↑|
The 2018 BraunPrize award ceremony in Kronberg celebrated the winners and 50 years of the BraunPrize together with 300 guests.

↙|
Jury discussions in the Braun Collection during the BraunPrize 2018. From left to right: Alexandra Deschamps-Sonsino, Duy Phong Vu, Rosario Costa, Lidan Liu and Chairman Oliver Grabes (jurist Stefan Schamberg not pictured).

↘|
Submissions poster in the atrium at the BraunPrize 2018 award ceremony.

BraunPrize
2018
DESIGN
FOR WHAT
MATTERS

STUDENTS
3087 SUBMISSIONS
FROM 107 COUNTRIES
CATEGORY STUDENTS: 2366
CATEGORY YOUNG TALENTS: 721

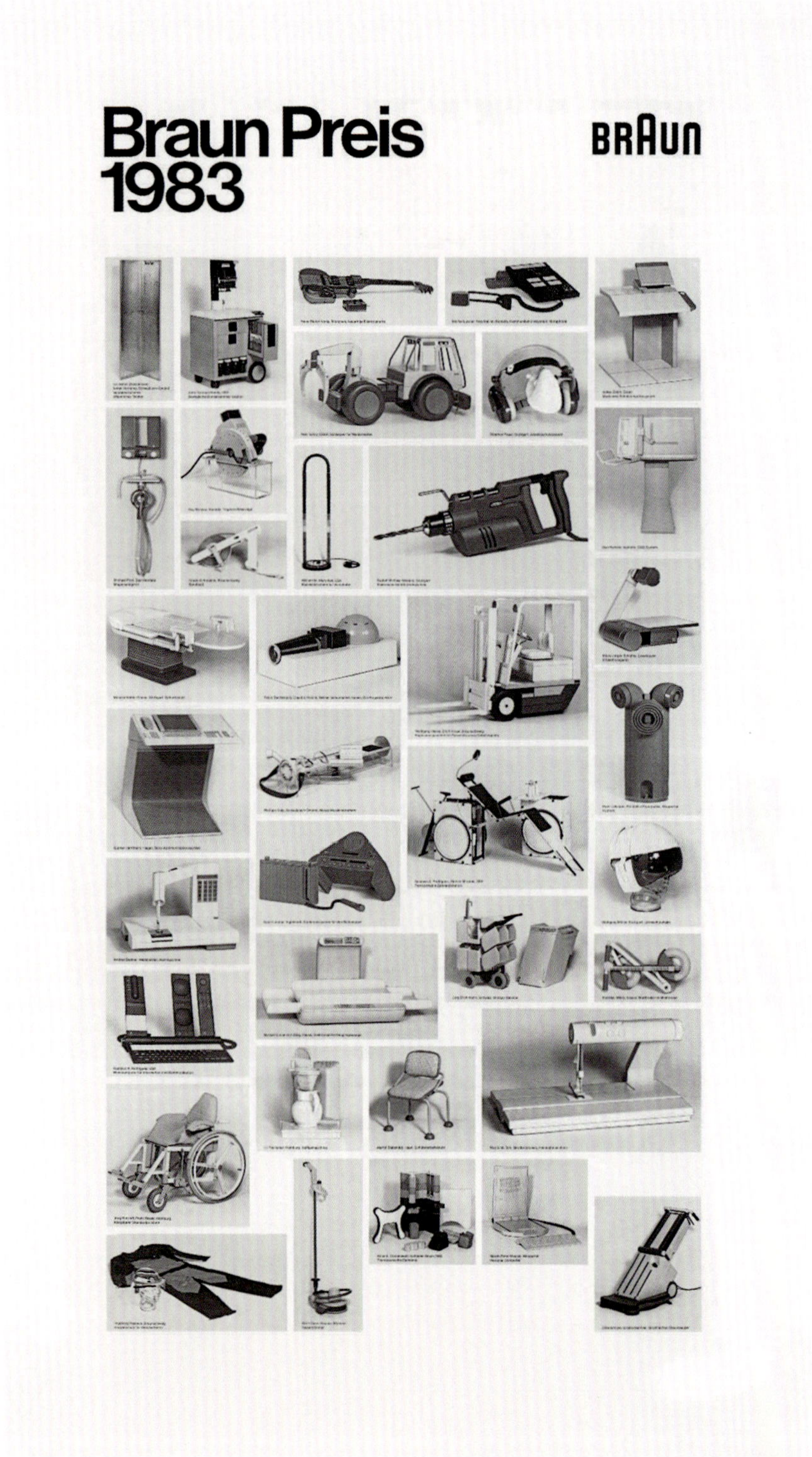

↑
Posters have been created for every BraunPrize, showing all awarded design concepts. In 1983 they were still in black and white but they later appeared in colour.

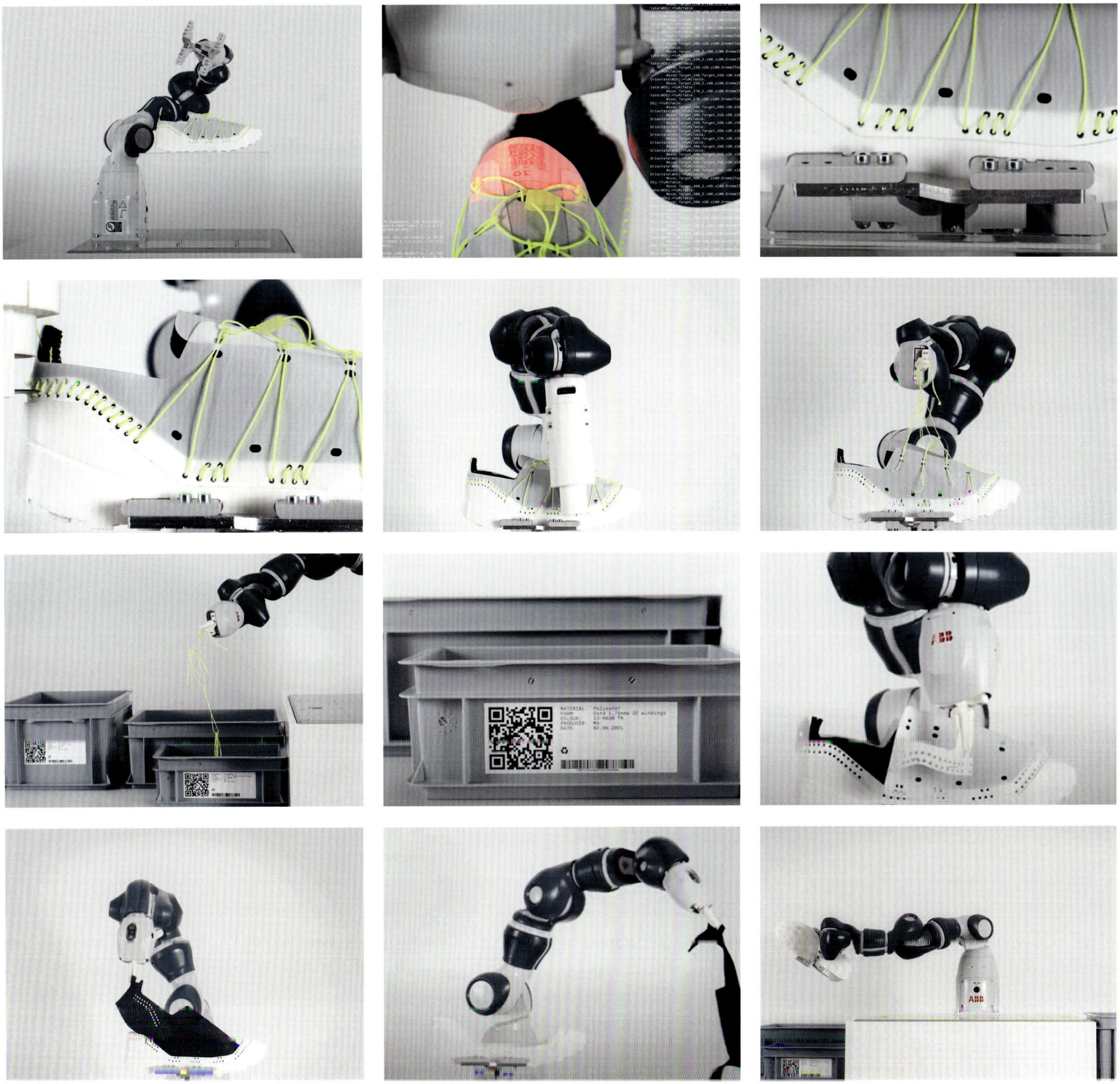

↑
RUEI-01 is a robotically recyclable shoe concept by Maxwell Ashford, a 2021 BraunPrize winner. Here, manufacturing, material and colour data is stored digitally in the product and read by the robot via g.code, enabling the robot to disassemble the shoe by type into the different materials and separate them.

KF 20

Coffee maker, 1972, Florian Seiffert

The tower-like structure with the water tank at the top, filter in the middle and jug below is logical and has a small footprint. Produced in six colours, it also paid tribute to contemporary Pop art.

KW
MW
LW
UKW
kanal
MHz
kHz
kHz
MHz
kanal
übernahme
stereo
monitor
band
phono
ukw
lw
mw
kw
afc
afn
übernahme
sendereinstellung
regie 550

regie 550

Receiver, 1976, Dieter Rams

Technically brilliant in its day and an icon of Braun Design. Over 18,000 units were sold, making regie a commercial success, too.

ET 66

Pocket calculator, 1987, Dietrich Lubs & Dieter Rams

The ET 66 is Braun's pocket calculator with the most coherent design and harmonious form. It also made design history as the inspiration for the calculator app on the first Apple iPhone in 2007. Since 2013, the device has been relaunched by the British company Zeon Ltd in both black and white.

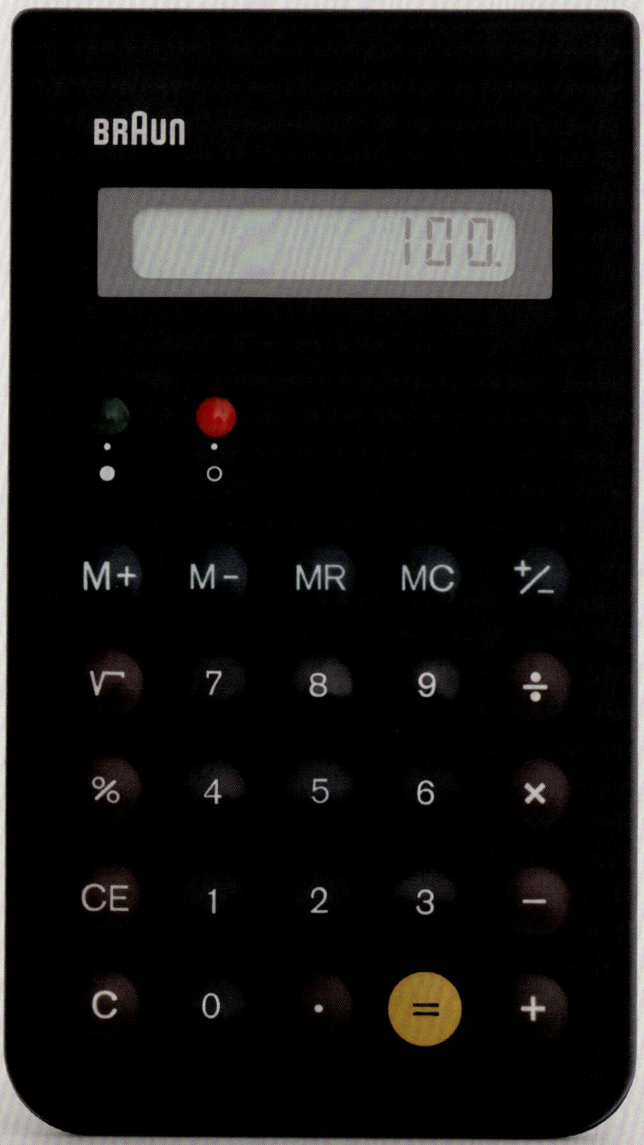

BRAUN
100.
+/-
MC
MR
M-
M+
%
CE
C

45
start
33
stop
PDS 550
fm3
fm4
fm5
fm6
tuning

slim line

Hi-Fi system with audio modules T 301 / TS 501, A 301 / A 501/ AP 701 and PDS 550, 1977–80, Dieter Rams

The last Braun audio units made in Germany. Only 6.5-cm high, the model is only half as tall as prior models and therefore gained the nickname 'slim line'. Slender, but powerful.

Lens made in West Germany
macro
UWL
ft
m
5
1,5
6
1,7
7
2
20
28
40
50

Nizo integral 7

Super-8 camera, 1979, Peter Schneider

The late cameras were given new technology and a new design. A characteristic feature is the adjusting bar that unites all switches and renders the functions visible at a glance. The new technology was not able to compete with the new video cameras for long and the last Braun camera came out in 1981.

functional

Digital alarm clock, 1975, Dietrich Lubs

0
1
48
-60

audio 308

Compact Hi-Fi set with record deck, 1973, Dieter Rams

The system angles eight degrees toward the user, hence the name 308. Plastic had become the defining element, but the visual elegance of Braun's devices was maintained.

dymatic

**Pocket lighter, 1980,
Dieter Rams**

Snug in the hand, you use your thumb to pull the activating button away from the hot flame.

micron plus universal

Electric shaver, 1982, Roland Ullmann

The hard metal shell is permeated by soft plastic grips, ensuring non-slip usage and also an appealing look. The grips evolved into a typical and unmistakable Braun detail that modern products are now referencing again.

KF 40

Coffee maker, 1984, Hartwig Kahlcke

The tower shape of Seiffert's KF 20 was retained, but Kahlcke places the water tank behind the jug so that only one heating element is required. This optimized concept became a bestseller and remains so today.

exact universal

Beard trimmer, 1986, Roland Ullmann

Lady Braun style

Women's shaver, 1988, Roland Ullmann

a/d/s/: Hi-Fi's Last Gig

Braun's core business since the 1930s, the radio and phonograph sector had already prompted a reorganization at the company in 1954 and 1955. At that time, it had grown into Braun's most difficult product area because competition from large suppliers, such as Grundig, Telefunken and Philips, had become overwhelming. Given its smaller size, Braun could not achieve real growth in the sector so long as it made devices that were formally and technically similar to those of its larger competitors. Braun thus shifted its focus into a niche: modern design. This was initially a successful strategy and allowed the company to establish itself as a Hi-Fi pioneer in the mid-1960s. While the return on investment had remained limited, Erwin Braun nevertheless knew that the company's image as one that sold high-quality Hi-Fi equipment would also benefit the lucrative shaver business. In the 1960s, the Hi-Fi sector established the Braun brand as a design role model, and it garnered a great deal of approval and won prizes, yet it was the shavers for the man on the street with a similar design ethos that actually made money. However, this situation gradually changed after the company was acquired by Gillette. From that point on, more attention was paid to the profitability of each individual product sector.

From BEL to a/d/s/

The regie units of the 1970s were still reasonably profitable, the elaborate slim line configurations less so. The atelier system issued in 1980 was a stereo system with modules consisting of a receiver, amplifier, turntable or cassette deck that could be combined in a range of ways. Initially designed by Dieter Rams and Peter Hartwein as a low-cost alternative to the predecessor models, it was also not very commercially successful. The more powerful and expensive later generations of the system did little to change this.

For this reason, in 1981 Gillette decided to spin off the consumer electronics division from Braun AG and transfer it to Braun Electronic GmbH (BEL) in Kronberg, whose shareholders were the Gillette Company and the American company Analog & Digital Systems (a/d/s/). a/d/s/ was an importer and exporter but also manufactured its own Hi-Fi systems and continued to produce the atelier systems in Asia. German American Godehard Günther, the company's founder, became the managing partner of BEL. From 1981 to 1987, the products were still sold in Germany with the Braun logo. After 1987, the company traded with the same shareholders under the name 'a/d/s/ Analog und Digital Systeme GmbH'. Yet even under the new brand, business was not great. Otl Aicher had created the corporate design for a/d/s/. The HfG Ulm archive holds his all his design work for the project, from initial sketches to the finished logo.

In 1989, Gillette bought back the partnership shares from Godehard Günther, making a/d/s/ a wholly owned subsidiary of the company. Ernst Ortmann became the new managing director, and products were again marketed under the Braun name. Ortmann wanted to launch two lower priced Hi-Fi systems for the

→|
Initially devised as an affordable alternative to the 'slim line' system, the atelier modular Hi-Fi system became ever more powerful across its four iterations. Designed in Kronberg and produced by various Asian manufacturers, at some point it could no longer compete technically or profitably with Japanese rivals. Gillette wanted to abandon the loss-making product and in doing so closed down Braun's most important design section.

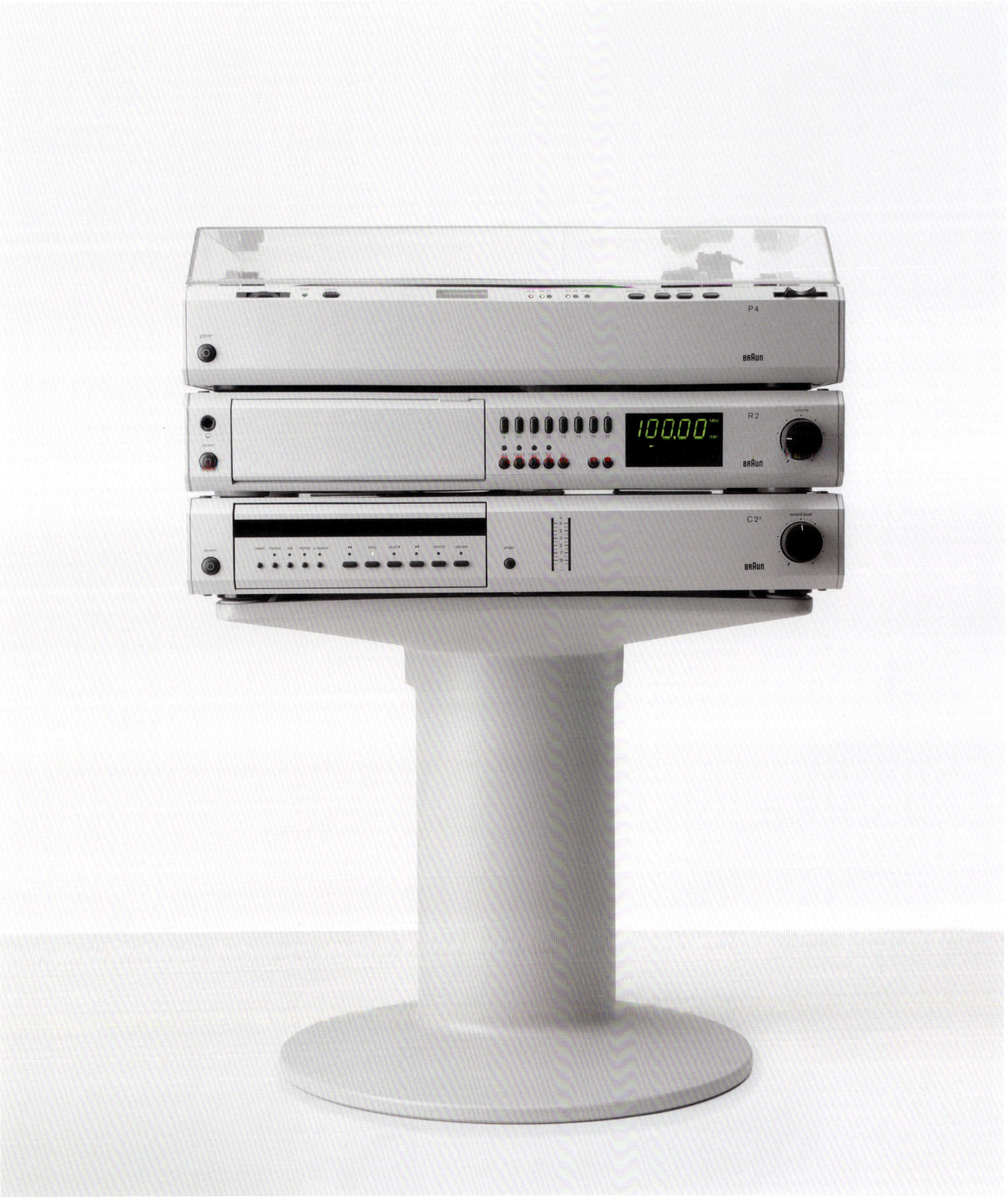
P4
BRAUN
R2
100,00
BRAUN
C2
BRAUN

lock
quartz
power
CD 2
BRAUN
record.
off
play
timer
Dolby nr
B/C
mpx
m. reset
reset
memo
set
repeat
a.space
stop
start
record
pause

atelier

Modular Hi-Fi system, 1979–89, Peter Hartwein & Dieter Rams

Details of the atelier system: slider on the cassette recorder and tonearm for the record deck.

1990 Christmas season. In May 1990, the difficult decision was made to discontinue the consumer electronics business altogether. Subsequently, in March 1991, a/d/s/ was dissolved. Unfortunately, the last business year resulted in a loss of 10 million Deutschmarks, despite a turnover of 45 million Deutschmarks.[24]

Last Edition

The end of Braun Hi-Fi nevertheless did not come quietly but with an advertising blitz that is now regarded as the textbook approach and which would cement Braun's image as a pioneer of design and technology. In 1990, the small Saarbrücken-based ad agency of Maksimovic & Partner was given the considerable budget of 2.5 million Deutschmarks to produce advertisements that were as humorous as they were original. Under the general motto of 'Braun Hi-Fi will be history on 31 March 1991', ads appeared with headlines such as 'How will you explain to your mother-in-law that you're spending the equivalent of a small car on a few black boxes that have just been discontinued?' Some 6,900 systems were sold almost exclusively in West Germany through 250 selected dealers. By October 1990, orders by the trade for all the systems had already been confirmed. It would probably have been possible to sell twice as many. More than 400 press articles, radio and TV reports contributed to the desired result of maintaining Braun's image.[25]

A gift was added to the 'last edition': a high-quality box covered in glossy black paper in the format of an LP album with minimalist typography – probably a reference to the legendary cover of The Beatles' 1968 White Album – designed by artist Richard Hamilton in a spectacularly reduced style.[26]

The inside of the Braun box[27] also refers to The Beatles albums released twenty years earlier that were as popular as they were artistically sophisticated. The inlay is completely white. The booklet, however, states: 'Braun design is much more closely related to the disciplined manual (and intellectual) efforts of the engineer than to the intuitive creation of an artist ... That's because Braun's atelier devices are conceived not as technical toys or pure design or art objects, but as high-quality and versatile Hi-Fi components.'

Der Spiegel published an article headlined: 'Stylish Exit' in 1990.[28] At a farewell reception in Darmstadt for the specialist press, Ernst Ortmann said: 'The initiative proved just how much potential there really is here. Maybe Braun will return with communication products in the future.'

And this did indeed happen, but not until twenty-eight years later.

↖
The book cassette – designed in 1990 by Jo Klatt, Ivica Maksimovic, Thomas Sauer, Michael Schneider, Dieter Rams and Peter Hartwein – was shaped like an LP and featured elaborate offset prints of Braun audio devices on cardstock and tracing paper. It was an added extra for buyers of Braun's 'Last Edition' of 1991.

↑
In May 1990, the Supervisory Board of Braun AG resolved to withdraw from the Hi-Fi sector. The units planned for that year were then issued as the 'Last Edition'. The focus was on an exit that would not be a detriment to the corporation's image as a whole. An agency was founded and commissioned for the sole purpose of managing this monumental task. Experienced ad-man Ivica Maksimovic, who had just been appointed professor at the Saar Academy of Fine Arts, brought other professionals on board: copywriter Ono Mothwurf and photographer Michael Erhart.

Daring to Be More Emotional

In 1995 on the initiative of Gillette manager Archibald Livis, a change in the management of the design department was instituted. At the age of sixty-three, Dieter Rams was 'asked' to hand over the reins to his deputy, Peter Schneider, and for the last two years until his retirement took on the role of 'Executive Director Corporate Identity Affairs' – a nice title that unfortunately did not come with a budget attached.

New products created during Peter Schneider's time that exemplified a new understanding of design at Braun were the AquaExpress WK 300 electric kettle by Jürgen Greubel and Ludwig Littmann in 1999; the Syncro system 7630 electric shaver with cleaning device by Roland Ullmann in 2000; and the Multiquick fresh system MR 5550 hand mixer by Ludwig Littmann in 2004. The design department had grown from sixteen to twenty-three people and was now also responsible for corporate identity and packaging graphics, as well as for awarding the BraunPrize, which the company has presented since 1968.

Braun opened itself up to the new zeitgeist of design theory and practice, otherwise known as the 'critique of functionalism'. The 'grand narrative', and more generally the primacy of reason, were being called into question. Design competitions had long since moved away from seeking the singular, correct and 'good forms' to which designers had previously aspired – the term 'good form' was now simply frowned upon. Many solutions, even if entirely contradictory, were the gold standard. While utility value remained at the heart of design at Braun, a greater 'emotionalization' of the products was required, with softer shapes and even some ornamentation being introduced. The design was now intended to appeal to everyone.

The changes in design at Braun provoked criticism in the world of design. Braun was under particular scrutiny since many designers felt a strong affiliation with the brand and its clear commitment to functionality and clarity. The many visual changes in both product design and visual communications from the 1990s onwards meant that Braun increasingly lost its significance within the design world.

The reasons for this were complex. Braun was responding to a confluence of these new currents in design along with utilizing the new technical possibilities for creating shapes that computer-aided design (CAD) offered, not to mention the phasing out of product categories that had brought forth a particularly large number of the most cherished Braun products.

At the same time, it is also remarkable how long the company was able to consistently maintain a reduced and clear design approach under Dieter Rams.

Against the Zeitgeist – With the Zeitgeist

A softer, rounder and more visually pronounced formal language became the spirit of the times. Philips, then under Italian head designer Stefano Marzano, successfully cooperated with lifestyle producer Alessi and, among other things, produced household appliances designed by Alessandro Mendini. Market successes by Alessi itself, by Swatch or Vitra, which collaborated with numerous well-known postmodern architects and designers, presumably also made the Gillette headquarters in Boston think twice about the previous 'Braun style'. Challenged by the new trends, Braun products were now designed to catch the eye, which opened the door for the 'superficially spectacular' from which Dieter Rams had always distanced himself. Peter Schneider, however, believed design had to become 'more democratic'[29] and regarded the fixed point-of-view as being anachronistic in a pluralistic and global design world. It was an era in which many companies, including Braun, were looking for new approaches to design. What is now subsumed under the label of 'postmodernism' was seen by many designers at the time as an expansion of the concept of function to include emotional appeals.

The same year Peter Schneider was made head of the design department, he gave an extensive interview on the future of Braun Design to Rudolf Schönwandt, a freelance writer who had previously worked in the Braun advertising department as a concept developer and copywriter. 'It is the quality of the design being experienced that makes product design credible and moves it forward' was his introductory comment. And this statement was certainly in line with the design discourse of the time: design was of course supposed to fulfil a function but also, and above all, stand for an experience. Schneider believed

↑
Braun product portfolio in the late 1990s. After Dieter Rams's departure, the design approach changed radically and swiftly. Braun seemed to turn its back on the once utterly clear designs.

design should not only be practical but also 'lovable'. Nevertheless, the products still needed to be durable, honest, unobtrusive, functional and easy to use. Innovations were to arise out of technological developments and ecological requirements. The term 'significance', which had played a role in design theory since Roland Barthes's essays of the mid-1950s, became important again, too. According to Barthes, 'every object is also a sign'[30] referring to other contexts, and humans are shaped by the objects that surround them. The design of postmodern objects was supposed to signify upbeat, light-hearted and even playful feelings and uses.

Psychological factors needed to be incorporated more strongly into design, Schneider stated. Furthermore, the new head designer explained, there was also the pressure of considerable cost restrictions. The demands made on design were intensifying: 'a sensitivity to "demographic trends" and a feel for the rapidly changing lifestyles and behaviours are key. Real-world experience and a true feeling for cultural peculiarities and self-evident facts will be necessary for our work in the future.'

These statements make it clear that Braun Design was now considerably influenced by marketing, which was also changing rapidly: consumer tests and 'global design' increasingly determined what shape and form products were supposed to take. Aspects of market psychology thus became more decisive when it came to design decisions at Braun, and consumer surveys took on greater importance in product development as part of the marketing activities of the time. In comparative tests, the company now wanted to stand out first and foremost and to present itself in a more colourful and emotional way. Compliance with the apparent wishes and acceptance of the market were the order of the day.

Closely connected to this was a preference for new design elements created in response to fashionable design trends. The former design approach was increasingly eclipsed by novelty.

New Technologies in Design

Technological innovations within design also gained importance at the time. The emergence of CAD processes accelerated the visual changes in Braun's products. Complex, curved shapes could now be manufactured easily with the aid of 3D software, and suddenly everything seemed possible. At Braun (and elsewhere) all manner of intricate shapes were now being produced. Devices were accordingly supposed to convey an ergonomic and intuitive use, making them appear more modern – a trend that many companies of the time followed.

Peter Schneider noted that:

> supported by modern technical equipment (CAD/CAM) in design model making, the 4 group designers, 2 program designers and 2 product designers work together with specialists in

↖
Dieter Rams's last visit to the Braun Design studio, 1997.

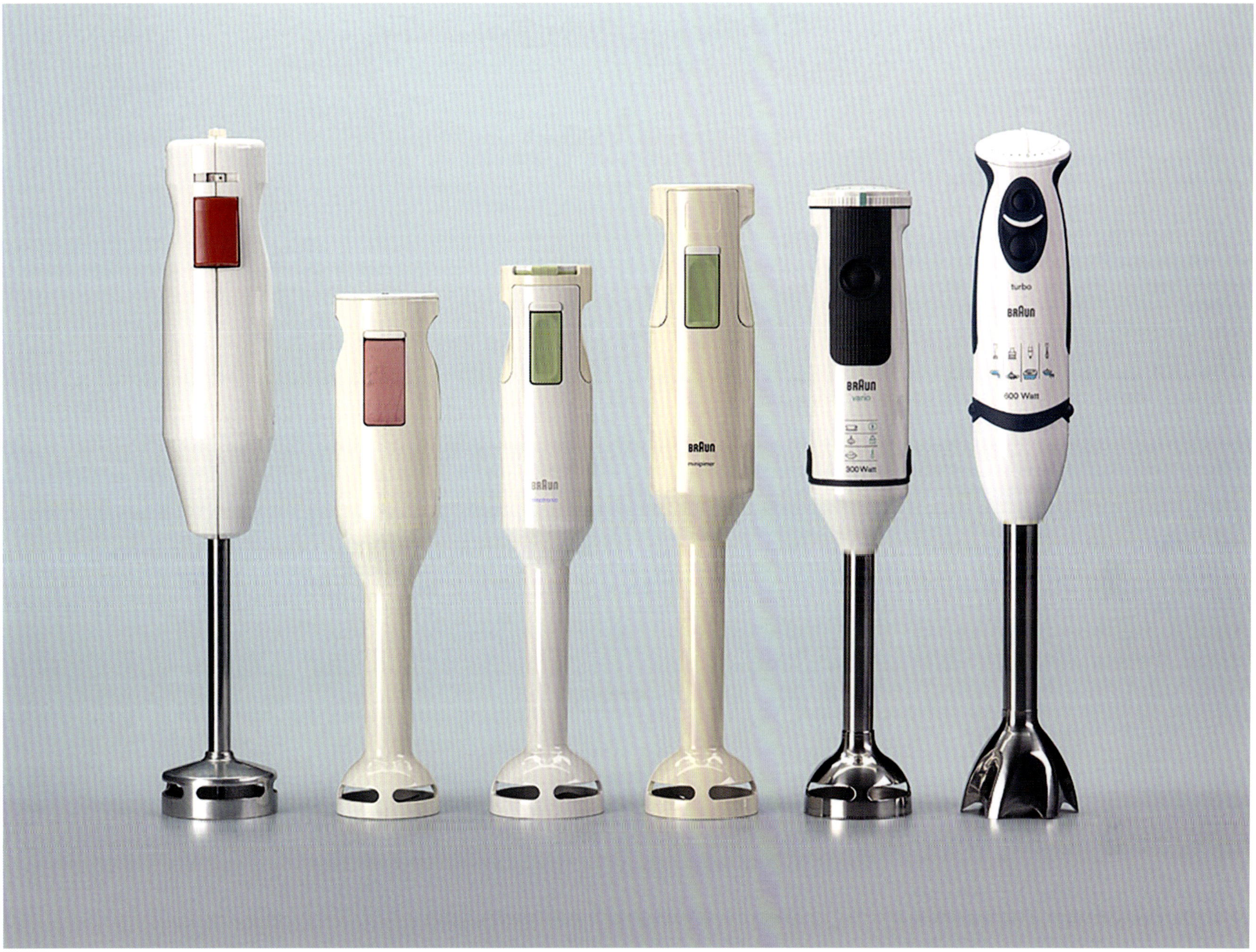

↑
The hand mixer was originally invented as the 'magic wand' by Swiss manufactures in the 1950s. Pimer, the Spanish household appliance manufacturer that Braun took over as 'Braun Española' in 1962, had been producing the Minipimer MR1 hand mixer designed by Gabriel Lluelles since 1959. In 1966, a new mixer was also designed in Frankfurt. Here, you can see the MR 6 from 1981 (left) to the MR 5550 from 2004 (right). All devices created in between were designed by Ludwig Littmann. The change in the design language at Braun in the 1990s and 2000s can be seen visually in this product line.

pro-engineering, graphics, computer design, corporate identity, model making, colours and finishes, as an efficient team capable of meeting the ever-increasing challenges of the design function.[31]

The transition from analogue designing and model-making took place from 1991 onwards and was driven primarily by Roland Ullmann and workshop manager Klaus Zimmermann. Computer programs for CAD and CAM (computer-aided manufacturing) were embraced by the design department and large drawing boards disappeared. CNC (computerized numerical control) milling machines, which were able to undertake five-axis simultaneous milling, changed and perfected the team's model-making capacities. The colour-mixing lab could now produce 14,000 different colour shades. This was something quite different to the packaging with seven RAL-colours for the Hi-Fi systems of the 1970s. The design process was changing for good and largely had to be reconsidered.

Because of these developments, designs reached the production process more quickly. This corresponded with the new company strategy under Archibald Livis, which consisted of shortening the product cycles and bringing new products to market ever faster. Many Braun collectors turned away in dismay at the time, but commercial success certainly continued, albeit with fluctuating sales figures.

From the Living Room to the Bathroom Cabinet

It was not just Braun's designs that changed in the 1990s but also its product categories: products that people needed but did not necessarily desire – such as a toothbrush compared to expensive audio equipment – took centre stage commercially in Braun's portfolio. Shavers had long been the most important output by far and were now joined by new product categories, including electric toothbrushes and epilators. These were all bathroom products that had utility, rather than luxury, value for consumers. Conversely, many of the categories that were once particularly relevant from a design perspective were gradually discontinued, and precisely those that had brought forth iconic Braun products, many of which were unrelated to personal grooming. Alongside Hi-Fi equipment this included slide projectors, film cameras and flashlights, table and pocket lighters, wristwatches and alarm clocks. Consequently, a younger generation of consumers who no longer used Braun's iconic products of the past or were not even aware of them, still viewed the company as a quality German brand, but no longer as a sign of exceptional design.

Past as Future

Not least due to the success Apple achieved with its iPod and iPhone – both conceived by chief designer Jonathan Ive, who decidedly referenced the model of Braun and Dieter Rams from the 1960s to 1980s – another turnaround in Braun Design came about at the end of the millennium's first decade. This was initiated with a realignment and a new Design Team head, a change to which CEO Bracken Darrell contributed significantly. At a meeting at the beginning of Darrell's tenure, he concluded that, while Braun was technically exceptional, it had lost face.

Another New Build

Built between 1998 and 2000 according to plans by the Frankfurt architects schneider+schumacher, Braun's new central administration building boasted 13,500 sq. m and an ecological orientation. The planning was supervised by Peter Hartwein from the design department, following a call for tenders from invited architects.

The winning architects described their concept as follows:

> The design consists of a three-story building arranged in a U-shape around an atrium. The newly developed facade, which functions like a box window and prevents heat loss in winter and overheating in summer, consists of an interior facade layer that can be opened manually and an exterior glass skin that is operated by a motor. Depending on the weather, season, time of day or individual operation, the facade fans out in different ways, creating an iridescent overall appearance. The central hall was covered using transparent membrane roofs, creating a suitable space for exhibitions and corporate events. The floor plan geometry offers maximum flexibility for spatial organization.[32]

The atrium, with its large water basin, forms a modern agora of sorts. Among other things, it is the place where the BraunPrize is awarded every three years, with the national and international design community gathering there for the occasion. The Collectors' Exchange is also held here annually, as are numerous internal company events. With its large canopy roof, the building is the central entrance hall from which the other units of the administrative buildings are accessed, which has great symbolic significance in terms of opening the company to the outside world.

→
The new Braun administrative building in Kronberg. Designed by Frankfurt architecture practice schneider+schumacher and built from 1998 to 2000, it boasts environmentally friendly and innovative air-conditioning technology and is still up to date more than two decades later. Among other things, the BraunPrize ceremonies take place in the atrium.

Syncro

Electric shaver, 2001, Roland Ullmann

The shaver can be placed upside down in a station that automatically cleans and charges the device.

0
1
2
3
automatic lock

WK 210

Kettle, 1999, Ludwig Littmann & Jürgen Greubel

Organic shapes and pastel colours became a part of Braun Design by the start of the new millennium.

Silk·épil eversoft

Epilator, 2002, Peter Schneider, Jürgen Greubel & Cornelia Seifert

Round, colourful and slightly translucent – the idea was for the new epilator to look friendlier. The ellipsis became a defining element of Braun Design.

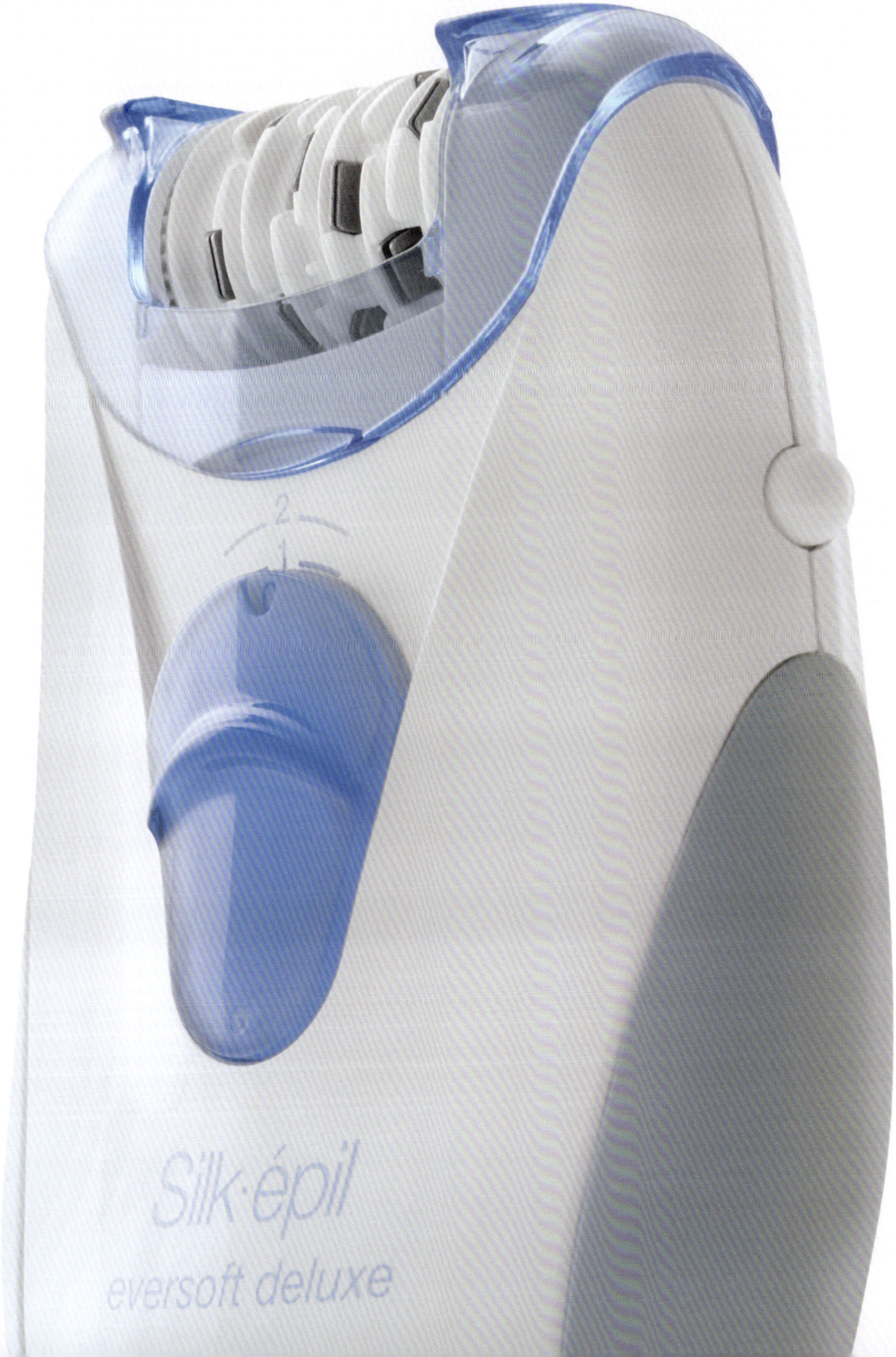
2
1
Silk·épil
eversoft deluxe

EP 100

Beard and hair trimmer, 1999/2005, Björn Kling & Roland Ullmann

Braun Design can even go asymmetrical.

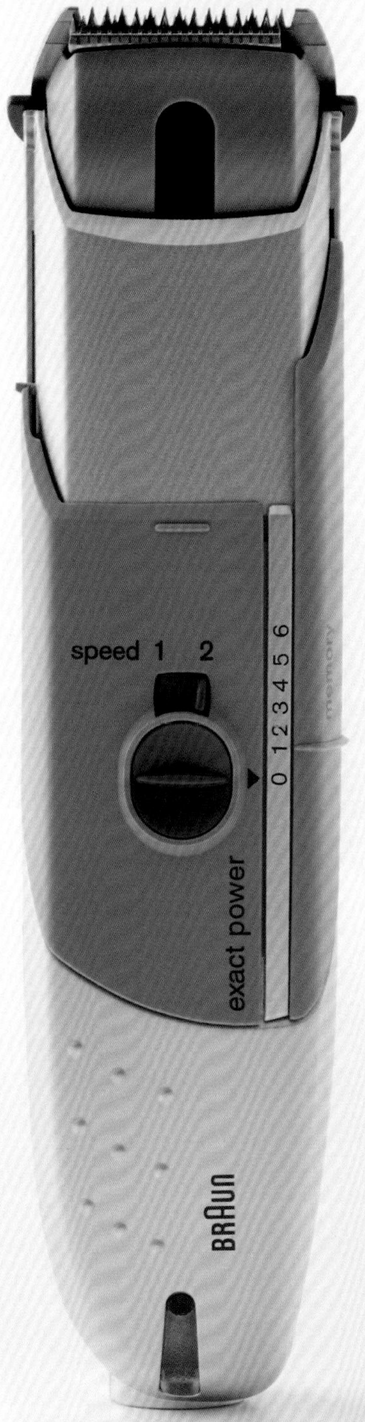

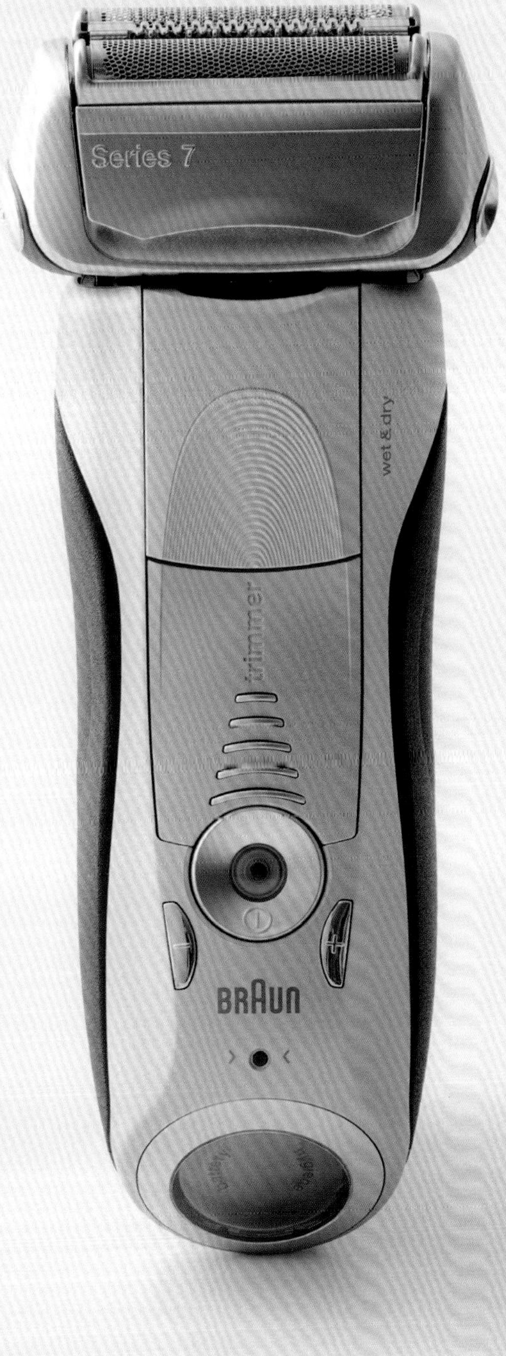

Series 7 Pulsonic

Electric shaver, 2006, Roland Ullmann

An adaptive head plus innovative sonic technology made for a new device. The slender, ergonomic handle and the free movement of the cutting elements have set the standard for the shaver product range up to today.

The 1980s and 1990s: The Era of German Reunification

In the 1980s, postmodernist, relativist and pluralist approaches to thought and action, critical of the primacy of reason, experienced their heyday in philosophy, art, music, literature, architecture and design.

An important prompt in the revision of product design in this sense came from the two Italian design groups, Alchimia and Memphis. Ettore Sottsass Jr, who had previously worked for Olivetti creating matter-of-fact designs for office appliances, became, along with Alessandro Mendini and Michele De Lucchi, the most important proponent of Italian postmodernism, in which symbolism, irony and humour took precedence over usability.

The German version of these emotionalized and communicative products was 'New German Design'. The term, coined by the Kunstflug group and design critic Christian Borngräber, subsumed the work of numerous designers and groups who often worked with unusual materials such as concrete, steel, iron or glass to carry out furniture experiments with the charm of the contradictory, the broken, the ironically ostentatious or the kitschy. Borngräber's colleague Volker Fischer nevertheless remarked that postmodernism had 'changed more in the minds than in the sitting rooms'. However, there had been no talk of 'good form' for a long time.

All the trends mentioned here were extraordinarily well suited for discussion in the media and museum exhibitions. Memphis products could be placed on a plinth in a museum as if they were sculptures. Yet postmodernism also contributed significantly to the popularization of design. Terms such as 'the designer as auteur' or 'star designer' were coined at this time. Some designers were celebrated in advertising or on stage as if they were pop stars.

Politics and Society

The 1980s and 1990s represent a major turning point in world politics. The communist command economies in central and eastern Europe went bankrupt, and with them the ideological economic system of socialism and communism. In 1985, Mikhail Gorbachev became the general secretary of the Communist Party of the Soviet Union. Under his leadership, reform processes were instigated in the Soviet Union. A new social and economic world order was developing. In China, economic liberalism emerged during the reign of Deng Xiaoping from 1979 to 1997, making the country the world's second-largest economic power. On 9 November 1989, the Berlin Wall came down after mass protests in East Germany and the two German states were reunited on 3 October 1990, when East Germany acceded to the Federal Republic of Germany.

Cheerful postmodernism suited the changing political situation and the affluent societies of the West grew and spread. People communicated with each other around the world by mobile phone, fax, text message and then e-mail. 'Anything goes', as the Austrian philosopher of science Paul Feyerabend had already claimed back in 1975.[2]

Technology

In 1984, Apple launched the first PC with a graphic user interface and mouse control on the market. The 'Macintosh' became the prototype of computer technology today. In 1981, IBM also shifted away from focusing solely on mainframe computers. Models compatible with IBM PC systems continue to dominate office environments worldwide today and have completely changed the way people work. Production processes without computers would be inconceivable nowadays. Private households of the 1980s boasted answering machines, microwave ovens, Walkmans and video recorders, and cars had anti-lock braking systems and airbags.

Culture

Music became fragmented into multiple genres: New Wave, Heavy Metal, Hip-Hop, Techno or House existed alongside one another and were equally successful. The musicians of Kraftwerk played only their computers on stage alongside some marginal vocals. The mechanically hard and sparse music has certain similarities with the 'New German Design'. Was this typically German?

With the Berlin Love Parade, Techno also took place in the public sphere from 1981 onwards and influenced graphic design, typography, photography and fashion. New types of lifestyle and music magazines such as The Face in the UK or frontpage in Germany appeared resplendent with new fonts, which were now computer-generated and focused more on visual appeal than legibility. And things were also loud, colourful and shrill at documenta IX in 1992.

The 1980s and 1990s were also characterized by the end of the Cold War and an easing of the global political situation, with new wars in the Persian Gulf and increasing globalization happening at the same time. Postmodernist thought and design led to a change in values in many respects and to the removal of taboos in both the personal and social spheres. The world was more colourful but also more complex. In terms of culture, things became diverse and experimental. It is not surprising that the characteristic attributes of simplicity, clarity and order in design were called into question at Braun, too. Nevertheless, they would be highly topical once more only twenty years later.

07
Braun Today and Tomorrow

Since 2005, Braun has been a part of Procter & Gamble, the international consumer goods manufacturer. There, the company plays a significant role as a technology brand and has revitalized its own design tradition: once again Braun products stand out for the simplicity of their design.

Now with the help of licensing partners, product categories that were once discontinued are being brought back to life – and the future of the brand is being shaped together under the slogan 'Designed to Keep'.

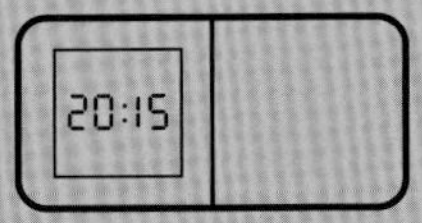

BNC010 2012

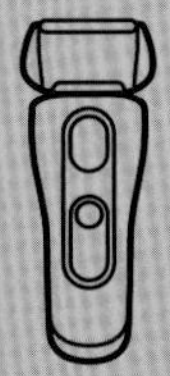

Series 5 2012

J 500 2014

JB 5160 2014

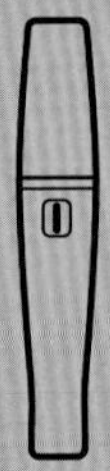

Face 2014

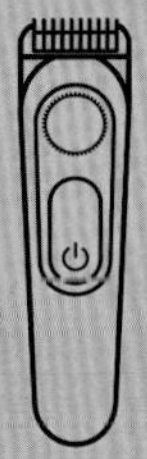

BT 7 2016

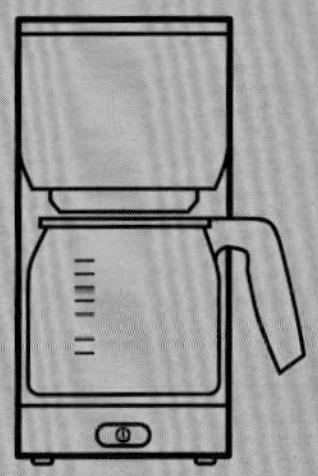

KF 3100 2017

AW10 EVO 2018

BC17 2019

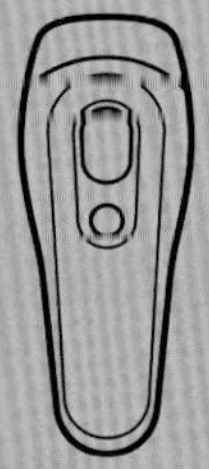

Silk·expert Pro 5 2019

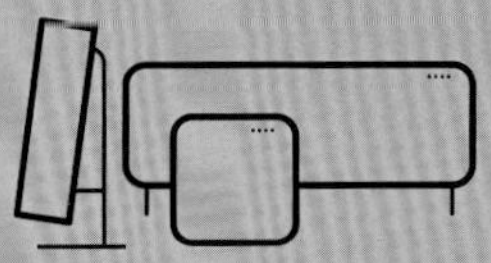

LE 01, LE 02, LE 03 2019

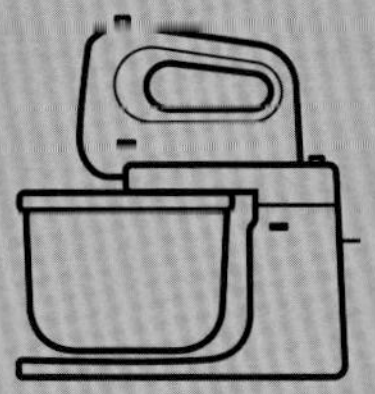

MM 1 2020

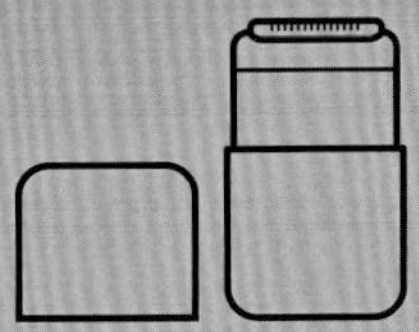

Pocket 2020

BC21 2020

XT 5 2021

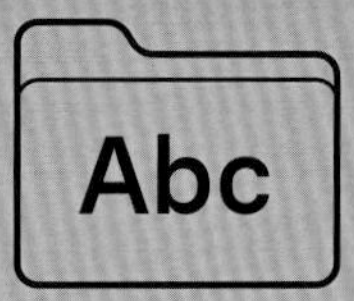

Braun Linear Font 2021

New Bigger Mother: Braun and P&G

On 1 October 2005, Procter & Gamble (P&G) took over Gillette for $57 billion and, with it, came Braun. The last annual report of Braun as a Gillette subsidiary noted Bernhard Wild's move from chairman of the executive board to the supervisory board on 1 December 2005. He was succeeded by Udo Milutzki. The integration into P&G led to, among other things, the former in-house IT team migrating into the system of its new 'mother', while some other departments were subject to outsourcing, since their function was already carried out by external service providers for P&G. This certainly eased the bureaucratic burden and led to higher efficiency.

Braun under P&G entailed an entirely new dimension for the company founded in a Frankfurt courtyard in 1921. Since the acquisition, and with the support of P&G, the brand continued its long-standing emphasis on design and product understanding that is highly relevant today. Long product cycles, repairability and a pared-down, use-orientated design focused on the essentials are Braun's offerings for a sustainable and responsible way of life.

Big Markets – Big Goals

In 2005, Gillette had about 29,000 employees and made sales of $10.3 billion. P&G had 97,000 employees in 2019, with sales of $67.7 billion. Today, Braun is one of over twenty billion-dollar global P&G brands and, together with Oral-B, complements P&G's product portfolio with technical devices and the necessary know-how required to produce them, which also flows into the innovative development of other P&G brands.

In the year of the acquisition, Braun celebrated '50 Years of Braun Design' since the 1955 radio exhibition in Düsseldorf, with shows in Europe, the United States and Asia, and opened its own collection within 450 sq. m at the Westerbach Center in Kronberg, right next to the company premises. Until the Braun Collection opened, a permanent exhibition had been on view at the Institut für Neue Technische Form (INTeF) in Darmstadt, with financial support from Braun.

Just as Gillette had done when it bought Braun in 1967, P&G also pursued a strategy of sharpening the focus of the brand. Braun was integrated into the new 'P&G Grooming'[1] business unit together with Gillette, while Braun product categories that were not in this core area were maintained and strengthened by being sold to competent partner companies or through licensing.

De'Longhi

In 2012, the Italian electrical appliance manufacturer De'Longhi acquired the trademark rights and associated patents for Braun household appliances from P&G, which had previously generated sales of around €200 million. The household segment was thus spun off. The purchase price was €140 million, plus a maximum future profit share of €74 million until 2017, depending on the business's success. The Italian electronics group with 8,300 employees is still owned by

↗|
Company sign at the entrance to the Braun campus in Kronberg.

the De'Longhi family and holds the trademark rights to Kenwood kitchen machines and other companies, thus pursuing a multi-brand strategy in more than 100 countries.

De'Longhi Deutschland GmbH is based in Neu-Isenburg near Frankfurt and has its own Braun Design studio there, which is headed by former Braun designer Markus Orthey. Initially, the head of design for all brands was Duy Phong Vu, who was also part of the Braun team in Kronberg and who first established the De'Longhi-Braun studio in Neu-Isenburg.

In 2017, De'Longhi's total revenue was €2 billion.

The product range now includes hundreds of different items from espresso makers to mobile air conditioning units. De'Longhi products are known for their ingenuity. Kenwood's food processors, for instance, feature an internally illuminated mixing bowl, built-in scale, touchscreen controls and a frying function up to 180 degrees. By comparison, the old Braun KM 3 food processor was still a technical dinosaur that offered only a few functions. De'Longhi's new food processors demonstrate a tendency towards complex convergence, or the ability to combine different appliances into one – a trend that can be observed in other areas of twenty-first-century product design, particularly after the 2007 introduction of the Apple iPhone, which combined a mobile phone, an MP3 player and a minicomputer with internet access into one device.

Zeon Ltd

Since late 2011, Braun brand analogue and digital alarm clocks and wristwatches have been distributed under licence by the company Zeon Ltd in London. Some Braun classics have been reissued, such as the first wristwatch AW10 by Dietrich Lubs in its original size, but also redesigned as AW10 EVO with a larger watch face. A number of new clocks and alarm clocks designed by the Braun Design Team in Kronberg have also been created, featuring a reduced, modern design language and Braun's typical ease of use.

A Multi-Faceted Family

Today, the Braun brand can be found in further subsidiaries and partner companies of P&G. Blood-pressure monitors and clinical thermometers are manufactured under the Braun logo by Helen of Troy. Due to the continued success, new products in the health sector are added regularly. Pure, a company in the United Kingdom, develops audio devices that employ new technologies under the Braun brand. In 2019, twenty-eight years after Braun's 'Last Edition', Braun once more stepped onto the stage of the highly regarded audio segment: with Pure as its new licensing partner, a series of modern smart speakers equipped with the latest technology was created by the Braun Design Team in Kronberg.

Braun as a part of P&G is now firmly integrated into the grooming segment with devices for both men and women. All of this is held together by the 'Braun spirit' of Kronberg and its team's long-standing design tradition.

Design at Braun now calls for a completely new focus and is no longer limited to discussions about the dimensions of gaps between adjacent parts or housing radii. The emphasis is now on designing products that are built to last, which includes tackling concepts like resource conservation, experiential functionality, durability and consumers' emotional and rational product acceptance. Over a century of corporate and design tradition certainly comes in handy in overcoming these challenges.

The 2000s: A New Plural and Global Order

Today, we no longer work with closed-off silos of styles. Adolf Loos's prophecy from the early twentieth century, namely that modernism would no longer require a 'style', has long since proven to be true. As a result of the huge expansion in media, we can now learn about the entirety of visual culture – digitally. We can at any time, for example, see a painting by Rubens or Rembrandt, Cezanne or Picasso, in a virtual museum (though this is no substitute for visiting an actual museum), we can listen on our phone to the Fifth Symphonies of Beethoven or Piotr Ilyich Tchaikovsky and, at the press of a button, the next minute switch to Hip-Hop. Today, there are arguably more artistic and music styles, and ways to interact with them, than ever before.

A World of Media

For most people, the smartphone and internet have become the most important media tools. The year 2001 saw the launch of Apple's iPod, followed in 2007 by the iPhone and the iPad in 2010. These were new types of products at each of their launches, whose designs, formats and uses are today offered by numerous companies. Today, the iPod has since merged into the iPhone, a typical phenomenon of the so-called convergence phase of a product in which the various functions come together in a new device. Though totally new, these processes now come about at a much faster pace and with greater complexity. The constantly growing number of smartphone apps, from home automation to health monitoring, have transformed what was once simply a mobile phone into a multi-purpose device.

The business world is no longer conceivable without digital communication. It enables ever more intensive globalization, especially as video conferences have become an essential aspect of international communication. Online stores are increasingly ousting their brick-and-mortar counterparts, proving tough competition for downtown department stores and changing cities' material infrastructure. Social networks such as Facebook, Instagram, WhatsApp, TikTok or Twitter have become omnipresent and in the next generation will surely be replaced again by new platforms.

In the last two decades, the media world has been radically altered. Millennials, or those born in the last two decades of the twentieth century, make up the first generation of 'digital natives', but it is the members of Generation Z who have been poised to use digital media for action.

Politics and Catastrophes

The 2000s began on 11 September 2001 with the terrorist attacks on the World Trade Center in New York, which marked violence and contempt for mankind on a new scale.

Neither has there been a scarcity of natural catastrophes in the first two decades of the new millennium. In 2004, 300,000 people died following a tsunami on the coasts of the Indian Ocean, and 70,000 perished in an earthquake in China in 2008. Three years later a tsunami in Japan led to the nuclear catastrophe of Fukushima. As a result, several countries altered their position on nuclear power. The year 2019 brought a global pandemic (SARS-CoV-2, or COVID-19), the economic and cultural impact of which still cannot be fully envisioned.

During this era, climate change has increasingly become a political topic, but world governments have come to no real agreement on the course of action we should take. Scientists have been issuing stark warnings for decades, but it took the efforts of Greta Thunberg, a Swedish school girl who skipped school to protest, to create a broader awareness for the topic. It would seem that promising young people can be more impactful than experts, inlcuding a brave fourteen-year-old environmental activist.

So, the world has more than enough problems today. Naturally, good design cannot save the world, but it can contribute to solutions in many areas. Product designers have long since ceased to be pure stylists. It is more a question of finding new concepts for a sensible and sustainable handling of the materials on our Blue Planet.

Braun Design Today

After the first decade of the twenty-first century, Braun set off in search of its 'lost face', which had been called for by Braun CEO Bracken Darrell as early as 2008. A special part of this quest was to look back at Braun's history, particularly the company's design history since the mid-twentieth century, to combine Braun's classic approach to design with the latest technology and innovative concepts.

On 1 September 2009, Oliver Grabes succeeded Peter Schneider as head of the design department at Braun and took up Darrell's mission. Grabes, an internationally experienced designer with roots in Hesse, had studied product design at the Offenbach University of Art and Design, in part under former Braun designer Richard Fischer. After working as a designer for many global brands in Germany, the Netherlands, the United Kingdom and the United States, Grabes became associated with Braun through joint projects during his time at a London design agency and in his capacity as Professor of Industrial Design at the University of Wuppertal. Several successful university projects for Braun were realized through Grabes's close collaboration with Jürgen Höser, the head of development for the electric shaver section at Braun.

After Grabes assumed leadership in Kronberg, the Braun Design Team sought to draw on the tradition that had characterized the company from the 1950s to 1990s. By means of contemporary shapes and concepts, the new Design Team wanted to echo Braun's original focus on a visually reduced and use-orientated design:

> Our new design language at Braun takes up the legendary legacy of Dieter Rams and his colleagues, above all by adopting their mindset and their search for new appliances that are easy to understand down to the last detail and are wonderfully functional. It is this mindset that inspires our design efforts today.[2]

Oliver Grabes himself writes below about Braun Design's fresh start after 2009 with an eye on the rearview mirror and another towards the future.

Moving Forward and Looking Back by Oliver Grabes

When I took over as head of the Design Team at Braun in 2009, the clear stance the company had once taken on design seemed to be miles away. There were very different approaches to design in the various product categories and, even within individual categories, you could no longer discern a consistent and distinct Braun Design.

This was confirmed by interviews we conducted at the time with international design experts. As a result, Braun management, led by Rainer Silbernagel (head of business) and Dirk Freund (head of product development), decided to change the orientation of the brand's design once again.

We onboarded young, talented designers and, in collaboration with a group of experienced

→|
In the Ceramic Watch BN0171, designed in 2015 by the Braun Design Team, classic Braun Design elements, such as the dial and the yellow second hand, are now combined with ceramic, a new material in the watch category.

BRAUN

↑ and ↗|
The new design language developed by the Braun Design Team references Braun's special design history and reinterprets it. Clear, visually timeless and archetypal product shapes are supplemented with geometric controls that are always round or oblong. Uniform design principles enable very different products to be linked by their forms and thus to be discernible as Braun devices.

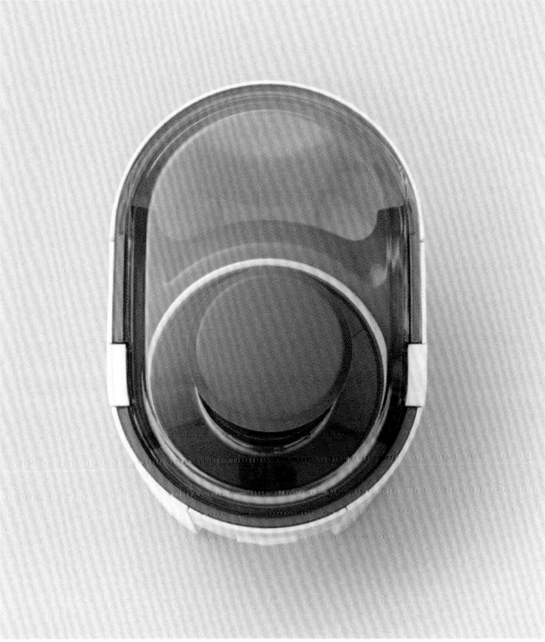

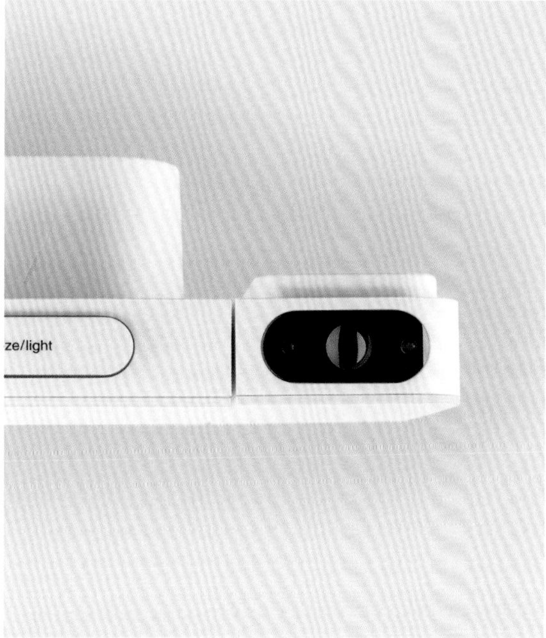
ze/light

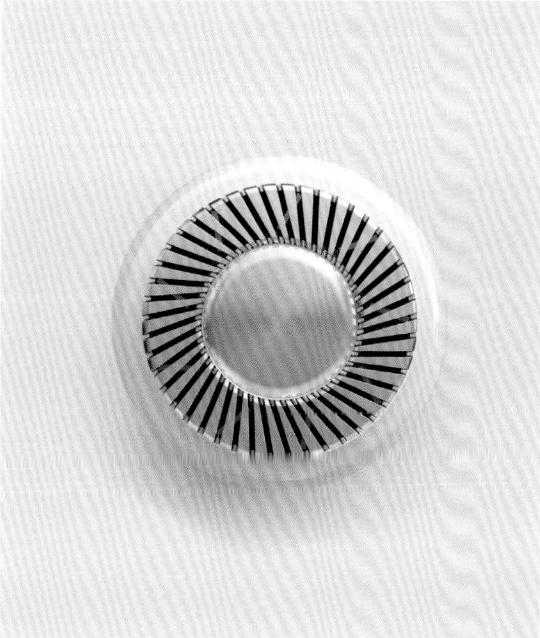

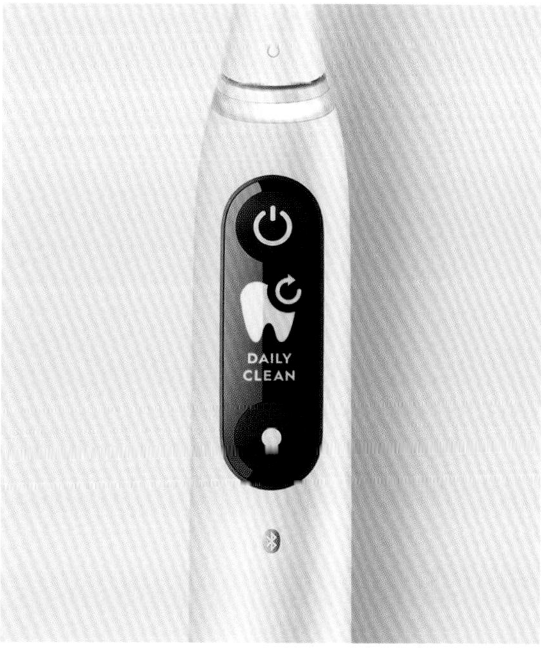
DAILY
CLEAN

Braun designers, worked on developing a new definition of Braun Design. A new spirit of optimism emerged in the studio, combined with the clear commitment of the entire team to return to the original values of a uniform design language that was in line with the Braun brand. The plan was to make certain this return to an appropriate shared design outlook was visible in all future Braun products.

The Strength of Pure: A New Braun Design Language with Roots

In early 2010, ten Braun Design principles were crafted that were presented as a commitment to this new focus. They were accompanied by the headline 'The Strength of Pure'. Unlike Dieter Rams's Ten Principles for Good Design (which were deliberately intended to be broad rather than specific), these new principles refer specifically to future Braun Design. In developing them, the aim was to translate the individual design interpretations created by different designers into a recognizable and consistent Braun Design language.

Despite, or perhaps precisely because of, the new diversity in the team, its members were remarkably quick to agree on shared design principles. These ranged from the distinct yet practical basic archetypal shape of a product to the geometric design of control elements: for example, switches and knobs today are round again or designed as oblong holes, replacing the era of ellipses, triangles and asymmetric shapes. In developing the design language, the designers drew on the company's own special design past and took their inspiration from the distinctive yet wide-ranging pool of geometric design elements.

In-house, we coined the term 'Past Forward' to express this strong recourse to Braun's own design history while simultaneously looking to the future. After all, Braun Design was never retro-design – it was, and still is, part of the Braun philosophy to always be on the lookout for new and better solutions and to reinterpret past designs to make them contemporary and advance them where needed.

That said, aspects of the 'Offenbach approach to product language' that influenced me during my studies at the Offenbach University of Art and Design can also be found in the new Braun Design principles, such as the emphasis on a product's functions that provide information on its use, the functional grouping of the control elements and the crucial importance of product graphics and pictograms that everyone can understand.

This return to Braun's 'design DNA' – a design language based once again on geometry and characterized by simplicity, intuitive use and functionality – motivated every single one of us to apply these principles in our day-to-day design work and consequently to test whether users would consider them valid and accept them.

Team Design: New Approach, New Procedure

Combining the 'outside views' of young designers who had recently joined Braun with the know-how of longstanding team members proved to be extremely productive for a change in direction, but it did also cause some friction. The experienced members of the team were asked to change their approach to Braun Design, which was not an easy task.

Jointly tackling this challenge proved to be important not only in defining the new design language, but also in designing many new products, which happened almost simultaneously. It was no longer the practice for individuals to design products and then discuss them with the boss; rather a team of two or more designers collaborated on several projects. Subsequently, the various concepts were presented and discussed at regular meetings of the entire Design Team. Working as a group enabled us to find the best possible design solutions, allowing designers to collaborate and build on each other's ideas. Moreover, the barriers between the various design functions were removed: while staff in packaging design had previously worked as a separate unit, they were now closely incorporated into the Design Team so that here, too, solutions could be discussed and developed as a team.

A decisive moment for this change in how people collaborated came when we regrouped as a team in the Braun Collection to discuss our path forward. Horst Kaupp, who managed the Collection at the time, explained to us the difficulty of naming certain designers as the authors of individual products that were introduced in the 1970s. Even back then, there was often not one particular designer behind a product form or idea. However, with the emergence of postmodernism, there was a call for personal authorship.

In 2010, we had to consider how to deal with the issue of authorship moving forward, precisely because we wanted to actively encourage other designers to contribute ideas and feedback. We were quick to reach a consensus that from this point onwards only the designation 'Braun Design Team' would be used and no individual would be named as the sole author of a

↑
The Series 5 electric shaver's burled case and indent for the thumb reference the hard/soft surfaces of the 1980s and ensure a secure grip when shaving. The buttons, both oblong and circular, form the centre of this shaver, which was designed in 2012. These elements became the norm for all new Braun products.

BNC010 / BNC020 / BNC008 / BC09

Digital alarm clocks, 2012, 2018 and 2020, Braun Design Team

Visual simplicity fused with intuitive controls. Typical Braun Design details, such as the green alarm button, are continued and reinterpreted in the digital alarm clock series. A flat, modern shape is enabled by relocating the battery to the clock's base. Gravity slants the alarm clock automatically, so it is perfectly legible.

alarm on/off

product's design. The aim was to make it visible both inside and outside the company that the entire team was responsible for a product's design and, ultimately, its success. This seemingly small detail was to have a major impact on the design process and the spirit of cooperation amongst designers.

15 in 1

Since Braun was expecting a change of direction in design due to the search for a new head of the department, many product developments were put on hold during the transition period between 2008 and 2009. There was still a push to produce technical innovations, but every project team wanted to see its product initiative as part of this new focus. In early 2010, as part of this new direction, the Design Team came under increased pressure to work on all these delayed projects at once. This period of intensive design work was a kind of litmus test for the new design language – and this work in turn helped to further hone the newly defined principles.

Fifteen new Braun products were designed in 2010 alone. It was a year marked by great intensity, but also a high degree of motivation on the part of the Design Team. This almost always entailed building on existing or already developed technical platforms and designing around them; in many instances, this required the team to compromise to achieve the new 'good design', but it also meant that the transition to this new design mindset was achieved relatively swiftly.

Series 5

The new design language first emerged in 2012 in the commercially important area of shavers with the Braun Series 5. The shaver's basic design is characterized by the ergonomically advantageous slim and longer handle from which the adaptable head clearly stands out both in shape and on account of its metallic look. In this regard, it emulates the Braun classics, the SM 3 and sixtant, as well as the Flex Control, the first electric shaver with a swivel head. Similarly, the grid dot pattern in the handle section, first used in the micron, was adopted from classic designs as a formal and functional element and reinterpreted in a contemporary way. The design concept for many subsequent Braun products was already visible here: centrally positioned, geometric controls integrated into an appliance that is also largely geometric in shape but includes soft formal transitions. The visual clarity of the product is created both through symmetry and by clearly separating colours and materials.

New Alarm Clocks and Timepieces by Braun

While product development for shavers, household goods and hair care devices took longer because of high production numbers and the more complicated automated manufacturing processes, the new thrust of Braun Design was visible as early as 2010 in the field of watches and digital alarm clocks. New products designed by the Braun Design Team in Kronberg were brought to market with the help of the company's new licence partner Zeon Ltd from London. This collaboration produced the new BNC009 and BNC010 alarm clocks, which both symbolize an independent and modern interpretation of Braun Design values while simultaneously reflecting Braun's earlier designs. Their operation is intuitive thanks to the green alarm button, and the slim product geometry is complemented by the battery compartment that functions as the base. Thanks to gravity, the display automatically tilts back to the right angle making it easy to read.

At the same time, a series of new wristwatches was also created that feature many design elements from iconic 1980s Braun timepieces, such as the basic geometric shape, a clear, functional face and, above all, the yellow second hand and red arrow on the date display – design elements that were included in both classic and modern sporty timepieces. Moreover, famous alarm clocks and wristwatches, as well as the ET 66 pocket calculator, were also relaunched in their original designs. Later, the iconic Braun AW10 watch was relaunched: in collaboration with its original designer, Dietrich Lubs, a modernized and larger version was produced and named the AW10 EVO. Braun's relationship with its new licence partner Zeon Ltd soon became one of the most important factors in making the new Braun Design approach more visible to consumers.

New Braun Household Appliances by De'Longhi

After P&G sold the trademark rights for the Braun household category to De'Longhi in 2012, more than a third of the designers from Braun's household appliances team went to work in the De'Longhi-Braun design studio in nearby Neu-Isenburg. This new team was headed first by Duy Phong Vu and then Markus Orthey, who brought with them many years of Braun Design experience and the new Braun Design language. Its deployment in De'Longhi's household

→|
The AW10 wristwatch from 1989 was redesigned in 2018 in collaboration with Dietrich Lubs, its original designer, and then relaunched in a larger version (39- instead of 33-mm in diameter) as the AW10 EVO.

BRAUN

department was demonstrated admirably in its IdentityCollection appliances, which once again boasted a timeless geometrical appearance.

Despite this significant change, Braun has maintained a consistent and recognizable design language, thanks in part to designers remaining in their positions for long periods and the close personal contact of both teams. There is a willingness to help one another to ensure a consistent design outlook despite the two teams working in different corporate contexts.

A Consistent Design Language

In the years that followed, many new innovative Braun products were created across categories, and the new design language was consistently applied. Geometric details united products as disparate as epilators and irons, wristwatches and beard trimmers – harmoniously designed without a narrow set of rules, meaning it was possible to use overall shapes that were generally standard and best suited to the function and task of the product while integrating them into a recognizable and unique canon of forms through common principles and design elements. The strategic design goal was not the radical new start that had occurred in the 1950s but rather the successful application and consolidation of the new design language across product categories. The idea was to prove that the approach to design could be implemented everywhere and contribute to Braun's business success.

The plan worked. Through their willingness to buy them, consumers have effectively demonstrated that practically designed products with modern, unassuming appearances once again resonate with the brand's understanding of design – one that complements and communicates the high technical quality and user-orientated focus of Braun's new products. The consumer receives a practical product, fit for its purpose and built to last. These are objects that are functional and visually long-lasting, rather than spectacular or highly emotional items that quickly become last year's fashion. A Braun product is designed to be with you for many years, like a good companion.

Moreover, combining good design, efficient and reliable technology and modern brand positioning in a single harmonious package makes for an excellent fit with the company's overall strategy to win over consumers by selling a few well-thought-through products in combination with good marketing. Contemporary Braun strategy also demonstrates the impactful influence of new P&G managers, who embraced the challenge of bringing to bear their marketing know-how, consumer understanding and efficient thinking in terms of platforms to a brand that was very different from P&G. This initial fusion of Braun's and P&G's approaches may have been challenging at times, but it clearly resulted in Braun's steady market growth. The Braun brand is once again globally competitive and economically successful, increasingly returning to its earlier values.

Braun as a P&G Family Member

In the years following P&G's acquisition of Gillette, leaders brainstormed how the Braun brand, with its unique understanding of product research and development (R&D), as well as its supply chains in Kronberg, could be meaningfully integrated into the group. It became increasingly clear that Braun's approach to products, markets and the development of technical appliances also represented a special opportunity. After all, although their production involved advanced technology at the time, there were no complex technical products to be found in the product portfolios of either P&G or Gillette. Growth through innovation is not easy in the highly competitive FMCG (Fast Moving Consumer Goods)[3] fields, so P&G hoped that both Braun's and Oral-B's technological knowledge in Kronberg would help give other P&G brands an innovative edge in their respective markets.

Braun was integrated into P&G Grooming under the management of Stefan Schamberg, a long-serving 'Braunian' who had previously headed Oral-B and was Braun's head of R&D from 2012 to 2022. An outstanding engineer and manager, he had become familiar with and managed many technical departments over the years. Simultaneously, the new 'Device GCO', or 'Global Capacity Organization', was launched to focus on the development of devices. As a new section in the former Braun headquarters and the now-new P&G Development Centre in Kronberg, the Device GCO carried out technology development not only for Braun and Oral-B but also for other brands within the group. Many managers and board executives were involved in this landmark decision, which, in addition to securing the company's location in Germany, was also intended to ensure the long-term growth of the Braun brand – a clear commitment from P&G to promote the traditional brand as part of P&G Grooming. This core business was supplemented by cooperation arrangements with specialized licensing partners who have successfully used the Braun brand in categories outside grooming, consequently generating a broad and strong brand presence recognizable to consumers.

|←
The J 500 juicer, designed by the Braun Design Team in 2014, features a clear geometric shape with practical design details for simplicity of use and cleaning.

Silk·expert Pro 5

IPL, 2019, Braun Design Team

The IPL (Intense Pulsed Light) technology sends light through the skin into the pigmented hair follicles, interrupting hair growth for a longer period than simply shaving. The device can be used with different heads, and the universal grip shape makes for easy use on any part of the body.

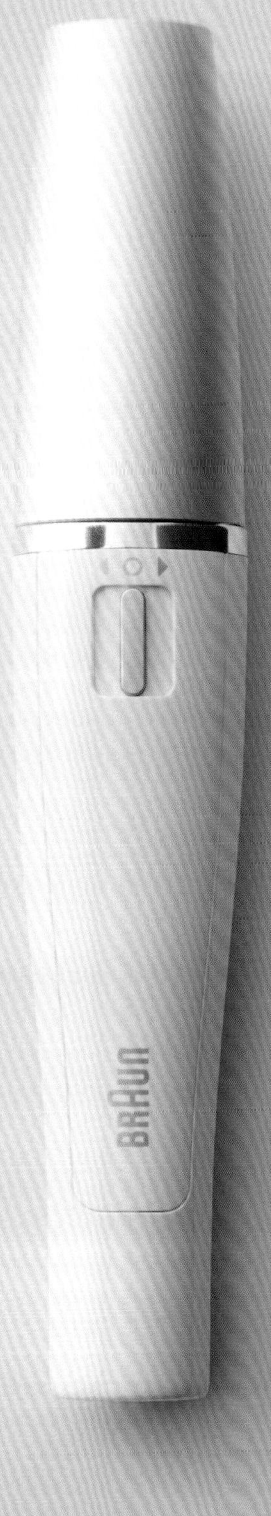

Braun Face

Facial epilator, 2014,
Braun Design Team

The Face removes even the finest of facial hair, especially around the sensitive areas of the chin, upper lip and eyebrows. To achieve this, a small precise head was developed that is attached vertically to the device to ensure full vision of the area needing hair removal. A metal cap protects the epilator head and creates a less-technical overall appearance.

Design and Brand Quality

The successful repositioning of Braun Design was confirmed not only by market success and positive customer feedback but also by Braun products winning international design prizes on repeated occasions. Specifically, from 2012 to 2023, Braun products received a total of 155 design prizes across all categories. This confirmed to all Braun designers the validity of their approach, and they gained recognition for the excellent quality of their work outside the company, as well as internally from colleagues and management.

Quality was also the aim in brand management. Set up in 2005, the Braun Collection moved to larger premises in 2013 so that, in addition to the 300 products already on display, it could exhibit more than 900 products. Staff handle loan requests from exhibitions and museums around the world for historical Braun devices.

The Collection proved to be an important meeting place for more than guided tours and lecture events by the Braun Förderkreis e.V.: it is also a seminar venue ideal for impressively conveying the continuity of Braun's brand values. Moreover, in 2019, an innovation hub was created in the entrance building of the Kronberg Development Centre where members of the public can familiarize themselves with contemporary products and their development process.

The Braun Design Collectors' Fair gives visitors the chance to experience the brand in a special way. A long-standing annual event held in May, it attracts hundreds of Braun enthusiasts to Kronberg. With an emotional intensity that is otherwise associated only with collectors of expensive vintage cars, old and new Braun electrical appliances are collected, cherished, cared for, bought and sold. Braun enthusiasts congregate in the atrium of the main Kronberg building to chat with the like-minded about their experiences as collectors.

While this event draws savvy members of the public, the BraunPrize, held every three years as an international design competition, addresses a different audience, namely young creatives. Through the BraunPrize, the Braun brand expresses its commitment to passing on the value of good design to the next generation of creative designers.

Back After Twenty-Eight Years: Braun Audio

The field of audio equipment has historically been a special product segment for Braun. Both the development of the company and the brand are closely connected to the subject of playing music. Over the years, Braun has developed different models for radios, Hi-Fi equipment, record players, tape recorders and loudspeakers. After the audio segment was closed in 1990, there was a surprise turn of events. In 2018, Braun Audio made a comeback after twenty-eight years.

To develop new Braun audio equipment, the company brought on board a skilful partner found

↖
Each year, Braun collectors and the likeminded meet up in the atrium of the P&G/Braun headquarters in Kronberg for a chance to buy, sell, swap and – above all – chat about Braun products together. It is a high point of the year for every kind of Braun fan.

↑
Founded in 2005, the Braun Collection relocated in 2013 to larger and newly designed premises in the immediate vicinity of the Kronberg campus. There, 900 products dating from 1921 to the present can be showcased in the exhibition and archive rooms, helping to visualize Braun's corporate and design history.

LE 01, LE 02 and LE 03

Smart speakers, 2019, Braun Design Team

The legendary LE 1 electrostatic loudspeaker by Dieter Rams, which came out in 1959, inspired a new series of smart speakers developed by Braun together with its audio partner Pure. As a high-end modular system with an anodized aluminium frame, it allows three different speaker sizes with different bases to be combined to fit a particular space.

↑
To mark Braun's 100th anniversary in 2021, a new corporate font was developed with the help of Stefan Dziallas of Iconwerk and typographer Ludwig Übele. 'Braun Linear' is a clear, geometric font that is based on the oblong form characteristic of Braun products and manages with little modulation in line width. Typical Braun details make the font precise and suitable for both product and brand language.

in English digital radio specialist Pure. A period of intensive collaboration began. The Kronberg Design Team was strongly involved in the strategic and creative design process to produce the right set of contemporary Braun audio products. The outcome of this collaboration was first presented in 2019 at the IFA (International Radio Show) in Berlin, where Braun had its own exhibition stand. Braun presented a series of smart speakers equipped with special technology and high-quality surround sound, designed as new classic models with sophisticated details in a timelessly simple yet modern Braun Design.

Inspired by the Braun LE 1 electrostatic speaker designed by Dieter Rams in 1959, the new LE 01, LE 02 and LE 03 speakers feature cutting-edge audio technology and can be controlled via smartphone or voice control. Designed as a networked product system with configurable standing versions, the low-key design blends easily into a wide variety of settings. The new Braun audio products have received the renowned Red Dot: Best of the Best Award and the iF Gold Award, as well as several other design and audio prizes.

One Hundred Years of Braun

In 2021, Braun's one hundredth anniversary year was celebrated with a range of activities, which took an illuminating and proud look at Braun's unique past but also ventured a glimpse into the future. The live events that had originally been planned were unfortunately cancelled due to COVID-19. Braun's anniversary, then, also marked its abrupt entry into the digital age: all celebrations were digital events and people learned how to log on to them, award hearts in video calls and communicate or applaud digitally.

The relevance, focus and realization of Braun's brand values were re-examined, honed and updated for the anniversary year. A global study[4] involving more than 3,000 young people confirmed once again how important it is to Gen Z that companies create sustainable and enduring products that can be understood and operated by everyone. The strong convergence between these convictions and the Braun corporate values and design principles as conceived and lived by Max Braun, the brothers Erwin and Artur Braun, Fritz Eichler, Dieter Rams and others, was a strong reminder to us all, especially in the anniversary year, to intensify our activities in the field of sustainable products.

Here, Braun can draw on its historical role and immense experience as a pioneer to push for a more critical handling of product design, ensuring that products are built to last, can be repaired and recycled and use sustainable materials. A decision was taken within Braun and P&G to prioritize sustainability, which is set out in detail in the '2030 P&G Sustainability Agenda'.

For today's fast-moving digital world this attitude has been expressed as 'Simple. Useful. Built to Last'. That is what Braun stands for today and it is vital that this brand outlook comes across to the younger generation. As part of the

→|
The design for the 100 Year Special Edition packaging celebrates the company's 100th anniversary and revisits the purist design of the 1960s. The boxes feature an illustration of the product, a few words of copy and the logo on a light grey background to create an eye-catching image for sales shelves.

BRAUN
100
years
Limited
Edition
Up to
5
Years
Warranty
Good Design
makes a product
understandable.
BRAUN
BRAUN
Styling Kit 7
100
years
Limited
Edition
Up to
5
Years
Warranty
Good Design
makes a product useful.

↑
American designer Virgil Abloh set out to open a portal into the future for Braun's 100th anniversary. Sixty years after it was first made, he reinterpreted Dieter Rams's wall-mounted Hi-Fi system as a 'Functional Artwork', fostering a 'dialogue on the role of design'. Completely chromed and featuring a modernized record deck, the primary function of this model was to attract attention. After the jubilee celebrations, the system was exhibited at the Brooklyn Museum, New York.

↗
Abloh turns the system's classical transparent Perspex cover designed by Dieter Rams into a violet shimmering piece of secular mysticism, and the tonearm features a contemporary high-tech system.

activities marking Braun's one hundredth anniversary, brand partnerships were created that reframe both current and classic Braun products so they appeal more to and resonate with younger people.

A special role in this context was played by the collaboration with Virgil Abloh, an architect and fashion designer who founded the cult brand Off-White™ and was the artistic director for Louis Vuitton men's wear. The Off-White™ × Braun Limited Edition BC02 alarm clocks in bright orange or blue, a re-edition of the AB 1 travel alarm clock from the 1980s by Dietrich Lubs and Dieter Rams, sold out within days. Braun's cooperation with Virgil Abloh in the company's anniversary year can essentially be seen as an experiment. This is especially apparent in his designs for the modified wall-mounted Hi-Fi set. The world-famous Braun audio system designed by Dieter Rams was altered by changing its material: satin aluminium was replaced with mirror-finish stainless steel, which resulted in a completely different kind of look. Abloh regarded it as an artistic experiment in translation to elevate the perfectly styled classic product to art, or 'Functional Art'. This did not sit very well with Dieter Rams's approach and had a highly polarizing effect, triggering a discussion about the symbolic function of products and what could be associated with the Braun brand. At any rate, the collaboration attracted attention from the media and younger age groups who had previously not been reached by the Braun brand. Abloh's acting as a juror for the BraunPrize 2021 would be one of his last public appearances before his untimely passing. His participation in the discussions remains unforgotten by the international members of the jury.

Other products celebrated Braun's one hundredth anniversary in a much less polarizing way. These special edition products were deliberately designed by the Braun Design Team to be restrained and refer to the past. They are high-quality and feature simple white or satin black colours and plain light-grey packaging that is no larger than necessary with minimal graphics. The aesthetic was reminiscent of the Braun packaging of the 1960s and 1970s, and this connection to the past through the packaging reflected the larger reference to the past in product design. Past Forward.

Moreover, to mark the anniversary year, the Design Team collaborated with typography designers Ludwig Übele and Stefan Dziallas to develop the new Braun Linear typeface. It complements Braun's corporate identity with a font that matches the brand, logo and design and expresses through typography the visual design characteristics of geometry, clarity, precision and order down to the last detail. At the end of 2021, Braun Linear replaced the previously used Neue Helvetica as the house font.

Continually Better

Rarely do brands radically alter their products from one day to the next. Radical changes only

BC02

Alarm clock, 2021, Virgil Abloh

The second anniversary piece by Virgil Abloh is the bright blue and orange version of the iconic black AB 1 alarm clock by Dietrich Lubs dating from 1987. As the BC02, it is also back on the market in both black and white, although here Abloh reverses the unobtrusiveness of the original design approach. He thus initiates a discussion on longevity in design, while generating interest and demand from a new generation of Braun consumers.

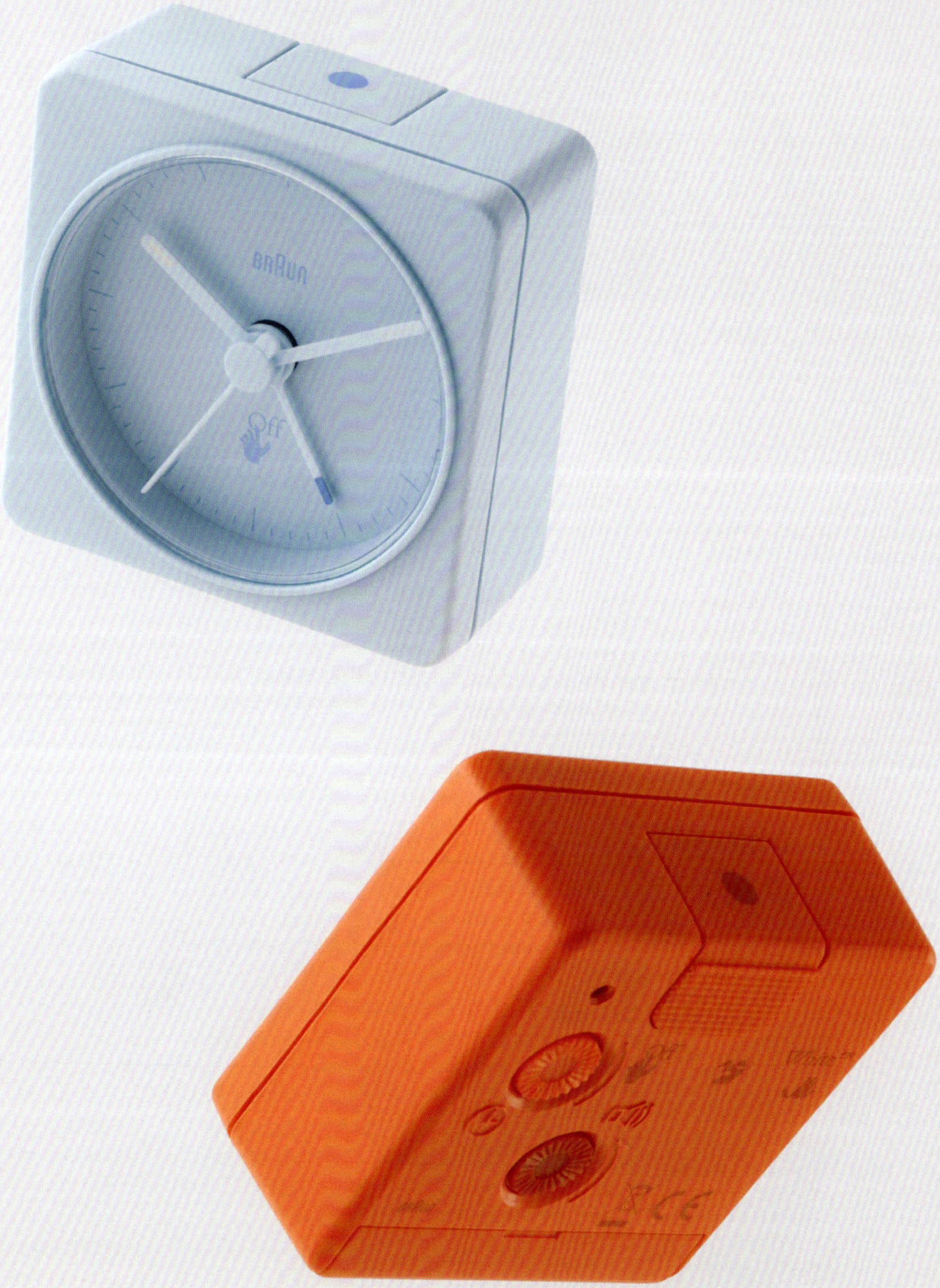

BRAUN
BRAUN

occur in the case of significant technological advances, like the ones we are experiencing in communications technology or electromobility, and even there the change might take years. More often it is a matter of continually developing valued, established brands and making them interesting and relevant for upcoming generations. Braun is ideally positioned to continually recharge the brand values that have been established for decades and the trust the brand enjoys with new innovative products and contemporary brand communications.

Braun's brand values today are more relevant than ever before. And the greatest danger for Braun would be to turn away from its own trusted values and core identities in search of the new and spectacular. Here, we can learn from our own history. And fortunately, successful brand management is one of the core competences of P&G, which places a strong focus on the long-term successful advancement of the traditional Braun brand.

This drive is particularly evident in the design. New products have been created in which technology and construction are integrated into the design process from the start.

It is still Braun's top priority to ensure its products are easily repairable and made of recyclable materials – or, in other words, that they are designed to keep and built to last. Braun products today stand out for their distinct simplicity of appearance. A simplicity that once again turns functional design into an emotional product experience.

↖
Joint design effort as a team: in 2023, the Braun Design Studio is located in the very same place as it was back in Dieter Rams's day, albeit in modernized offices. Teamwork remains at the heart of the design process.

→|
The Braun iCheck 7 is a blood pressure monitor designed in 2018 by Helen of Troy and the Braun Health Design Team. Firmly in the Braun tradition of intuitive user interfaces, the device itself shows you the correct way to use it. The colour coding of the heart enables swift and simple readings on your wrist, while a detailed assessment can be viewed in the relevant app.

BRAUN
SYS DIA PUL
120 80 68

KF 3100

Coffee maker, 2017, De'Longhi & Braun Household Design Team

Here, the successful tower shape of the KF 20 and KF 40 is applied in a modernized and simple form. The reinterpretation also boasts a favourable price tag.

MM 1

Mixer, 2020, De'Longhi & Braun Household Design Team

The universal, entry-level kitchen appliance system features a mixer that can be used by hand or as a standing unit.

BRAUN
BRAUN

Pocket

Electric shaver, 2020, Braun Design Team

A straightforward design resembling those from Dieter Rams's day. The chargeable, waterproof shaver with its protective cap is distributed in various colours in China and Japan in cooperation with the Xiaomi brand. The grip made of soft plastic with its diagonal fluting reinterprets classic Braun Design elements.

BC17

Clock, 2019,
Braun Design Team

The iconic design of this large clock dates back to the ABK 30. The clear clockface with the satin casing makes it very simple to read the time.

BC21

Digital alarm clock with wireless smartphone charger, 2020, Braun Design Team

The tradition of Dietrich Lubs's alarm clocks, designed in the 1970s and 1980s, has been rejuvenated. These clocks can now be used to charge a smartphone, while the daily alarm function can be simply and visibly activated by pressing the green alarm button.

Oral-B iO9

Electric toothbrush, 2020, Oral-B Design Team

Functionality in a minimalist design and with innovative magnet technology. A smartphone app helps optimize cleaning performance.

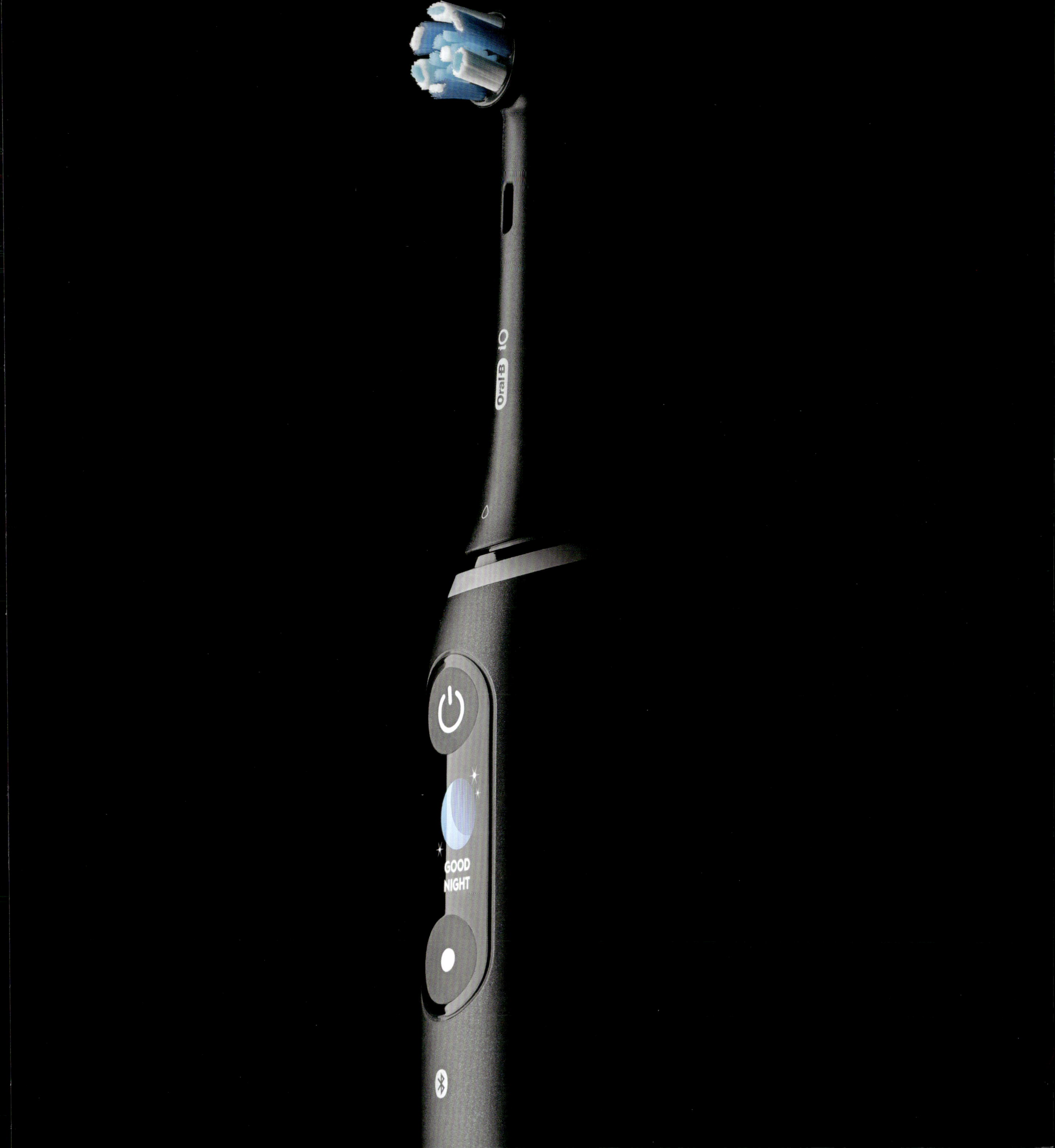
Oral-B iO
GOOD
NIGHT

BT 7

Beard trimmer, 2016, Braun Design Team

The precise dial allows for intuitive adjustment of the comb and easy setting of the desired beard length.

XT 5

All-in-one trimmer, 2021, Braun Design Team

This hybrid trimmer for face and body combines trimming, styling and shaving functions all in one device. The handset is rubberized all around for optimal grip.

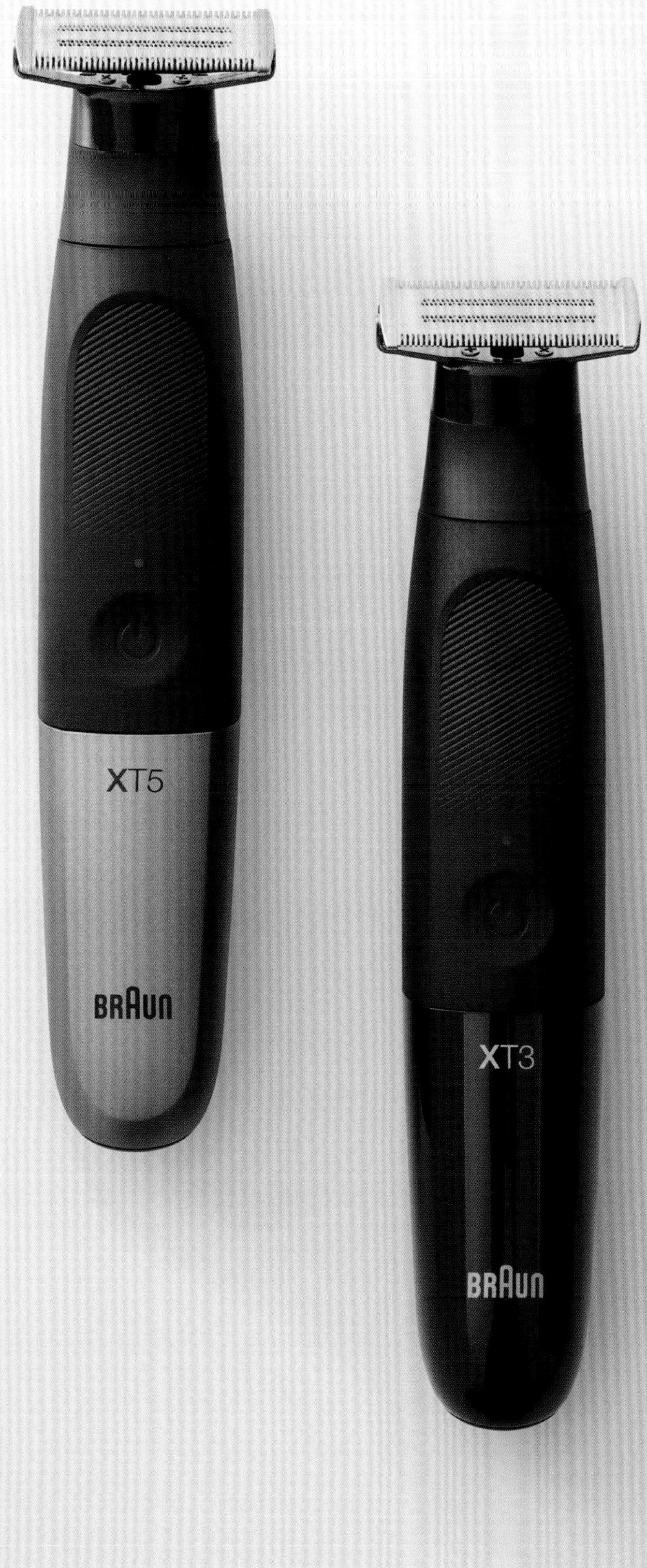

Part 2

Core Identities

What Braun Is About

Braun stands for not only economic success but the creation of useful and sensible products. What is at the core of such an organization? How might you describe its 'DNA'?

The following section will discuss eight 'core identities' that run through more than a century of Braun's corporate history and that have shaped a unique approach to products and design. In this context, the concept of 'core identities' stands for characteristics that distinguish the products and the particular way they are created that makes them entirely unique and unmistakable.

The first five core identities – simple, useful, aesthetic, long-lasting and better – characterize requirements that Braun products have always had to meet and still do. The final three describe the conditions under which these products come about: responsible, together and international.

The eight core identities were developed in collaboration with Braun's current Design Team. These designers see the basis of their work in the attitude and mindset that these core identities express, which is perfectly described in German by the word 'Haltung'.

Simple
Useful
Aesthetic
Long-Lasting
Better
Responsible
Together
International

Simple

Braun products are characterized by the greatest possible simplicity and clarity in use, construction and design. Reduced to the essentials, they avoid superfluous forms, functions or decoration and instead focus on what is important to users. The emphasis is on functional simplicity that is visually expressed in a design that is as clear as it is individual.

T 3, pocket radio, 1958, Dieter Rams. No chrome trim, no colourful plastic and no printed ornamentation of the kind that could frequently be seen in contemporary competing products can be found here. Instead, this pocket radio is defined by a clear, light grey box. The visual quality of simplicity results from the circular openings for the loudspeaker and the dial, forms which essentially originate from the HfG Ulm's school of design. The long- and medium-wave frequency ranges do not have to be set with station names, but only with the frequency numbers, making it easier for the user to find their favourite station quickly.

15
0
150
24
110
22
90
20
75
70
18
65
17
60
16
55

Simple

So, what does simplicity actually mean? Perhaps the most accurate historical definition of simplicity in relation to design comes from Le Corbusier in 1929: 'as such simplicity is a choice, a discrimination, a crystallization. Its object is purity. Simplicity synthesizes.'[1] Design orders and simplifies so that people find objects easy to handle. Complex technical products should also be designed so that they can be used by everyone and not only by experts.

To achieve real simplicity in handling something, users have to be involved in the design process. Observing and understanding how people use things leads to meaningful product improvements even if sometimes it is only small details that make things easier to handle.

Since Braun products are sold around the world, they must be easy to understand in different cultural contexts and be simple for many different people to operate. Moreover, this simple handling must also be reflected in simplicity of design. Simplicity thus becomes a democratic design concept that can fit into the lifestyles of many.

Looking Back

Company founder Max Braun was passionate about construction. 'The simpler the construction, the better it is', was his credo. And, he would add, 'A good device consists of a nail and a board.' He called complicated devices 'machines that prevent milk from boiling over'.[2]

With his commitment to efficient construction that was as simple as possible, Max Braun made a major contribution to how the company saw itself – an image that still holds true for Braun today. Initially he was not that interested in design, and his tool machines and radio components can best be described as looking technical in appearance. Braun's early radio appliances tended to rely on conventional designs so that the devices would fit well into the living rooms of middle-class central Europeans. With very few exceptions until the 1950s, Braun also treated radios and phonographs as decorative 'sound furniture'. This was especially evident in the portable radios whose design featured only functional elements without any decoration – such as the Braun portable radio Kofferradio BSK (Braun Super Koffer) 238, designed in 1937.[3]

We do not know what Max Braun thought about the large-scale New Frankfurt project, which sought to revolutionize design in the 1920s. Certainly, like its proponents, he was in favour of simple, functional design; after all, the modernist New Frankfurt programme was all about simple forms and lines, cube-shaped architecture, boxy furniture devoid of decoration and practical luminaires. Its architect Ernst May also envisioned electric kitchen appliances for the new, modern apartments, but they did not materialize as there were no inexpensive devices on the market back then. Conceivably, Ernst May might have cooperated with Max Braun, another man who made simple yet crucial devices, but this never came about.

After 1955, the collaboration between Max Braun's sons with the Hochschule für Gestaltung (HfG) Ulm led to a radical new understanding of design at Braun that was reflected in updated designs for appliances. Braun's audio devices, with their polished dark wooden chassis and decorated fabric and switch elements, were replaced by housings with hardly any embellishment, made from light, almost white, maple wood. At first, these box-like housings with their large wooden surfaces seemed almost too plain. It was only with the combined radio and record player, the SK 4, also known as 'Snow White's Coffin', that a new design emerged: a combination of white lacquered metal, wooden sections at the ends and a transparent cover that made technology visible in the living room. Though simple it was convincing in its simplicity; here form followed function and, though the design was guided by practical considerations, it still had a semantic poetry.

From the late 1950s onwards, this simplicity, which was in no way mundane but rather harmonious and precisely conceived, was continued by the company's own Design Team under Dieter Rams in all Braun products: in kitchen appliances, flash units and film equipment, as well as in electric shavers, audio equipment and TV sets. However, the idea was not simply to apply this Braun style to every item but rather to rethink simplicity each time. An early example of this is the 1958 T 3 pocket receiver, a hybrid of design elements conceived by the HfG and Dieter Rams. Equally simple but also reimagined are the lighters T 2 / TFG 2 cylindric from 1968 and dymatic from 1980, both designed by Rams. Today, the body of the elongated device with rounded edges is repeated in the Pocket travel shaver, where simplicity has been reinterpreted to suit the context.

This refined simplicity also made its way into product communication. In 1955, Otl Aicher and Hans G. Conrad designed a modular exhibition stand of great visual clarity that was used for many years. The advertising section under Wolfgang Schmittel developed appropriate graphics and fonts for products, with texts that were factual but easy to understand and made even complex topics, like Hi-Fi technology, comprehensible for everyone.

Outlook

As the twenty-first century progresses, product design is no longer only about form. It increasingly addresses the broader compatibility of products with the demands of everyday life – and this life has become complex to say the least thanks to technological changes and new consumption cycles. Here, good design can and must break down the barriers created by complexity; it must enable or facilitate people's access to and handling of (technical) products. However, simplicity in form and handling requires a high level of innovation. The development of digital technology demonstrates this vividly: while initially computers took up entire rooms and could only be operated by specialists using punch cards, today, accompanied by exponential growth in performance, they have become increasingly simple to use. While baby boomers grew up with desktop PCs, millennials grew up with portable laptops and zoomers have never known life without smartphones.

With each of these steps, information technology has become faster, smaller and more versatile – to the point where there is an almost unlimited supply of information today, available everywhere around the clock. A device's ability to sort, categorize and discover this information determines whether people feel it is easy to operate. Voice control makes things even easier today: it is now possible to ask Siri, Alexa or Google almost anything at any time without the hassle of typing, scrolling or swiping. But simplicity also comes with the price of complete transparency. Technical progress, automation and networking will result in even larger amounts of data being available and greater complexity in information.

Going forward it will remain the task of design to order functions and information and make them usable in a simple and understandable way – just as Braun did back in the 1950s, say, in designing the SK 4 with a self-explanatory user interface. Similarly, in the recent design of the new LE smart speakers, Braun not only makes a visual reference to the legendary LE 1 loudspeaker from 1959 but creates a link to earlier Braun appliances through their design simplicity and self-explanatory display language. Fittingly for the times, Braun's smart speakers can be controlled from a smartphone via Bluetooth or an app, and it also boasts voice control. In addition, the smart speaker has a special privacy button: when pressed, it shoots up and its light turns red, signalling that all microphones have been physically disconnected.

True, such a design approach might in the future be applied to immaterial rather than material things; however, the principles will remain the same. What we refer to as simple, whether for a virtual or physical product, is often the most difficult thing to achieve – and arguably it will continue to be the most important task facing product design for years to come.

Simple

T 2 / TFG 2 cylindric, table lighter, 1968, Dieter Rams.
Three geometric elements define this table lighter: a cylinder, a circle and an oblong hole comprising circular arcs and straight lines. These suffice to create a unique product that not only looks good but is ergonomic.

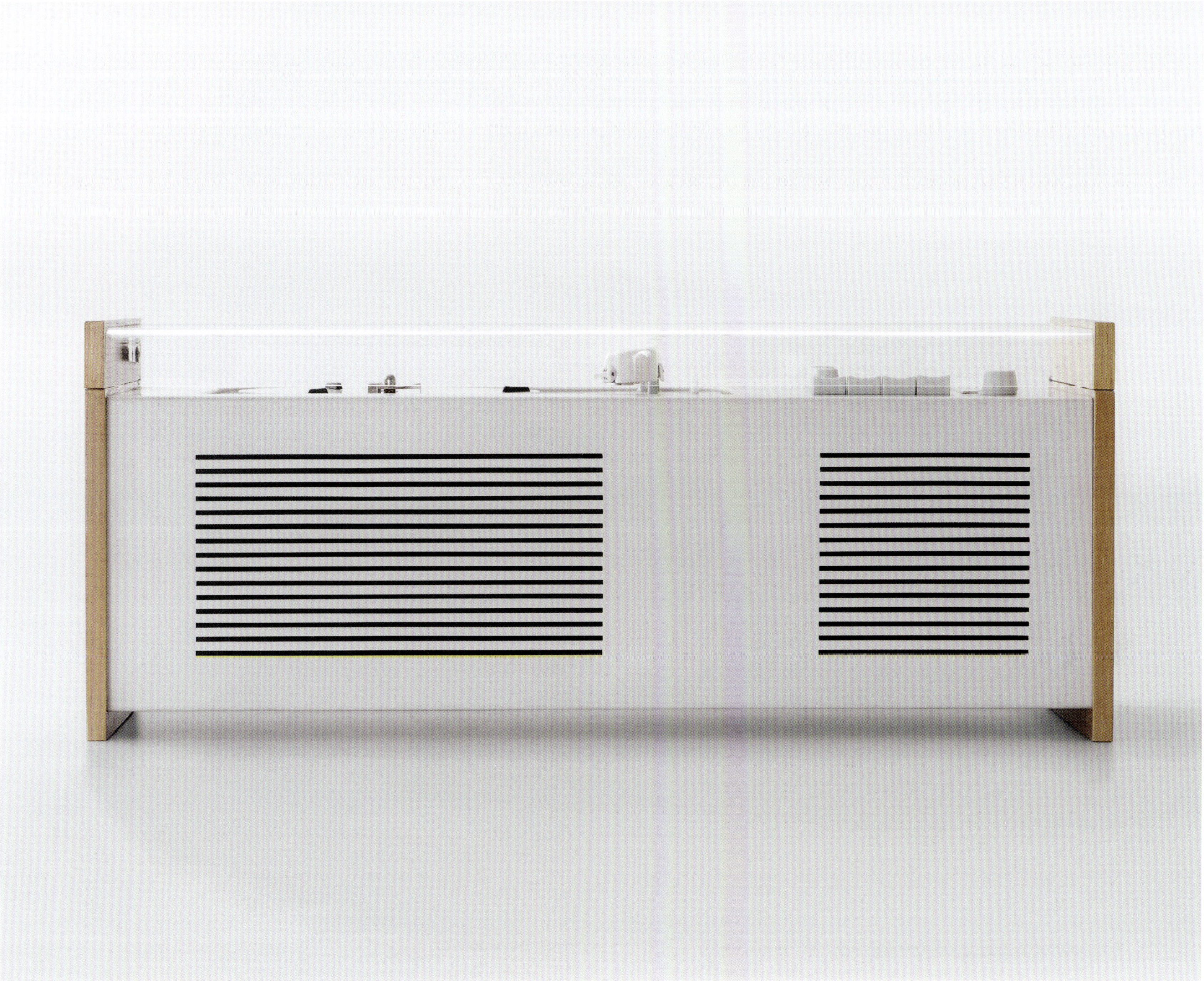

SK 4/1, radio-phono combination, 1956, Dieter Rams / Hans Gugelot / Gerd A. Müller / Werkstatt Wagenfeld. The SK 4 has a clear shape, consisting of a horizontal rectangular block that is transparent at the top to reveal uncluttered and clear controls. Thanks to this clear lid, the controls are always visible, and the device is thus utterly simple and self-explanatory to use.

Simple

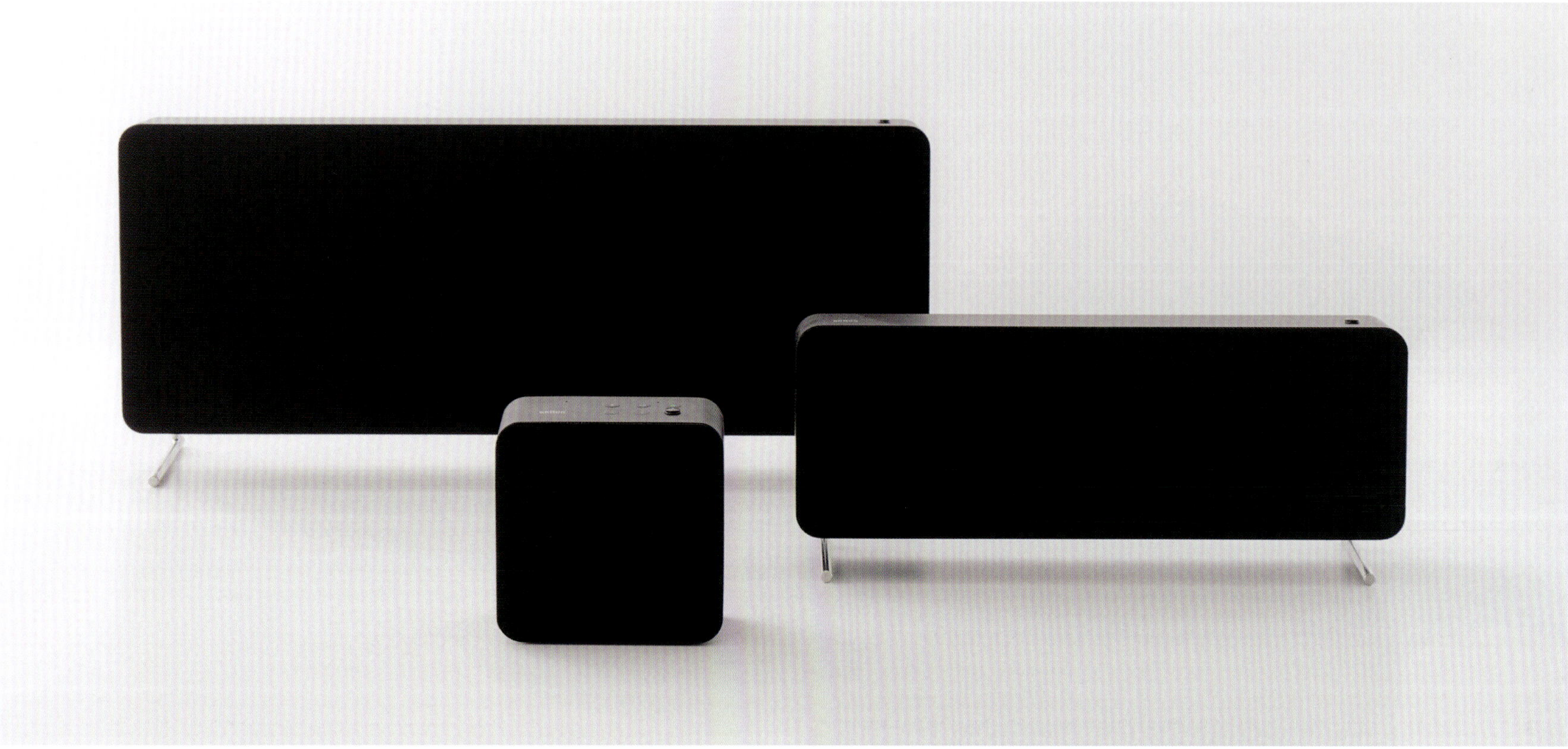

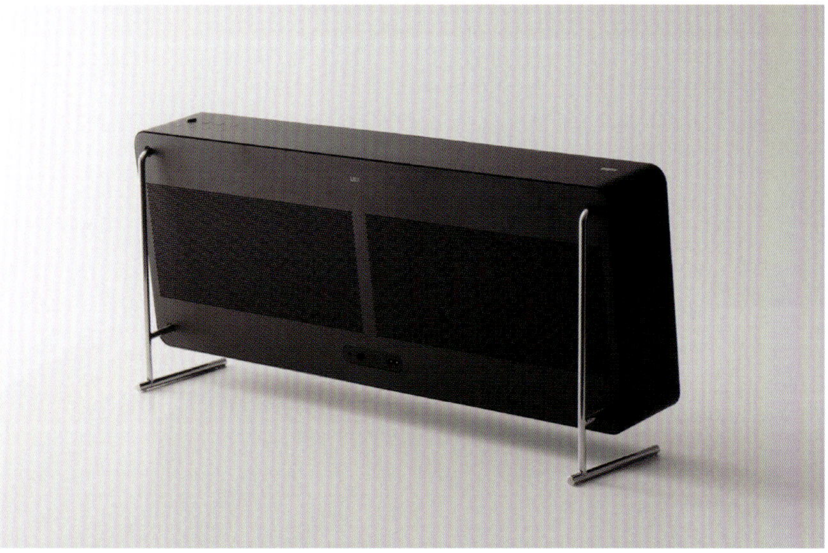

LE 01, LE 02 and LE 03, smart speakers, 2019, Braun Design Team. The reduced design of these smart speakers in combination with their different stands gives them a high-quality yet unobtrusive appearance – a quotation of the LE 1 speaker's form from 1959. They also seem to float. These speakers can be controlled using a smartphone or the human voice. To prevent the device from unintentionally turning on, its microphones can be physically turned off with a push-button switch – whether they are on or off is indicated directly at the switch position by a coloured ring of light.

Pocket, electric shaver, 2020, Braun Design Team. The Pocket combines visual and functional simplicity on several levels. The rectangular but soft shape is reduced to a compact design, while the rounded corners make the shaver comfortable to hold. A large cap protects the shaving elements during travel and eliminates the need for a case. Once removed, it reveals the rubberized grip zone. The shaver has a flat base where the battery indicator and the charging port are located, allowing it to stand and enabling space-saving storage. Superfluous elements are nowhere to be found.

Useful

Braun products are highly functional and useful. With a deep understanding of human behaviour, they are designed down to the smallest details to meet users' needs. Their design makes technology easily accessible and intuitive to use. Usefulness is also the result of versatility. That is why Braun products are often systems that unite various functions in a single product.

KM 3/31 and 32, food processor, 1957 and 1964, Gerd A. Müller.
The KM 3/31/32 food processor, first launched in 1957, is a prime example of practicality through conceptual design and technical longevity. A variety of attachments and inserts, which allow for the performance of a wide range of tasks with just one appliance, make it suitable for universal use in the kitchen. For almost four decades, this food processor has been manufactured nearly unchanged, and the clear design by Gerd A. Müller has endured as fashions come and go.

BRAUN

Useful

Things that you want to use to achieve a certain goal need to function properly. This may sound obvious but sadly it is not. Many products promise to provide a certain function and then fail to do so. They are technically complicated or of poor quality, unreliable or awkward to use. It is not immediately apparent how to use them – to do so you often must study lengthy operating instructions before you are able to handle them.

Braun products explain what they are and what they do by the way they are designed, from the overall shape to the detail of a switch. A high degree of functionality, versatility, simple usability and a self-explanatory, intuitive user interface are part of the Braun tradition and still make up the DNA of its products today.

Looking Back

This DNA was already evident in the very first devices Max Braun produced: both the 'Trumpf' belt connector and drum connector stood out for being practical and easy to use. Max Braun and his successors have always sought to improve on existing products and systems. Throughout the company's history, one of Braun's real strengths has been to improve the technology and design, and hence the usefulness, of electronic devices.

Fewer Control Elements

'You press the button – we do the rest' was the advertising slogan for the Kodak No. 1 camera, also known as the Kodak Box, which was first produced in 1889 and made photography possible for everyone. You cannot get more self-explanatory than that. With the HF 1 television set, the 1958 brainchild of Herbert Hirche, Braun also produced a device that seemed to manage with a single button. In the middle of the television's front, there is a push-button that promises: 'Press me and you've got television.' The relationship between the button and screen (the main components of a television set) is therefore obvious to all. Hirche relocated all the other switches that were not constantly needed, such as the volume control or the channel selector, beneath a flap at the top of the set. The idea was to persuade those people still sceptical about television's technology to see just how simple it would be to operate. At that time in Germany, however, there was only a single television channel.

Another highly revolutionary device would later take on this one-button symbolism: the iPhone by Apple. Pressing the power button opens a wealth of possible applications that are also typically self-explanatory. Apple's chief designer Jonathan Ive repeatedly drew on the Braun model for this idea.

Intuitive Operation

At the start of the new digital age, intuitive operation was already a key concern at Braun and not least of all an important factor for the economic success of its products. Although the first Hi-Fi modules had numerous adjustment options, they were clearly arranged. To reduce the number of rotary dials on the stereo equipment, Dieter Rams developed a dual control consisting of rotary dials and a flanged tongue for the left and right channels, which was first employed in the CSV 13 amplifier of 1961. His idea halved the number of dials in this area of the product.

Moreover, the use of colour coding for the switches to indicate the many functions became a typical design feature at this time. This was illustrated very clearly in the audio 308: the on/off switch was green – something people are familiar with from traffic lights – while other important functions were visually highlighted using coloured switches. The colour coding was restricted to a few colours that were set out on a simple Leitz cardboard box with RAL colour numbers and information on the respective function. This simplified operation using different coloured switches can be found in many Braun devices over a long period of time. Consequently, this usefulness, alongside simple operation, became a distinguishing feature for Braun products.

The last Braun Hi-Fi set, atelier, had control elements on two levels. Frequently used switches were located visibly on the device's surface, while those needed less often were concealed behind flaps ensuring an excellent layout for the entire control panel. The design was inspired by the idea Herbert Hirche originally used for his HF 1 television in 1958. This ease and efficient mode of operation, even when a product offers a wide range of functions, is typical of Braun Design.

Self-Explanatory Functions

Looking back, Dieter Rams says, 'in the products I designed it was always important to me to ensure the functions were as self-explanatory as possible ... It is still a big problem today designing the exterior of a device in such a way that the user understands the various functions available. It's a big challenge for designers.'[4] It was not only Rams himself who took up this challenge, but the entire Design Team. Take the first food processor KM 3, designed by Gerd A. Müller in late 1955 or early 1956 and launched in 1957. It simply had one blue rotary dial with Roman numerals from I to III and a (zero) dot – you cannot show how to increase the mixing speed more intuitively than that. Additionally, the blue colour also cropped up on the top of the beaters so that an association was created between switching on the machine and the function this triggered.

Usefulness through Versatility

What sets the KM 3 apart is not only its intuitive operation but also its versatility in the kitchen. Thanks to a host of different attachments for cooking and baking, users benefit from a wide range of applications in just one device – a complete system that does not take up much space. This concept that products which are versatile are also automatically useful continues to define Braun products today, and not only its kitchen appliances.

In the case of the BT 9 beard trimmer, the basic archetypal shape is complemented by a central, precise rotary wheel that clearly shows both the setting and the various comb length options. The hold switch located directly beneath the dial prevents the setting being altered involuntarily. What makes the trimmer especially useful is not only its precise trimming but also the many combs and trimming attachments that can be used for a wide variety of tasks. Once again here clarity of use is complemented by versatility.

Outlook

Simplifying the operation of technical devices will continue to be one of the biggest tasks facing designers in the future. While in the past designers relied solely on the design of actual switches and product graphics to create pathways for intuitive use, devices will increasingly be used via a broader interaction of gestures, including touch, virtual display, movement or voice control. The almost unlimited freedom and flexibility this offers when it comes to representing information for users requires design principles and an exact understanding of the way things are used to enable the simple handling of products despite their great complexity. Moreover, product functions are increasingly automated. Even our coffee maker tells us in good time, 'I want to be descaled now.' As early as 1992 the Braun AB 40 vsl alarm clock used new technology to switch off the alarm via voice control, while the AB 50 rsl relied on reflex control technology. The concept behind what was quite simple in technical terms back then is not too different from contemporary scenarios.

Take the Braun iCheck 7 blood pressure monitor from 2018. Physical interaction with the device is connected to the digital measuring and monitoring of vital signs in a logical and intuitive fashion. The product's display and gyro sensors, then, help the user to adopt the correct arm position for measuring their blood pressure.

Digital systems are penetrating more and more areas of our lives. For designers, this means developing suitable solutions that are attractive to look at and, above all, fulfil the 'Aesthetics of Use'[5] criterion. It is vital to think in terms of enduring product systems that unite various functions and are characterized by clarity of use thanks to displays, apps or voice control. Good, intuitive design is of crucial importance in every form of interaction with users.

Useful

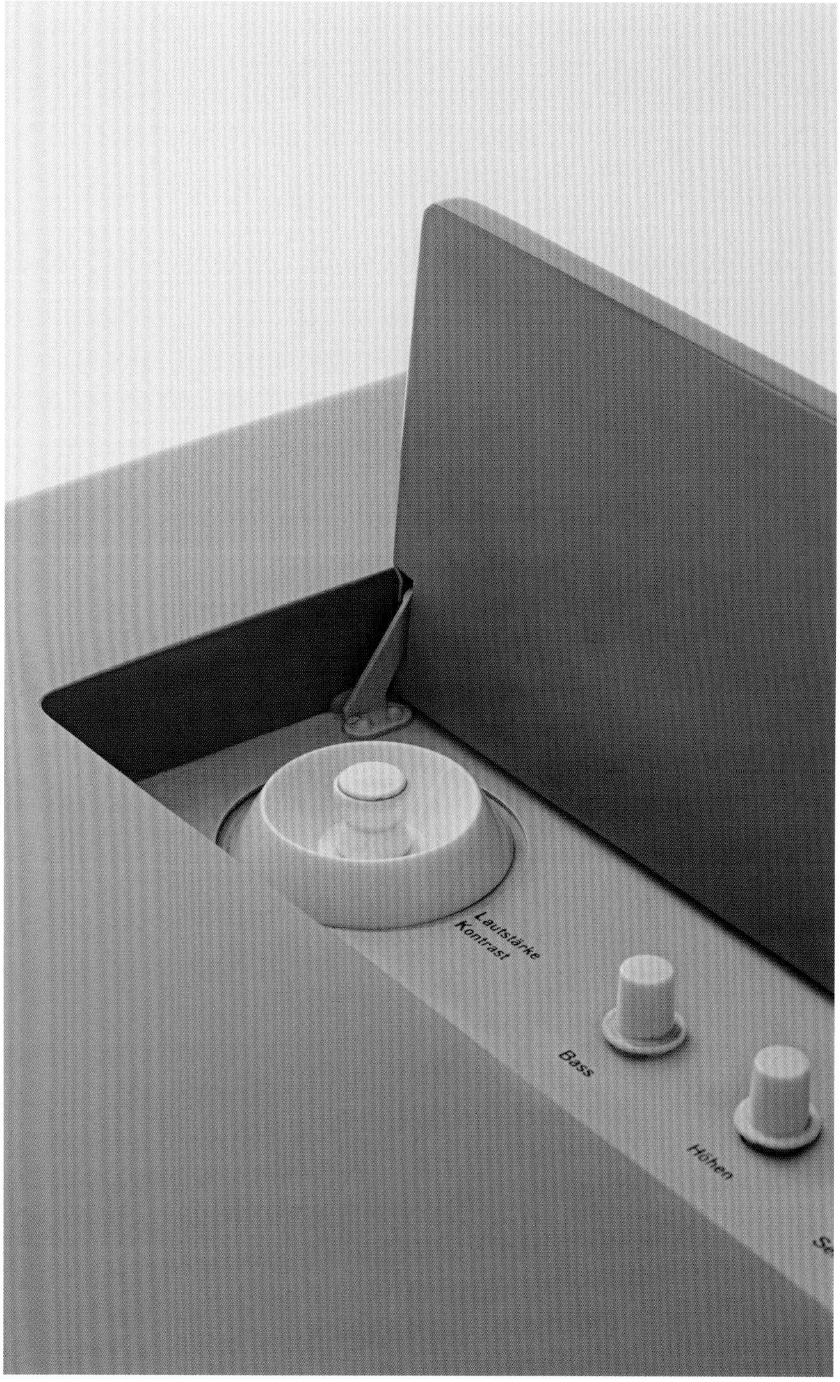

HF 1, television set, 1958, Herbert Hirche. Until 1958, there were no television sets with only one visible button. With the HF 1, Bauhaus student Herbert Hirche creates clear semantics: 'Press the button to watch television'. All other functions, which are required less often, are hidden under a flap on top of the television. This self-explanatory user interface becomes the standard for later Braun devices as well, long before the topic of interface design became a subject of more general interest.

BT 9, trimmer system, 2021, Braun Design Team. The BT 9 is a multifunctional trimmer system with interchangeable cutting elements and combs. With its high-quality execution, the various product parts together form a long-lasting, practical system for various applications on the face and body. A precise wheel allows the exact setting of the desired hair length in 0.5-mm increments, while the different blade widths enable precise execution of the various cutting tasks.

Useful

studio 1000, Hi-Fi system, 1965, Dieter Rams. This Hi-Fi system, which bears the suffix '1000', is top tier in terms of its technology and design, forming part of Erwin Braun's project for a 'Grand Design'. All the switches are precise and clear not only from a technical perspective but also in their design. They consist of circles and lines only, which are sometimes combined in a special way, as is the case for the dual-function switches.

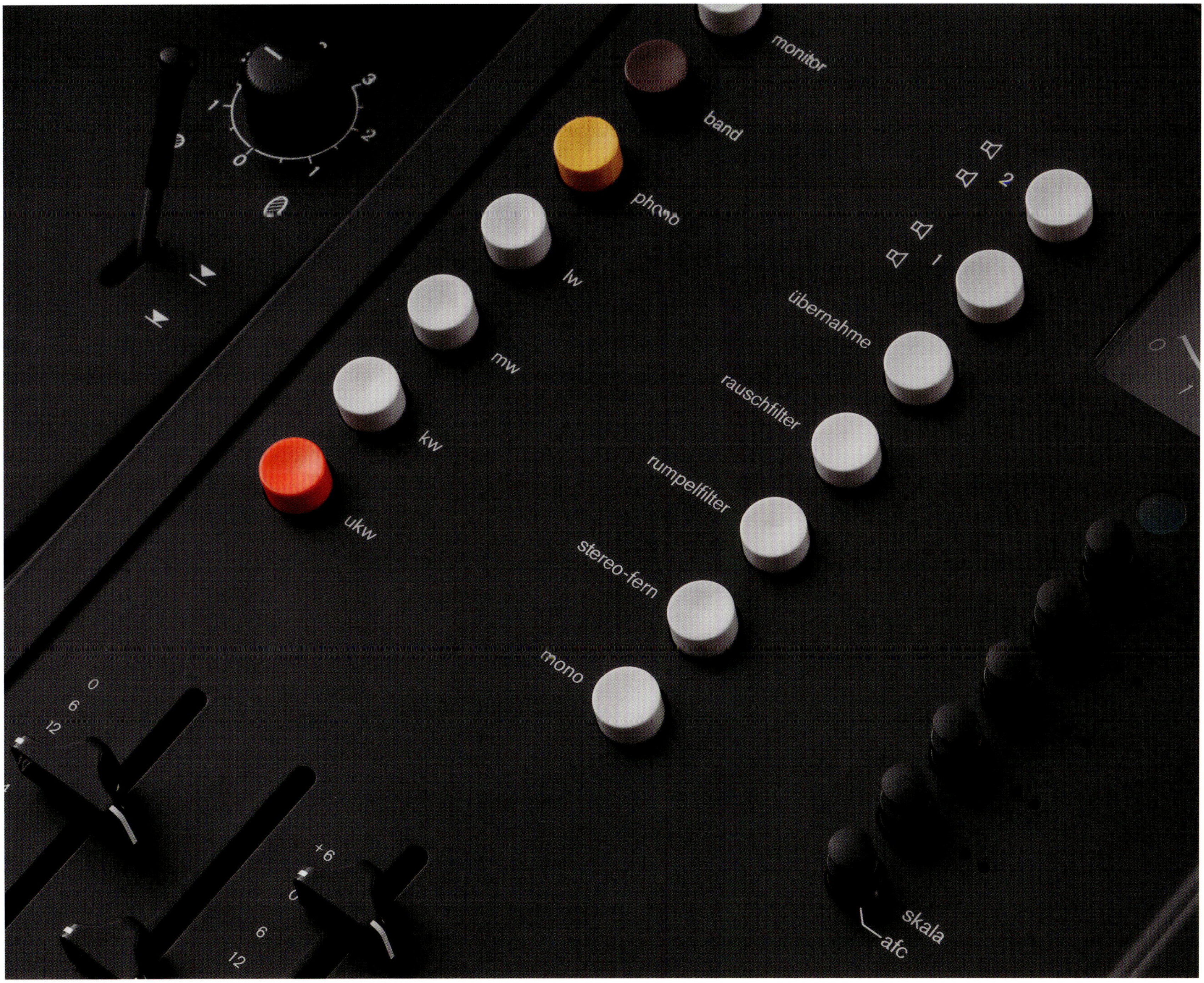

audio 308, radio-phono combination, 1973, Dieter Rams. The plastic age reached Braun, too, in the early 1970s. With the audio 308 from 1973, however, the functionality and value of the design remain at the same level as before in the large radio-phono combinations. The coloured highlighting of individual switches simplifies use and becomes a typical design element for Braun appliances.

Aesthetic

Braun products combine functionality and simplicity to create visual harmony. The individual design components are carefully selected and arranged with a strong eye for aesthetics, reducing them to only those that are truly necessary. The design follows a precise understanding of proportions, order and balance. This results in products that are functional and stand out for their simplicity, clarity and harmony; products that have a universal appeal while remaining unique.

ET 66, pocket calculator, 1987, Dieter Rams / Dietrich Lubs. This pocket calculator is a testament to the fact that plastic can equal durability and precise operability. But something else is also essential here: a beauty that results from harmony and contrast. The convex keys create highlights on the polished surface, while the function keys in brown and dark green are only slightly offset in colour from the black numeric keys. The on and off buttons are in the complementary colours of red and green, and the sum sign strikes a dominant final chord in yellow.

BRAUN
100.
M+ M− MR MC +/−
7 8 9 ÷
4 5 6 ×
% 1 2 3 −
CE C 0 · = +

Aesthetic

People have appreciated the beauty of simplicity for centuries in Japan, long before it was employed in art and design by European modernists. The fifteenth century saw the emergence of the Japanese room style known as a 'washitsu' (or tatami room), and the corresponding tea ceremony that rejected lavishly decorated Chinese imports in favour of more minimal decoration. 'As soon as we recognize emptiness our mind attempts to fill it with the content it is missing,' writes Japanese designer Kenya Hara in his book White.[6] It is telling that the Japanese word for white ('shiroi') is derived from the word 'ichijirushi', which means clarity.

Aside from focusing on designing useful products at Braun, the company's other aim is to always produce something attractive. Apart from their usefulness or logical, economic, ecological, functional and haptic qualities, Braun products always stand out for their lack of ambiguity: they do not refer to something else but solely to themselves and their function.

The product semantics of Braun since the mid-1950s can be compared to a painting by Piet Mondrian or Kazimir Malevich. The less visual information a picture contains, the more observers can add their own personal associations. Consequently, the actual image is only created in the mind, becoming even more striking. Canadian media theorist Marshall McLuhan coined the terms 'hot' and 'cool' media to describe this phenomenon and compared it with Hot Jazz and Cool Jazz. While 'hot media' completely engage the senses through their many details, the 'cool medium has so few details that it requires greater attention and involvement on the part of recipients in order to be able to understand it'.[7] It is this subtle interactive process of focusing on what we see and our own associations that makes us look at it again and again.

Richard Fischer was a former designer at Braun and, after 1968, a professor at the Offenbach University of Art and Design, which at the time favoured rather lavish product semantics. He would later say, 'although precision rejects everything that is human and emotional, interestingly it is precisely those products that stand out for their great precision which have a high emotional value.'[8]

Looking Back

Visual quality plays a huge role in product design because every item of use is looked at much more often than it is actually used. And who wants to be reminded of an industrial machine every time they look at their kitchen appliance? It bears considering that into the 1950s the appearance of the great majority of household appliances was still inspired by heavy machinery. However, Braun succeeded in also establishing an elegant look for the kitchen.

Back in 1971, Fritz Eichler commented on this topic:

> Anyone who looks at our devices more closely will ascertain that every one of them appears 'neat and tidy' – there is no superfluous decoration – and that everything has its place. Every detail and be it no more than the head of a screw fits in with the other elements and is coordinated with them – the separate parts of the device are organically arranged to form a whole. We try to achieve this as economically as possible, but not for reasons of economy. It has to be one of the most difficult tasks there is – using the fewest and most simple means to arrive at a harmonious design ... But when designing our products, we not only have its function in mind – we also try simultaneously to make it as attractive as possible. We go to a lot of effort to ensure that aesthetically too, they are styled down to the last detail – in material, form and colour. And it is precisely in the details that the quality becomes apparent – all the more so the more similar things are.[9]

The combined radio and record player SK 4, dating from 1956, is a prime example of this. It broke from the traditional phonograph design that people were accustomed to seeing back then by establishing a completely new look for audio devices – not least of all thanks to the transparency of its acrylic glass cover. Furthermore, the appliance's rear sides were carefully styled, so that it became a standalone design object of sorts, designed to be viewed from all sides. The SK 4's unusual loudspeaker slits are another striking feature. While they are assuredly functional, they are realized with the utmost care and became a trailblazer for the design of future loudspeaker openings at Braun. Since the SK 4, many of the Braun audio devices have rather unique appearances. The RT 20 portable radio particularly stands out here: its loudspeaker openings consist of horizontal slits arranged in a circle that are strikingly interrupted at the centre by a vertical bar.

This unusual aesthetic quality continues to be a hallmark of Braun's technical appliances. The XT 5 beard trimmer, inspired by the look of the sixtant shaver of 1962, likewise features two contrasting materials for a clean-cut appearance. We find a similar trend with the wristwatches Braun has produced up until today. The carefully selected thicknesses of the minute markers and the ever-present yellow second hand continue to reference the original design of the AW10 wristwatch from 1989.

Outlook

Dieter Rams set this out in one of his Ten Principles of Good Design: 'Good design is aesthetic. The aesthetic quality of a product is an integral aspect of its usefulness.'[10] After all, devices that we use every day colour our surroundings and influence our sense of well-being. But something can only be attractive if it is well made.

This could already be a formula for the future, since it explains why design that is felt to be attractive can enter our lives and stay there – aesthetics remains one of our basic needs and a lasting one at that.

Aesthetic

RT 20, table radio, 1961, Dieter Rams. Next to the table radios that, until the early 1960s, tended to be made of polished wood with gold and silver trim, this Braun RT 20 tube radio is an aesthetic exception. Both the arrangement and design of the controls are essentially taken from the SK 4 radio-phono combination and set at an angle to the front. Instead of the turntable, a 20-cm round speaker from Isophon is used behind the front panel, which also delivers decent bass. The most striking design element, however, is the sound openings: horizontal slits inserted into the circular shape with a vertical centre bar.

atelier, modular Hi-Fi system, 1979–89, Peter Hartwein / Dieter Rams. The atelier Hi-Fi systems of the 1980s represent a final high point in the design of audio equipment at Braun. The bevelled edges of the housing make the modules appear particularly slender, while switches that are rarely used are hidden behind flaps to keep the control panel clear – an element that recalls Herbert Hirche's HF 1 television. All visible switches are carefully designed from clear geometric elements.

Aesthetic

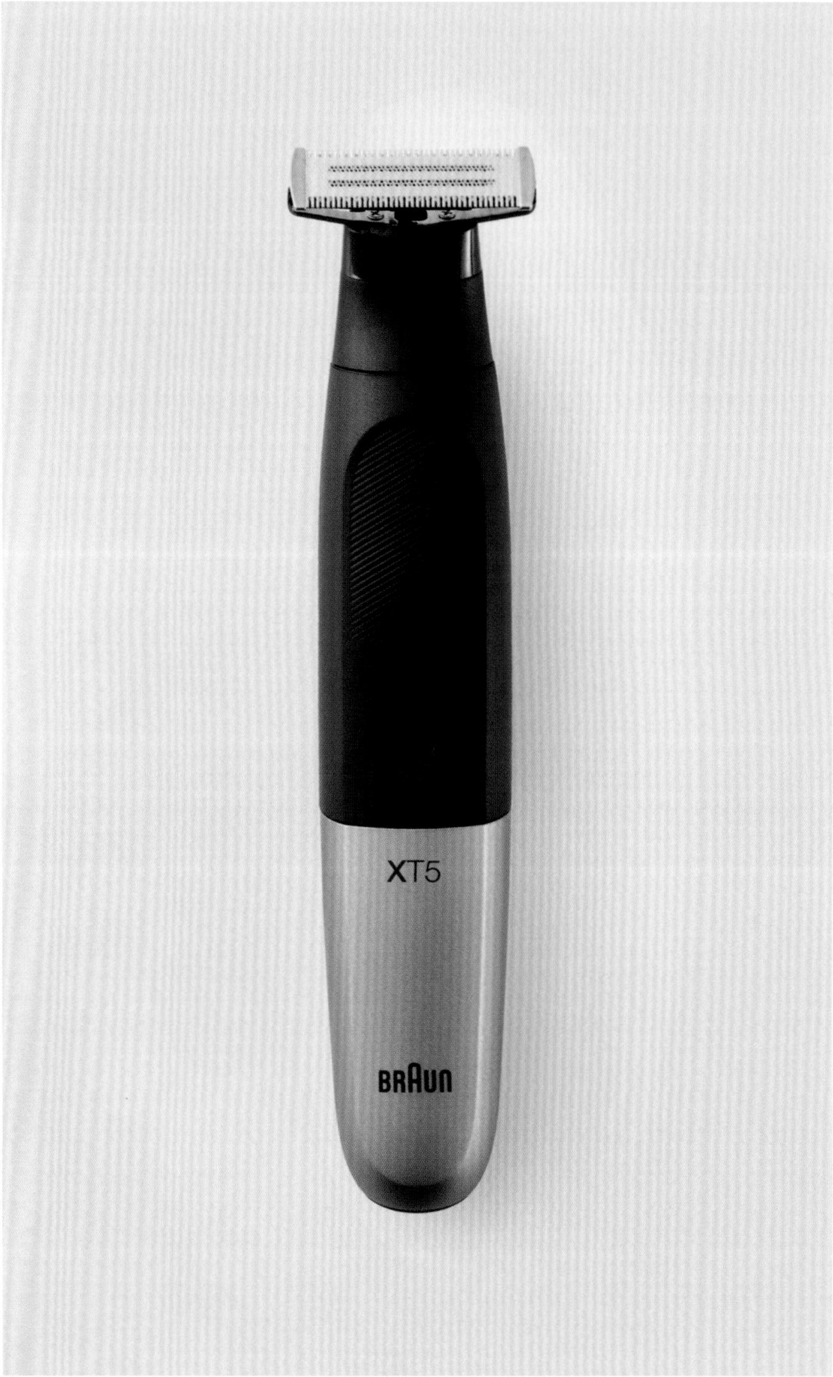

sixtant SM 31, electric shaver, 1962, Gerd A. Müller and Hans Gugelot / XT 5, all-in-one trimmer, 2021, Braun Design Team. The sixtant shaver from the 1960s is visually unmistakable thanks to its simple form combined with the striking colour and material design of its brushed aluminium head and matte black housing. The vertical separation of metal and plastic is also taken up in the XT 5 all-in-one trimmer, which reinterprets this typical Braun aesthetic of simplicity that has characterized many other modern products. As with the sixtant, the XT 5's interchangeable cutting head is formally integrated into the simple overall shape.

AW10, wristwatch, 1989, Dietrich Lubs / BN0211, wristwatch, 2016, Braun Design Team. Braun originally planned to make analogue wristwatches only as promotional gifts. With the AW10 from 1989, however, Dietrich Lubs succeeded in creating a form so self-evident and convincing that the watch became the basic form for numerous subsequent models and has lost none of its appeal to this day. It is still in production in its original size, as well as in a slightly enlarged version (the AW10 EVO), and has inspired newer models, such as the particularly thin BN0211.

Long-Lasting

Braun has always been committed to creating durable products to counter excessive consumption and swift obsolescence. Braun products are technically advanced and built to be repaired for extended life and use. Its products are of a high quality and work for many years – sometimes for an entire lifetime.

Another core characteristic of Braun Design is that it is visually enduring. This is achieved by omitting anything fashionable. Combining useful features, sound technology and an enduring design results in products you can use for ages and enjoy having around you.

citromatic, citrus juicer, 1972, Jürgen Greubel / Dieter Rams. The long-selling citromatic has been in production since 1972, so it is safe to say that it is a true classic. Art history lovers will see an echo of Doric column capitals in its top, and a select few might even relate the squeezing cone to the Duomo in Florence. The device does not require a switch – simply pressing the halved fruit onto the cone activates the juicing motion. An elegant cover made of transparent acrylic glass can cover the device's top for when it is not in use.

BRAUN
citromatic

Long-Lasting

Long before the words 'ecology' or 'sustainability' became such buzzwords, Braun focused many of its efforts and ideas on creating enduring products. And this also means that a product must be visually appealing to its users for many years.

As early as 1985 in his Ten Principles of Good Design, Dieter Rams pointed out the need for longevity in products:

> Good design is enduring. There is nothing fashionable about it that quickly seems outdated. In this respect well-designed products differ radically from the short-lived trivial products of a throwaway society that can no longer be justified today.[11]

And elsewhere Rams wrote:

> The keyword is 'built-in obsolescence'. A product with a highly fashionable design loses its appeal quickly and ages. I'm convinced that today design must actively contribute to slowing down the cycles of product renewal and our throwaway habits. The era of lavishness and waste has come to an end. We need our resources for much more important tasks. You can no longer justify fashionable, short-lived products.[12]

Today, considering the global shortage of resources, dramatic scale of climate change and environmental destruction, these ideas articulated almost forty years ago are discussed in all sections of society and political parties but without any consensus on how to respond. Braun was truly well ahead of its time.

Looking Back

It was back in the 1920s that mechanical engineer Max Braun insisted on building things that were above all sound and enduring. His drum detector is not only technically better but more durable than the crystal detectors that preceded it.

Then, in the second half of the twentieth century, when consumer society was already well established, Braun began to apply this durability concept to the outer appearance of a product. However, not every technical advancement at Braun automatically got a new design simply to motivate people to buy it. On the contrary, designs were retained until a truly technical innovation made a new design necessary. For example, from its S 8 (1965) model onwards, the Nizo film cameras featured a body made of an anthracite-coloured die-cast core with sides of sheet aluminium – a design that would be retained for a decade. Even older models do not seem outdated. Braun would arguably have continued to make highly attractive video cameras had it become involved in new camera technology after the Nizo.

Another well-known example of enduring Braun Design is the KM 3 food processor from 1957, whose design would remain largely unaltered for thirty-six years. In 2022, the MPZ 2 citrus juicer even celebrated its fiftieth birthday and is still being made today. Similarly, in 1972 Florian Seiffert created a space-saving coffee maker, the KF 20, with an innovative vertical design. Twelve years later, it was advanced by Hartwig Kahlke to become the KF 40 and indeed continues to be manufactured by Braun as the Aromaster Classic. Braun has realized several such archetypes, such as acrylic glass covers for its record players, the HT 1 toaster with its long slots designed by Reinhold Weiss and the wall-mounted Hi-Fi system designed by Dieter Rams.

Enduring products also need to be easy to repair. In 1984, Albrecht Schultz, a Braun board member, declared, 'Repairs and maintenance must be easy to carry out and require little effort. In terms of reliability and service life, Braun devices should also be well above the average of comparable devices on the market.'[13]

Schultz touched on a topic that is highly relevant today. Making products easily repairable to extend their life calls for innovation. This includes being able to replace rechargeable batteries. Replacing the battery in Braun shavers, for instance, has been possible for many years thanks to an easily accessible screw that is hidden from sight in a cable duct.

Outlook

Consumer products with their short lifecycles and limited replaceable components will undoubtedly have to be rethought if we want to save the environment. Economic growth cannot be achieved through the increasing use of disposable products. Accepting this and adapting our behaviour has to be one of the greatest economic challenges of our times. Braun has recognized this global need, and since its products are built to last and designed to be visually appealing for years, it will continue to stand out amongst its competitors.

Today, Braun is deliberately returning to a design approach based on simplicity and geometry that visually conveys the imperceptible product values of precision, quality and durability. Underlying this approach is an understanding of design that stands a good chance of being well suited to future product requirements – where a screw for simple and understandable disassembly of the product may also be part of the design.

dymatic, lighter, 1980, Dieter Rams. The dymatic is a flat lighter with a large switch that is pulled down to keep the thumb safe from the open flame. With its calm, round shapes, the dymatic is easy on the hand and the eye – then as now.

KF 40, coffee maker, 1984, Hartwig Kahlcke. A tower-shaped coffee maker takes up little space. Hartwig Kahlcke integrated the water tank, pot and filter to give the appearance of one single cylinder. This coffee maker has been a constant in a great many kitchens for forty years and is still available today.

Long-Lasting

AB 1, alarm clock, 1987, Dietrich Lubs. Displaying as much clockface as possible and only as much housing as necessary, the AB 1 alarm clock from 1987 shows the time and wakes you up with the legendary Braun signal. And that is it. Because it is so reduced, it does not seem to age.

BC10, BNC010 and BC09, digital alarm clocks, 2021, 2012 and 2020, Braun Design Team. The clean, visually reduced design of the digital alarm clock series matches the devices' functions: the simple display of the time and uncomplicated activation of the daily alarm. With different implementations of size, display and colour combined with sound, radio or projection features, the Braun alarm clock finds its permanent place next to the bed.

Better

Braun products stand for innovation and perfection. At Braun, it is not so much about creating spectacular inventions as it is about taking the difficult task of perfecting something, both technically and in terms of design, so that users intuitively feel it is consequently 'right' in every regard. This understanding of constant improvement defines the way Braun products are conceived, designed, constructed and manufactured.

T 1000, world receiver, 1963, Dieter Rams. When radio was still the most important electronic mass medium, a world receiver was considered the ultimate in radio equipment, enabling the user to receive shortwave stations from all over the world. The first portable devices were developed in the US shortly before World War II. In 1963, Braun launched the T 1000, which was supposed to be able to do everything that a normal world receiver did but better. Indeed, it did – in both reception as well as design.

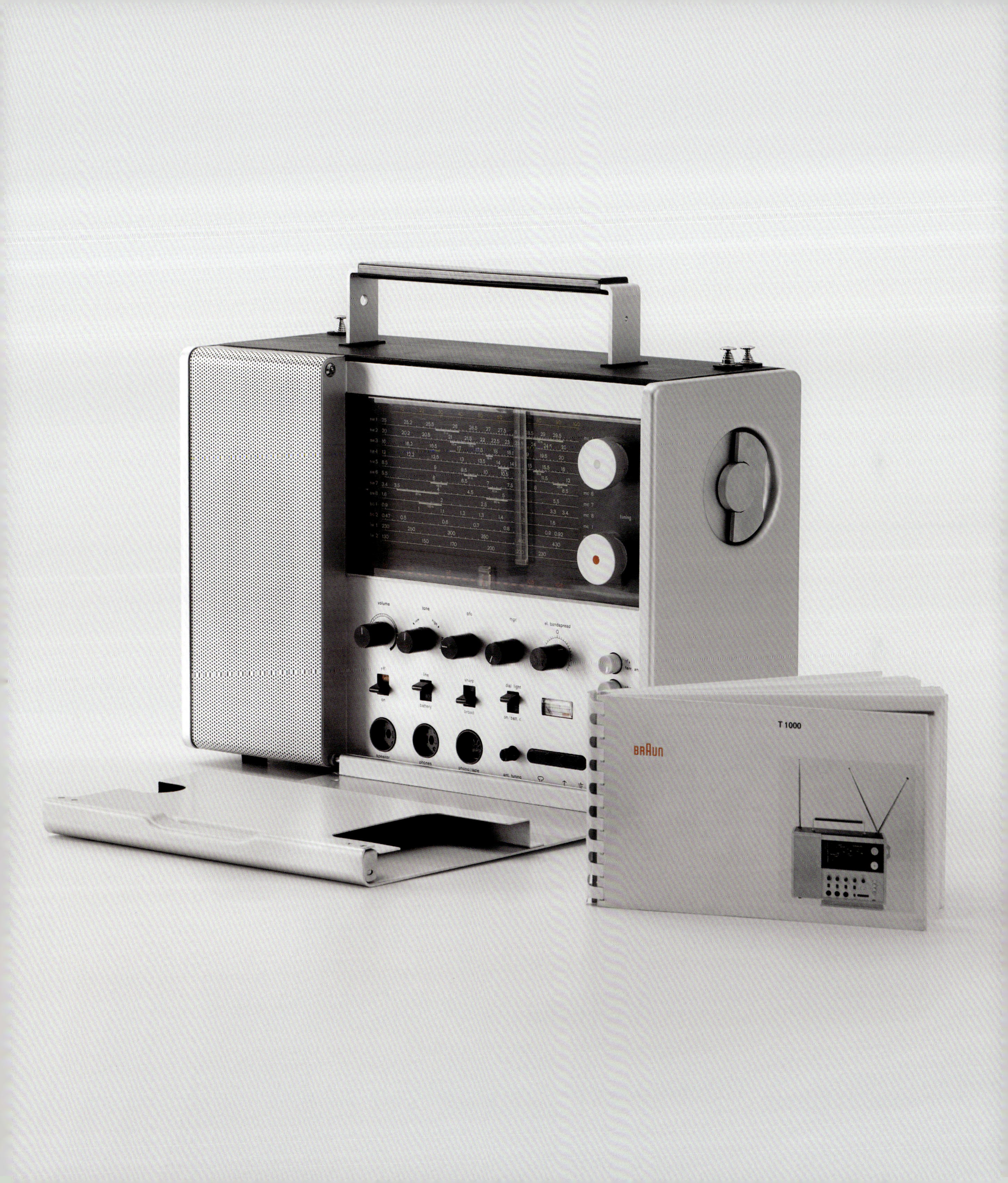
BRAUN
T 1000

Better

From the very start, the desire to always make good products better was reason enough for many employees to join Braun, and it was not unusual for someone to spend their entire working life with the company. Today, designers and engineers still focus on crafting products with the best possible technical properties. This does not necessarily mean inventing new products but rather rethinking and improving existing ones.

Needless to say, Braun did not invent shavers, radios, cassette players, film cameras, slide projectors, hair dryers, lighters, clocks, calculators, toasters or coffee makers. However, Braun products were often better than those of competitors, and that is often still the case today.

Looking Back

We can go way back in Braun's history to trace this desire to realize the better product. For Max Braun, 'doing it better' than others was the motivation for his actions. However, the drive to do and make everything better than the competition started in earnest when the second generation of Brauns took over. Erwin Braun was primarily driven by the aspiration to have a superior company: he improved the firm's internal structures, encouraged staff to take responsibility for their actions, promoted a constructive culture of debate and created a sales structure characterized by close relationships with retailers and fixed sales prices. Another innovation was the direct delivery of the early flash units to photographic dealers – in return for payment in advance and avoiding wholesalers as middlemen. This ensured the units were delivered to customers fast and at Braun's low cost. That said, Erwin Braun also cared profoundly for the well-being of his staff. In the 1950s, the company's lavish health service even included physiotherapy, a sauna and dietary cuisine. What Braun provided was anything but standard.

The company's most visible metric of 'doing it better' was the revolution of Braun Design. Thanks to the early involvement of the Hochschule für Gestaltung (HfG) Ulm, Braun was able to present its new design approach boldly at the International Radio Trade Fair in Düsseldorf in 1955. This was followed by milestones of better design such as the SK 4 Hi-Fi deck, known as 'Snow White's Coffin', various kitchen appliances and the modular audio elements by Dieter Rams.

Launched in 1963, the T 1000 World Receiver might not have represented a new product type, given that comparable devices had been around since the 1940s in the United States, but it was superior technically and a design masterpiece, created by Dieter Rams when he was barely thirty years old.

Not least of all, this 'doing it better' at Braun can be detected in innovative solutions that improve the usefulness of products. Since the late 1970s, its hair dryers have featured a handle that tilts inwards, making them much more ergonomic to hold – a feature that was subsequently copied by many other manufacturers. Its coffee makers with vertically aligned elements take up much less space in the kitchen. Similarly, with the new CareStyle Compact steam iron line, professional ironing technology is skilfully arranged so that the iron sits on top of a water tank, again following a vertical design. Thanks to its highly compact footprint, the unit, which otherwise normally takes up a lot of space, can be accommodated on a normal ironing board.

The various aspects of 'doing it better' can be illustrated especially well by the development of shavers at Braun. In simple terms of unit sales, electric shavers have represented the most important product area at Braun for more than sixty years. Beginning with the first model, the S 50 by Max and Artur Braun developed in 1949, there has been a constant improvement in shaver performance and design. The sixtant SM 31 dating from 1962 represents a milestone: it has a highly effective, extra-thin shearing foil and an iconic design. And indeed, it swiftly became one of Braun's most successful products. In fact, it was Gerd A. Müller who two years earlier had provided the basic shape with the SM 3, but it was its makeover by Hans Gugelot that resulted in a perfect design comprising a housing of brushed black plastic and a brushed injection-moulded head clad in chrome.

The micron plus from 1979, designed by Roland Ullmann, is another milestone that offered better grip and a new look. Since the first shavers, attempts had been made to improve the grip using ridges or grooves, but now a new hard-soft technology came into play that involved combining soft grips with a hard shell. The idea was workshopped in lengthy trials by Ullmann and Klaus Zimmermann. A multi-component injection-moulding process developed by Bayer AG made it possible to produce this development in a way that was justifiable in terms of production technology.

Compared with these shavers, today's models from the series 9 rely on a much more advanced technology. Rather than one large foil, the S 9 has several narrow, flexibly suspended shearing foils and long-hair trimmers that are mounted on an oscillating and laterally pulsating shearing head, which enables large areas to be shaven but is still gentle on the skin. The basic concept of shearing parts that are easy to clean, as seen in the sixtant, has been retained to ensure an excellent shave over a long product life.

Outlook

Today, improving products means considering not only their usefulness and aesthetics but also the ecological impact of their materials or production process, not to mention their repairability. It goes without saying that a long period of use is favoured over rebuying or replacing. Products and the materials they consist of must be part of recycling processes that have as long a lifecycle as possible. To achieve this, it is crucial to have high-quality products with functions that meet the real needs of their users.

Dieter Rams articulated this concept very fittingly with the catchphrase 'Less but better'. In the future, better design will also mean persuading people to behave more responsibly as consumers.

Better

micron plus de luxe, shaver, 1980, Roland Ullmann.
Shavers usually had a housing of either metal or plastic, but the micron plus de luxe from 1980 has both: a perforated metal housing with inserted plastic grips, giving it a non-slip grip even in wet hands. Even before that, Braun was already producing its micron shavers with a hard plastic shell and soft grips. Today, almost every Braun shaver and epilator boasts this hard-soft technology.

Series 9, electric shaver, 2014, Braun Design Team.
The Series 9 shaver stands for engineering excellence. Decades of experience in the development of electric shavers have gone into the technical design and manufacture of the Series 9 to optimize cutting performance and shaving comfort. The replaceable shaving head not only allows for many years of use with consistently high shaving results, but the user can also benefit from technical innovations to the cutting system of newer product generations.

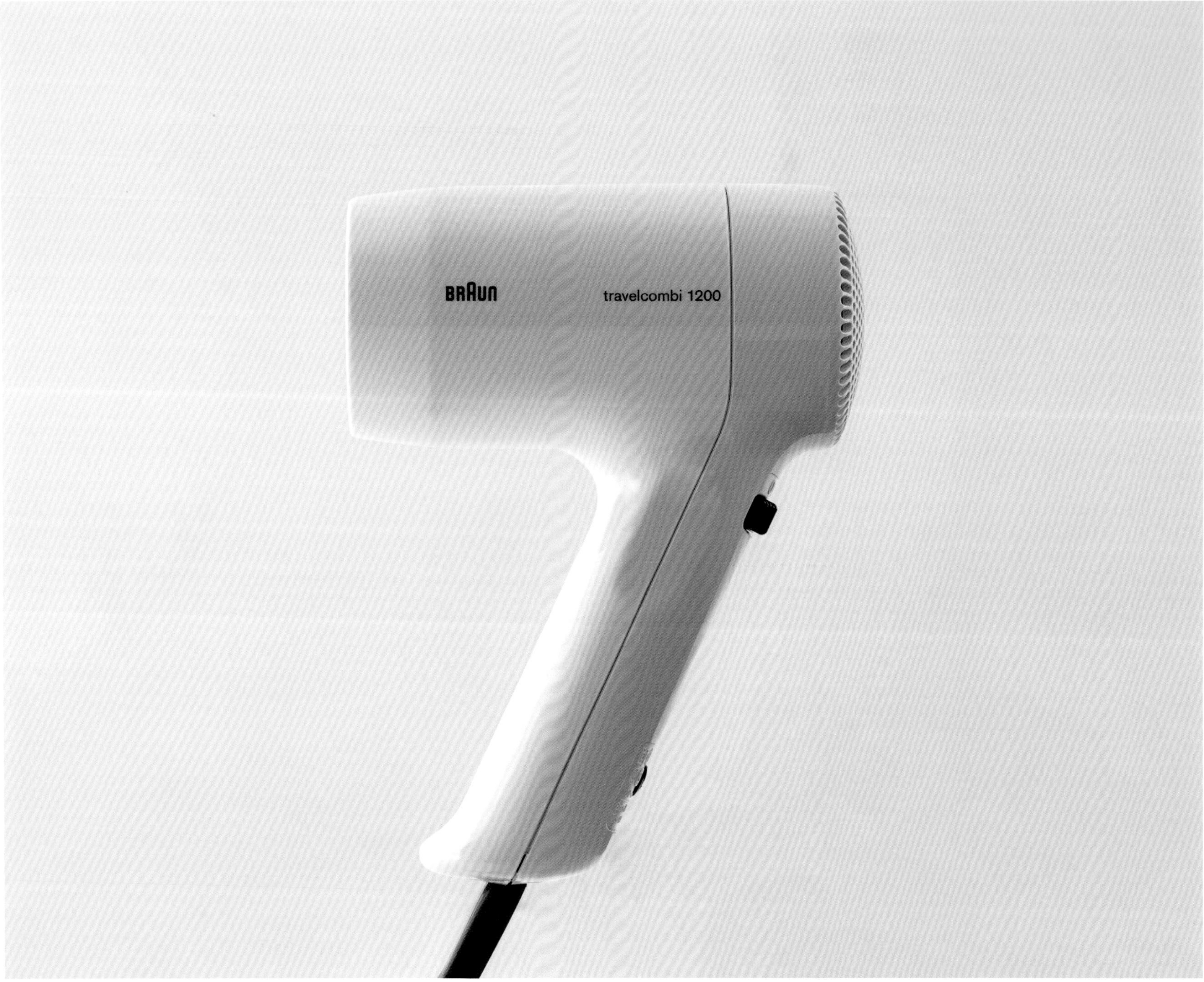

travelcombi 1200, hair dryer, 1982, Braun Design Team.
The handle of a hair dryer had historically been at a right angle to the blower. Around 1960, however, Braun's competitor Moulinex introduced a pistol grip angled backwards on its model. In 1978, Braun's design department put the handle in the other direction, making it much easier to dry hair at the back of the head. The travelcombi 1200 is also remarkable because of its compact size.

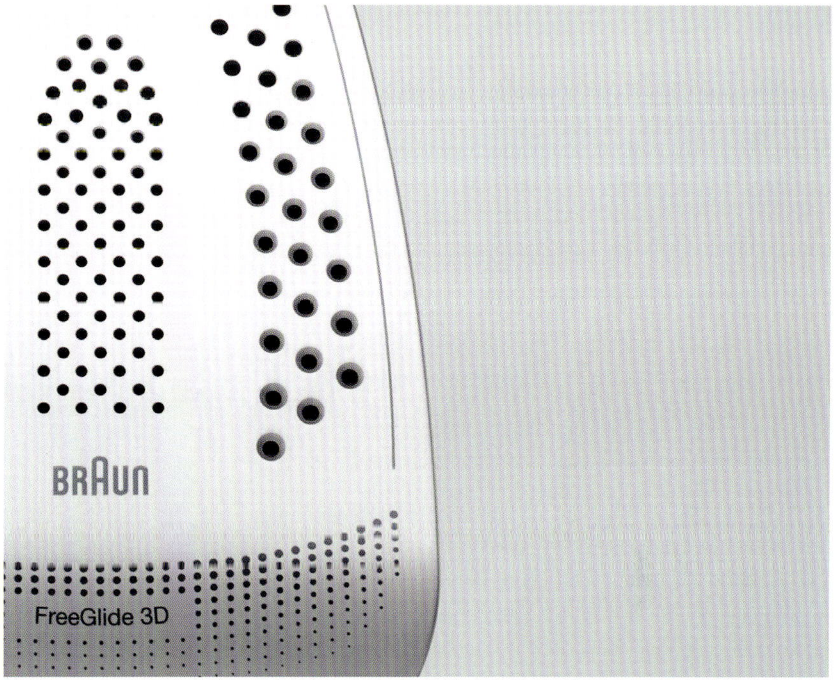

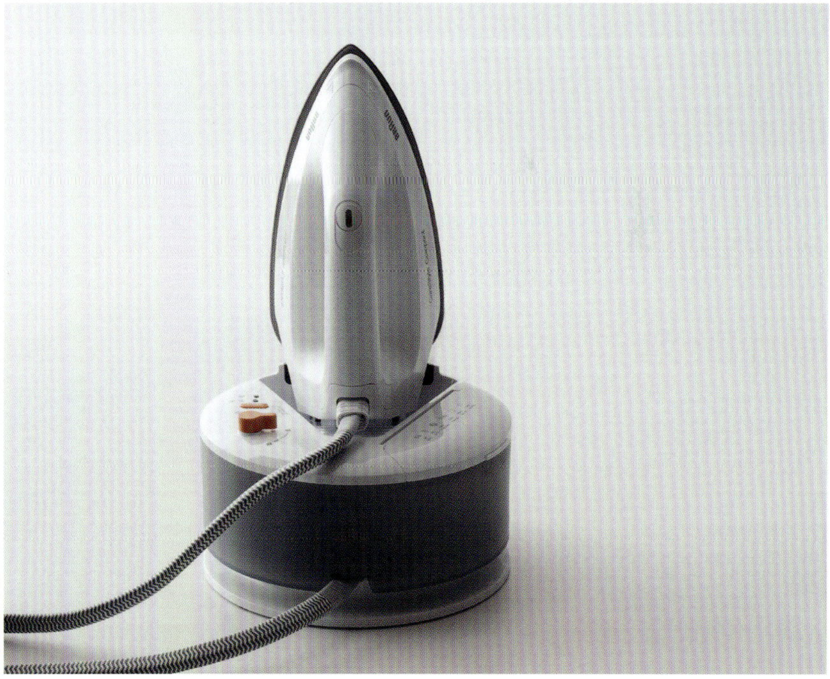

CareStyle Compact, steam iron, 2018, Braun Household Design Team. In the CareStyle Compact iron, the steam technology has been improved to such an extent that an entire steam ironing station can effectively fit in the smallest possible space. This means that the new technology can also be used in confined areas. In addition, the bent iron soleplate on the back prevents shirt buttons from snagging – a seemingly small improvement with a big effect that makes ironing work considerably easier.

Responsible

Do the objects a company produces have a purpose? Are they good for people and our environment? And how does a company treat its employees? Corporate responsibility has many faces. It involves how a company sees itself and how it wishes to be seen, as well as how it reconciles social and economic responsibilities. These are precisely the topics that preoccupied Erwin Braun, the son of the company's founder, and his ideas have shaped corporate values that continue to have a lasting influence on the brand today.

Series 7, electric shaver, 2019, Braun Design Team. Braun shavers allow easy replacement of the external shaver parts over a period of many years, but components hidden inside can likewise be easily replaced. In the new generations of products, care is taken to ensure that they are simple to disassemble: screws positioned in the Series 7 shaver's charging plug, for example, make it straightforward to open the device and change its rechargeable battery.

Responsible

Protecting our planet has become a major challenge for all of us. Product design today is therefore concerned with answering questions about how we want to live with our objects in the future and, above all, how we can live with them. The answers to these questions inform how design needs to adapt, and so the slogan 'form follows function', coined by Louis Sullivan in the late nineteenth century, has gained a new and wider meaning. Design no longer only gives shape to a form but rather shapes a concept. This idea has long been the case at Braun.

The need for a new concept of design has now been recognized, at least in some industries, and can be seen in debates on the need for a circular economy[14] or the 'cradle-to-cradle' principle.[15] In some cases, these ideas are even implemented. If this transformation from a linear economy, or a 'one-way street' leading from resource removal to waste disposal sites, towards a regenerative system of energy and materials is to succeed, we need designers and engineers who are not content to merely use environmentally friendly materials and have them recycled. Specifically, we need to insist on longer product lifecycles, options for repairing, reusability and the ability to disassemble objects for recycling.

Looking Back

When Erwin and Artur Braun took over the company in 1951 after the death of their father, they envisioned creating a modern company, one that would distance itself both from the Nazis' management principles (that demanded blind loyalty) and from purely profit-orientated thinking. Their model for the future was inspired by the concept of a social market economy as articulated by Ludwig Erhard, the German Minister of Economics at that time. With this in mind, Erwin Braun sought to create a model company that not only saw its employees as staff but also as members of a community with shared values.

As early as 26 April 1951, company founder Max Braun had proposed a company pension fund, although it was only realized after his death. Erwin Braun took up and expanded this idea of corporate welfare. Not only was a pension fund introduced but also a range of other benefits, including a moderate share of company profits, a total healthcare package and the provision of affordable housing for company employees. Erwin Braun's vision was of a comprehensive welfare system for all Braun staff. This resulted in a large majority of employees identifying as being part of a modern corporate family and feeling filled with a sense of purpose. For many staff even today, being a member of the Braun family, or a 'Braunian', is more than simply a professional identity.

From the 1950s, the company health service played a large role in employee well-being, especially since medical care in post-war Germany was still inadequate in many areas. A company that not only employed its own doctor and dentist but also offered sport, physiotherapies and even a sauna was a great exception at the time. Werner Kuprian, who built up the health service at Braun starting in 1954, had a holistic concept of preventive medical care. When the company moved to Kronberg in 1973, an impressive health centre was realized that is still used by many staffers today and serves as a model for setting up corresponding centres at other Braun locations.

Today, Braun's parent company, Procter & Gamble (P&G), also strives to operate as a socially responsible company. It runs comprehensive social and healthcare facilities for its employees and promotes a way of living and working in line with today's social values. In doing so, P&G continues the corporate ethics of Braun in areas such as diversity, inclusion and equality. The proportion of women and men in the entire workforce is roughly equal across all levels of the company – a rarity worldwide. Just as Braun considers social responsibility for its own workforce as hugely important, the company also thinks carefully about the social responsibility of its products and technical production conditions.

In his manuscript Zum Gestalten bei Braun (On Design at Braun), Erwin Braun wrote in 1961: 'The term "design" is given a broad interpretation. Designing our devices includes everything relating to function, construction and material. The impact of the appliances on people's senses is seen as particularly important ... The design of appliances is related to their use.'[16]

And, in a lecture he gave in Canada in 1975, Braun designer Dieter Rams explained:

> We are in the process of destroying the biosphere and are now starting to worry about getting poisoned, falling ill or suffocating. But what impact do high-voltage pylons, skyscrapers, motorways, satellite towns, street lighting, car parks or elevators have on our psyche, and our relationships to one another? ... I can imagine that centuries from now people will get goosebumps when they consider the lack of thought we gave to filling our homes, our cities, our countryside with a chaotic mix of paraphernalia of all kinds ... The age of mindless design that could only flourish in combination with a mindless production aimed at mindless consumerism is coming to an end.[17]

A social orientation based on values such as tolerance, equality and the notion of sustainability still plays an important role at P&G today. These values are not only embedded in the company's culture but are firmly written in its policies. Climate change, renewable raw materials, wind energy and zero waste are firmly on the agenda of this company with a payroll of almost 100,000 that reaches roughly five billion people worldwide with its products. Since 1999, P&G has published an annual sustainability report that sets out its efforts and measures and, since 2020, this information is supplemented by clearly defined sustainability goals for 2030 and 2040.

Outlook

In many companies today, managing work-life balance plays an important role in reconciling our work and private lives. However, the two are fluid, and younger generations no longer place such a strong focus on ensuring a strict division between them. For many, working from home has long since been routine – more so since COVID-19. As companies with a social responsibility, Braun and P&G also must reinvent themselves in view of shifting work conditions and adapt to not only those changes but the expectations of talented specialists. How important will it be in the future to get together to do work when even today many people perform their work virtually? Will we only meet in the Metaverse? Can you still feel part of a company if you no longer see your colleagues in person? In a world that is constantly changing, both companies and employees will have to find ways of enabling a meaningful balance of work and life. We will need to create a sense of community as an expression of responsibility for others but also as a fundamental prerequisite for corporate success.

Responsibility is demonstrated above all by our attitude to environmental protection. Climate change and environmental policy will continue to have a stronger impact on manufacturing firms and will require a change in both thinking and action, either voluntarily or through taxes and legislation. Braun was early to take a stance on protecting the environment through its corporate values. For decades, small technical appliances by Braun have stood out for their longevity and repairability. Braun shavers can be opened using a screw located in the plug socket to replace the battery, for instance. Cutting parts and shearing elements can also be replaced.

Other important goals of product development now and in the future include ensuring that components are not only made out of recycled materials but that they can easily be separated for reuse or recycling, even for future regeneration processes that might not currently be supported. Only by operating in a climate-neutral and environmentally friendly manner can companies do justice to the high level of responsibility they have in relation to people and the environment.

Braun's health service in Kronberg even had its own swimming pool, with swimming seen not just as a leisure activity but part of Werner Kuprian's preventive healthcare.

Artur and Erwin Braun inspect the canteen, which opened in September 1960. Since as far back as 1957, the company's cafeteria only served organically produced food.

Together

The process of creating a product is influenced by the range of internal and external conditions under which a company operates. These include technical innovations, business parameters, how a company sees itself, cultural contexts and ethical issues, not to mention general social and political circumstances. Ultimately it is the people involved who have the greatest influence. Much depends on how they interact and whether they succeed in consistently developing a convincing shared attitude.

Discussions at the big drawing board take place frequently and are part of the design process at Braun. Here, the team can discuss not only the big picture but also the smallest details, or even correct them right away with a pencil. Left to right: Rams, Pechstädt, Schneider and Oberheim, mid-1970s.

Together

A company's ability to create good and responsible products also depends on the atmosphere in which those products are made. This collective atmosphere has a lasting impact on the cooperation of those involved in the development process and ultimately on the products and their use.

To this day, an environment of respectful collaboration, mutual motivation and great trust plays a decisive role at Braun. Indeed, without such an environment, Braun's products could not become what they are. Therefore, the Braun Design Team and technical development team are not only closely connected in terms of location at the Kronberg Development Centre but also closely aligned organizationally. Frequent meetings ensure that all issues can be dealt with thoroughly yet swiftly, leading to a joint solution. The heads of the design and technology departments are part of the Braun executive team and have a direct influence on future product strategies and their realization – the joint goal being to develop better products.

For the Design Team, this means that workers can trust in each other and the larger organization. Only in an atmosphere of certainty and mutual trust is it possible to inspire creative competition and encourage opposing ideas. Effectively, this means people must be prepared to develop ideas collaboratively and always be open for new perspectives to achieve the best possible design. Such an iterative process is more successful than an adherence to dogmas. After all, an ongoing joint process of reflection and communication does not block creativity or the realization of designs but rather encourages them.

Looking Back

Arguably the first Braun workshop, located as it was in a building courtyard, was not particularly inspiring. However, all that would change in 1928, when the company opened its own factory building on Frankfurt's Idsteiner Strasse. With its large strip windows, the building thoroughly reflected the modernist design outlook of the 1920s. The functional, austere architecture seemed to propel the firm into a new era. At any rate, source material reveals that it led employees to identify more strongly with the company. It was the people much more than the architecture who then defined the company's working environment and spirit of cooperation. Max Braun was the typical patriarch, a man who was strict with others and with himself but who also took care of everything. Such a hands-on, authoritarian leader with active experience of the war was not atypical for post-war Germany in the 1920s. Without a doubt Max Braun could be very direct both in manner and language, but he was also a team player and sought his employees' advice.

When his two sons took over at the helm in 1951, the atmosphere and manner of collaboration altered considerably. It was Erwin Braun above all who established a different kind of corporate structure and culture. Though Erwin was not unlike his father in terms of his style of management and wish to shape things, his aim was to create a model firm based on teamwork, more benefits for employees, a quieter tone of voice and a willingness to listen and discuss things. Until 1967, he developed a working environment of mutual recognition between the diverging mentalities of those concerned with the business and the employees from fields of technology and design. Fritz Eichler, who from the mid-1950s onwards was responsible for design at Braun in a management capacity, was also successful as a mediator. It should be noted, however, that Erwin Braun did not have a master plan for reorganizing the company: he instead gathered people around him whom he trusted and who contributed their expertise so that any changes could be implemented smoothly.

In 1969, Fritz Eichler wrote that the prerequisite for the new corporate concept was the 'ever closer interaction of everyone involved – an interaction of marketing, technical development and design – and from the very start: from the idea and the conception for a product via the development through to its final formal design and its presentation to the public.'[18] Discussions between equals, mutual respect and trust are more important and more efficient than some of today's 'design methods'. Then as now, Braun consisted of one large communications network. The 'conversations in corridors' are often more important than formally arranged meetings.

When so many people work together who then makes the decisions? In a lecture he gave in 1977 about the design and decision process at Braun,

Dieter Rams describes things as follows:

> Useful design must always be revised from the very start. You have to be aware of what is possible both technically and in production. You have to understand the wishes, the requirements of the consumers or perhaps more accurately the 'users'. You have to understand the rules of business – what makes it possible, and what – to put it harshly – would be cheating ... So, the decision to produce good design has to be a company decision. It can't only be the designers or the design department that tries to establish it or is responsible for it. It must be one of the fundamental goals of the company, and it must be consistent with all the other objectives of the company.[19]

Outlook

Since communication and collaboration in such a company can be complicated, it remains important for future decision-makers to share a joint location and remain in close contact. New ideas are often the result of intensive talks where different thoughts are swapped and help to move things forward. Simultaneously, there will be new ways of communicating and working digitally. Video conferences have been firmly established in the working world. For years, it has been the practice for employees of global companies to forward their projects to colleagues in different parts of the world. Such collaborations result in time and space being experienced differently. Hybrid collaborative working in the office and remotely from home across languages and time zones requires cooperation in an atmosphere of trust if excellent results are to be achieved.

Discussing a design concept with someone and asking for their feedback is quite different from truly collaborating as a team to create a new design. Thinking and designing together combined with the willingness to advance another designer's idea, or to allow someone else to advance your own idea, results in a quality of design that one person alone is hardly capable of achieving – no matter how talented they are. To further strengthen this type of creative collaboration, the decision was taken at Braun to no longer attribute products to single designers but always to name the 'Braun Design Team' as a product's author. It then follows that responsibility is also shared across the entire team.

With increasing geographical distances between staff members, clearly defined corporate values with which employees can identify and which define the context of work will become ever more important. Not only do they foster a sense of community they also make it possible for new product developments to stand out from those of the competition. Moving forward, collaboration will increasingly take place in virtual spaces under higher creative and temporal pressures and be challenged with transferring the special atmosphere and culture of a company into the digital realm.

In the future, the principles of trust and mutual acceptance, as well as a culture of open discussion that is as non-hierarchical as possible – long championed at Braun – will doubtless remain the basis for working on and designing better things.

Together

A device also requires labelling when it comes to communicating numerical values. This philosophy applies just as much to the analogue dial of a clock as it does to the settings on Hi-Fi systems or kitchen appliances. For many years, this was the responsibility of Dietrich Lubs, assisted by Waltraud Müller and always in contact with colleagues working in design, model-making and technology. After 1955, the house typeface was Akzidenz Grotesk, a sans serif typeface that dates to 1898 and itself had many co-designers.

Conversations and discussions are an essential part of the design process at Braun. This collective action takes place not only at the drawing table but, perhaps more importantly, during the model stage – as seen here in 1977 with a model for the Hi-Fi system with the PC 1 and RA/RS 1 devices.

International

The world of the twenty-first century is often described as a global system of networks, international supply chains and diverse yet connected cultures. German philosopher Wolfgang Welsch asserts that the differences between cultures need not only be divisive but that they can mingle harmoniously. According to Welsch, when two cultures come together, they can find mutual points of interaction that overcome divides without this convergence producing a bland global sameness.[20]

However, ongoing wars and a resurgence of nationalist thinking today seemingly fly in the face of this optimistic view, and both globalism and the desire to overcome borders are once more being put to the test.

VW buses from a Braun advertising campaign in front of the Eiffel Tower in Paris around 1970 – modern marketing for modern design.

BRAUN
BRAUN ELECTRIC FRANCE

International

Products intended for sale in many countries must be highly compatible with different cultures and societies. If this can be achieved, such products can help promote the connection of cultures.

From the start, Max Braun was active in international markets, rubbing shoulders with people whose mindsets also impacted his own ideas and work. This early openness paved the way for Braun to become a company whose products are globally relevant, comprehensible, attractive and consequently enjoy acceptance worldwide. As a subsidiary of Procter & Gamble (P&G), Braun belongs to one of the largest global corporate players.

Looking Back

Braun products and their design are often described as being 'typically German': they are orderly, tidy and functional. That might be correct to a certain extent, but Braun has always been anything but nationally orientated. This holds true for the company's history, for the design of its products and for the company's founder and his sons. Already as a young man living in Berlin before World War I, Max Braun attended English courses; in the days of the chauvinist Reich, that was most certainly not customary practice and would at least seem to indicate his transnational mindset and interests. Just a few months after he had set up what was initially a small firm in Frankfurt, he attended the Frankfurt Fair and paved the way for a trade relationship with an English partner. Several years later France would become the firm's most important export country for its record players. Max Braun had even set up companies in Europe and North Africa early on in Braun's history. In other words, relatively soon after it was established, the company was already doing business with international customers.

A European People's Radio

After 1933, all German radio manufacturers produced the so-called 'Volksempfänger' (People's Receiver) that sold at a fixed price of 76 Reichsmarks, making it affordable for everyone. Back then Braun standard radios cost roughly twice as much as this, and the best devices with a large range had a price tag of more than 300 Reichsmarks. The idea for an inexpensive device initially came from Joseph Goebbels, minister of Nazi propaganda, who wanted to use radio for the dissemination of Nazi ideas to the masses. Indeed, over time, the new medium would become the most important propaganda instrument of the new ruling party.

Goebbels might have wanted everyone to listen to the radio, but he did not want them to hear just anything, especially foreign programming. The radio he championed was therefore a very simple model lacking shortwave bands and intended mainly to receive German programmes. Until the late 1930s, however, when advertising its radios, Braun often described them as enabling 'European reception'. For example, in 1937 the tagline for the Phono-Super 738 was 'Overseas reception via the perfect short-wave band is a matter of course'.

To convert the Volksempfänger into a receiver of European stations, a special attachment was developed at Braun in Frankfurt to improve reception. The advertisement explains: 'The Braun attachment ... converts the Volksempfänger into a powerful dual-circuit receiver and consequently enables clear reception of almost all European stations.'[21] Many radio owners continued to use it during World War II when those caught listening to so-called 'enemy broadcasters', such as the BBC or Radio Moscow, faced imprisonment or even the death sentence.

Internationality since 1951

Max Braun's sons shared his views on internationality and took them on board wholeheartedly. In 1956, both sons visited America: while Artur Braun studied factories and production methods in the eastern United States, his brother Erwin met up with the designers Ray and Charles Eames in Santa Monica. Even though they did not find models for their own production in the United States as hoped, the trip did widen their horizons and ended with a visit to Olivetti in Italy. With its new approach to product design, Braun was in a similarly awkward position to Otl Aicher, who had been commissioned to create the corporate design for Lufthansa. Anxious not to overemphasize the company's Germanness, Aicher decided to use Swiss typography and forms, avoiding anything national that would consequently identify the brand as 'typically German'.

From the 1960s onwards, the world market was opened to Braun thanks to its functional design that made its products compatible with diverse cultures. Within just a few years, the company transformed itself from a regional player to a supplier of electrical devices across Europe. Aside from the excellent technology of its products, their appearance was also a decisive factor: neutral designs that suited every setting and fit in anywhere, making them truly transcultural. Slick Braun shavers, for instance, are valued the world over because consumers everywhere find them just right.

Braun on Tour

Braun's international outlook and cosmopolitan attitude are also revealed by the company's early interest in the topic of mobility. Both in the radio age and later years, the mobility of its devices has always been important. The development ranges from portable radios that became more and more compact over time, starting with the portable record player TP 1 and progressing to the T 1000 World Receiver, which offers every single radio station no matter where you happen to be, enabled mobility through both space and media. Braun's timepieces also reflect the company's emphasis on global usability. The AW24 wristwatch from 2003 can display two time zones simultaneously. Small alarm clocks come complete with a map of the world in the lid. The compact radio alarm clock ABR 313 sl from 1990 was the perfect travel companion in hotels without a radio or wake-up service. And it was even small enough to fit into a pocket or handbag.

Outlook

Braun continues to deliver devices for a life on the go that will not come to a halt, whether in person or virtually. What will arguably change is the form mobility takes and the devices it requires. It is already possible for people to be connected via digital media no matter where they are, providing they have internet access. Likewise, digital media serve to control a whole range of devices across geographies, from 'connected home' drones to virtual operations performed on a patient in California by a robot controlled by a surgeon in Frankfurt.

What will remain vital moving forward is the need to work on design that crosses and connects both cultures and spaces and is seen by many people the world over as being compatible with their lives. Given that life on our planet is becoming ever more complex, this can only be achieved with an international team working in all the areas involved in product development. Influenced by P&G, whose workforce is much more international than Braun's staff ever was, Braun has accepted multiculturalism as the norm.

But accepting multiculturalism by no means guarantees that a company's products will truly reflect consumers' wishes or bring economic success in the future. For the following generations of millennials and particularly Gen Z, for whom transcultural communication has long become quite natural, new areas will assume greater importance. One such area will be overcoming outdated gender models. While the functionality of Braun Design secured global acceptance early on in many product areas, especially in the more gender-sensitive product categories such as shaving and hair removal, other products have not always been as functional and neutral in design and have tended to play towards stereotypes. Today, this is an area that benefits from the progressive attitude of P&G, for whom acceptance has long since taken in different cultures, as well as different lifestyles and identities.

Diversity and inclusion as clear corporate guidelines enable an open and proactive handling of these issues. The underlying interpretation of diversity as enriching is comparable with Max Braun's cosmopolitan attitude and global outlook. Then as now, these corporate values impacted products and shaped their design. Today, they are part and parcel of every discussion on design at Braun, so that future generations of consumers will also perceive the brand as remarkable and relevant for them in the same way that young people once did back in the 1950s and 1960s.

International

Brochure from Wolfgang Schmittel's advertising department on Braun's successful participation in exhibitions in Berlin, Milan and Brussels. Such successes – including those in the media – are still extensively communicated internally and externally because they promote acceptance of Braun Design at a fundamental level. The company's most important exhibition success in this early period was undoubtedly the 20th Century Design exhibition at the Museum of Modern Art in New York in 1958.

There are private collections of Braun products not only in Germany but all over the world. South Korean collector Jason Park has amassed a brilliant collection of over 2,000 pieces relating to Braun and Dieter Rams, which also includes other function-orientated design objects from the 1950s to the 1990s. The collection was open to the public as the '4560 Design Haus' from 2020 to 2022, taking the form of a small design museum in the south of Seoul, and has since become a travelling exhibition.

Part 3

Design Is Made By People

The following sixty biographies show the wide range of qualifications, activities and personalities among the individuals who have worked for Braun over the last one hundred years.

They serve as representatives of many employees who cannot be named here. Design is made by people – even by those in the company who are not directly involved in the actual design process. In those instances where the biographies are about people who are still alive, they were naturally consulted, and sometimes personal emphases were made. We have indicated when this was not the case. Materials from archives or previous publications were used for the biographies of those individuals who are no longer alive.

Otl Aicher

Artur Braun

Otl Aicher
born 1922 in Ulm
died 1991 in Günzburg

Otto (who went by the name Otl) Aicher became interested in reform Catholicism and philosophy at an early age. Aicher met Werner Scholl in school and went on to befriend Werner's siblings Inge, Hans and Sophie, who would later join the resistance against the Nazi regime. After the war, Aicher and Inge founded the Volkshochschule Ulm, an educational centre. The centre gave rise to the idea of a private university that would promote democracy in Germany through design, politics and individual education, a 'school of tomorrow', as they put it. Otto and Inge married in 1952 and had five children: Eva Aicher (b. 1953), Florian Aicher (b. 1954), Pia Aicher (1954–75), Julian Aicher (b. 1958) and Manuel Aicher (b. 1960). In 1953 they founded the Hochschule für Gestaltung (HfG) Ulm. Considered to be the most important post-war design college in Germany, it was initially headed by Swiss architect, artist and designer Max Bill, and organized as an independent foundation. Otl Aicher adopted a rational approach for the field of visual communication at the HfG that made no reference to the fine arts. His own design was grid-based, modular and as objective as possible. When contacts between the HfG and the Braun corporation were established in December 1954, Aicher was, along with Hans Gugelot, the most committed promoter of Braun's new approach to design. As a result, he developed not only a working relationship with the Braun brothers and Fritz Eichler but also lifelong friendships where they exchanged ideas about a new democratic and humane society. Otl Aicher provided significant initial concepts for Braun's corporate design, which Wolfgang Schmittel then took up and elaborated. The Braun exhibition stand was designed by Aicher as a collaborative effort with Hans G. Conrad for the radio exhibition in Düsseldorf in 1955. It was used at trade fairs for another thirty-two years and decisively shaped the style of Braun.

Aicher created definitive corporate designs for Lufthansa and the 1972 Summer Olympics in Munich, as well as for companies such as the lighting manufacturer ERCO, fittings manufacturer FSB and fitted-kitchen manufacturer bulthaup.

For Otl Aicher, the world existed 'as a design', and he claimed: 'We do not live in a world of being, but in a world of projection. We are not content with what is, we are concerned with what should be. Our problem is not nature, nor technology, nor culture, but what we do with nature, what we do with technology, what we do with culture.'[1]

Inge Aicher-Scholl
born 1917 in Crailsheim
died 1998 in Leutkirch im Allgäu

Inge Scholl was the oldest daughter of tax consultant and liberal politician Robert Scholl and his wife Magdalena. When she was a teenager, Inge and her younger siblings organized a discussion group on literature and religion in their parents' apartment. The siblings' initial enthusiasm for the Nazi party gave way to the realization of its status as a criminal regime, and her brother Hans and sister Sophie set up the 'White Rose' resistance group; both were executed by the Nazis in 1943. Her experience of totalitarianism laid the moral foundations for Scholl's professional activities after 1945. In 1952, her book Die Weisse Rose (The White Rose) about the resistance group was widely printed and published in several languages.

In 1946, she and Otl Aicher founded the Volkshochschule Ulm educational centre, which she directed until 1974 and where she organized high-profile lectures, primarily on a new democratic society. In 1950, she established the Geschwister-Scholl Foundation, which later became the sponsor for the HfG Ulm, the college of design that she cofounded with Otl in 1953.

In 1952, she married Otl Aicher. The couple had five children: Eva Aicher (b. 1953), Florian Aicher (b. 1954), Pia Aicher (1954–75), Julian Aicher (b. 1958) and Manuel Aicher (b. 1960). Manuel has authored an extensive but as yet unpublished biography of Max Braun.[2]

The Aichers' first meeting with Erwin Braun took place in Ulm at the beginning of December 1954. From then on, Inge remained in close contact with the Frankfurt-based entrepreneur, who not only became a client but a friend who provided extensive financial support for the HfG.

Werner Apel-Dube
born 1921 in Weimar
died Heiligenmoschel, Magdalenenhütte

Werner Apel-Dube became friends with Erwin Braun during his school years. After passing his exams in 1940, he was drafted as a soldier during World War II and, after being captured in 1944, became a Russian prisoner of war for five and a half years. After the war, he joined Braun in February 1952, initially working in customer service and advertising. In autumn that same year Erwin Braun instructed him to set up a section for phototechnology. In 1956, he embarked on a world tour of twenty-two countries for Braun, with the goal of familiarizing himself with the Asian markets. After June 1960, he was briefly chairman of the executive board, whose other members included Dr Kurt Georg and Ferdinand Simon.

In the early 1960s, Apel-Dube played an important commercial role in founding Erwin Braun's Swiss companies and in 1960 he was made president of Braun Electric International S.A. in Baden, Switzerland, where he was seemingly destined to further expand foreign business. He left Braun in 1963 to become general manager at General Electric and, until his retirement in 1984, acted as a member of the executive board of Brown Boveri & Cie AG (BBC) in Mannheim.[3]

Artur Braun
born 1925 in Frankfurt/Main
died 2013 buried in Kronberg-Schönberg

Artur Braun, the younger son of Max Braun and his wife Mathilde, began an apprenticeship as an electrician in his parents' company after completing high school in 1943 but was then drafted as a soldier and returned to Frankfurt in September

Erwin Braun

Mathilde Braun

1945 after a stay in hospital. On 1 October 1945, he started as an engineer for the company.

In 1951, after the death of his father, Artur took over the technical management of Braun. In 1952, he married Elfriede Lautner and they had a son, Thomas. Having previously played a key role in the development of his father's shavers, Artur was now tasked with the design of the company's new kitchen appliances. In the spring/summer of 1955, he and Fritz Eichler conceived of and designed the SK 1 small receiver. After 1954, he was also responsible for Braun's collaboration with Wilhelm Wagenfeld.

Organizing the company's operations was another aspect of Artur's brief. Additionally, he was in charge of constructing factories in Walldürn in 1954 and Marktheidenfeld in 1961. After the company was converted into a stock corporation, Artur became chairman of the executive board in 1962 and moved to the supervisory board a year later. He visited the United States on three occasions – in 1953, 1956 and then again in 1966 – to study business operations there. He was also largely responsible for choosing Kronberg in the Taunus region as the new company headquarters. This attracted other companies to the town and is why the city council decided to name a Kronberg business park after Artur in 2017.

After Braun AG was sold to Gillette in 1967, Artur founded his own engineering practice in Königstein and worked for Braun in shaver development until the mid-1970s. He also had his own office on the Braun premises in Kronberg with a staff of around ten employees.

Erwin Braun
born 1921 in Frankfurt/Main
died 1992 in Wilen (Sarnen), Switzerland

'Erwin Braun was the most imaginative and at the same time most restless soul I have ever met in my life.'

– Hans Horst Semmel, CEO of Braun Española[4]

Erwin Peter Richard Braun, the elder son of the company's founder Max and his wife Mathilde, completed high school in Frankfurt in 1939 and was drafted into the Wehrmacht shortly afterwards. There, he met Fritz Eichler, as well as many other people that he would later employ at the company. Erwin Braun was a networker who relied above all on personal contacts, and these were mostly individuals he had encountered during the war.

After a childhood characterized by serious illnesses and medical complications that were potentially life-threatening, his health remained fragile throughout his life. This probably also explains his later interest in well-being and the introduction of nutritious food in Braun's cafeterias. As a soldier, he also became seriously ill and, after appropriate training, spent the rest of the war teaching radio engineering at an army intelligence school in Halle. On 19 January 1945, he married Erika Ludwig (1921–2024), with whom he lived in Halle until the end of the war.

After returning to Frankfurt, he would have liked to become a doctor, but his domineering father insisted he study business administration instead, which he did at Frankfurt's Goethe University from 1946 to 1949 while completing commercial training in his father's firm. Max Braun wanted his son to take over 'the business' later together with his brother Artur.

During his studies, he took courses in American marketing, which would play a significant role in his work at the company. For Erwin, market analyses and a systematic approach pointed the way to the future for businesses. He developed his ideas during long walks in the Taunus hills, typically accompanied by business partners or employees. This constructive practice of conversation in nature became central to his strategic management: walking and talking as a means of thinking.

Undoubtedly, Erwin was the driving force behind Braun's success in the 1950s and 1960s. He was 'the powerhouse' who moved the company forward. He was bold enough to completely reorganize the company and was quick to recognize the importance of good and new design. In 1967, however, it was also Erwin who pushed ahead with the sale of the company so that he could devote himself entirely to his interests in medicine.

At that time, he already spent part of the year on a farm in Switzerland. With the millions realized from the sale, he founded new companies (Maxon Motor AG in Switzerland) or bought existing ones (Ritterwerk GmbH near Munich). Most importantly, however, he took over the 'Friedenfels' naturopathic spa in Wilen (Sarnen) and in 1980 established an institute for preventive medicine in nearby Engelberg. In 1981, he became a Swiss citizen, and the Academy of Postgraduate Medical Education of the German Democratic Republic in Berlin-Lichtenberg awarded him an honorary doctorate in 1988. Erwin died in 1992 in Switzerland, where he had been active for so long.

Mathilde Braun, née Göttelmann
born 1891 in Armsheim, Rheinish Hesse
died 1987 in Frankfurt/Main

Mathilde Göttelmann, later wife of company founder Max Braun, came from a wealthy merchant family that was forced to emigrate from Flanders to Rheinish Hesse in 1728 because of its Protestant faith. Her father ran an extensive agricultural business and owned his own vineyard. Mathilde and Max Braun met at the wedding of Max's brother and married on 6 November 1920. The couple then moved to Frankfurt, where Mathilde bought a house at 30 Robert-Mayer-Strasse. In 1930, Max Braun acquired a plot of land at 17 Cretzschmarstrasse in Frankfurt, where he built his own home, which still belongs to the family today.

Mathilde supported her husband in many ways from 1920 until his death in 1951, both financially but also through her involvement in the administration of the company, at trade fairs or on business trips. She always spoke of 'our company'. Mathilde's sister Annie (1883–1976) also contributed financially to Max's investments in France and Frankfurt.

Max Braun

Julius Calzaferri

In 1921 and 1925, sons Erwin and Artur were born, but this by no means prompted Mathilde to assume the role of housewife. She was more than just 'the supportive wife'. Together with her family in Armsheim, she helped shape the young company and contributed to its success – and not only financially. After the early death of her husband, she spent part of her long life in Switzerland and on the lakes of Northern Italy.

Max Braun
born 1890 in Schillgallen, East Prussia
died 1951 in Frankfurt/Main

Wilhelm Max Braun was born in 1890 in a village with just 300 inhabitants in what was formerly East Prussia. He was the sixth of seven sons of Friedrich August Braun, a farmer and sailor, and his wife Karoline.[5]

After attending school, he completed an apprenticeship in mechanical engineering. His first entrepreneurial activity demonstrated his business clout: at harvest time he toured through the villages of a region growing large amounts of grain with a steam-powered threshing machine. In 1910, since there was no longer anything to keep him in the provinces, he took up a position with the engineering company Wilhelm Fette in Hamburg. However, he was soon conscripted for six months' military service in Berlin. The big city was the right environment for this ambitious young man, who worked there for Siemens, R. Stock & Co. and AEG after his military service. Max Braun wanted to be more than just an industrial worker. Before World War I, Germany had become the leading industrial nation – with all the opportunities this opened for ambitious young people.

Alongside his professional work, Max Braun attended a private polytechnic that specialized in the trades of mechanical, electrical and iron engineering, run by the engineer L. Barth, and took English language lessons. His years in Berlin were formative. Here, before World War I, the technical and cultural modernity that would later characterize the Weimar Republic prevailed. He was initially drafted during World War I in 1915, but as soon as 1916 AEG claimed him as an employee important to the war effort and he returned to his job in the capital. At AEG, he rose from the position of machinist to that of technical draftsman. He also came up with his first inventions, for which he received a bonus of a handsome 1,600 Reichsmarks from AEG. If he had stayed in Berlin, he could have conceivably had a career as a machine designer, but things turned out differently.

In May 1918, Max Braun met and fell in love with Mathilde Göttelmann at the wedding of his brother Richard in Wiesbaden. The relationship marked a turning point for Max Braun in the form of a change of location, since his future father-in-law insisted that his fun-loving daughter remain in the Rhine-Main region. Certainly, concern for her safety also played a role, as violent barricade battles were still raging in Berlin in 1919 over the future of the new state. From 1920 onwards, Max Braun lived with his wife in Frankfurt where, in 1921, he set up his company with just four employees. On 28 January 1922, he passed the Wiesbaden Chamber of Crafts' master craftsman's examination, which meant he was also able to take on and train apprentices.

'What's that fellow doing there?' might have been a typical remark of Max Braun. His somewhat blunt manner and succinct and direct way of speaking was no doubt due to his East Prussian origins and the time he spent in Berlin. He was the typical self-made entrepreneur of the 1920s, who rolled up his sleeves, got involved and was constantly busy, full of ideas and enthusiasm. He saw obstacles as challenges to overcome. In his company, he was the uncontested authority and strict patriarch who attached great importance to punctuality and thriftiness and could also raise his voice to make his point.

According to many sources, he is also noted as being fair, helpful and trusting, something that was not infrequently abused. He countered hollow statements by claiming that some people were 'all mouth but no substance'. Fine manners and conventions were not for him. Max Braun did not think much of war and nationalism, and he did stand up to the Nazis. Unlike his wife, he also had no deep relationship with religion. He travelled to England, France and the United States and quickly realized that exports were essential for his business, so at an early stage he established an international orientation for the company, which Braun has continued to develop to this day.

Max Braun died unexpectedly of a heart attack when he was sixty-one and his sons then took over the business.

Julius Calzaferri
born 1907 in Frankfurt/Main
died 1981 in Frankfurt/Main

Julius Calzaferri was taken on as Max Braun's first apprentice on 1 August 1921. He was a 'real Frankfurt lad', as they say in the city. His having an Italian surname was not so unusual in Frankfurt, it being a central European trading city that has always attracted settlers from abroad.

Calzaferri became an enthusiastic employee at Braun's company and in 1927 was involved in the planning of the new workshops. From 1927 to 1933 he also worked in several other Frankfurt companies as a mechanical engineer. Perhaps Max Braun even encouraged him to gain a range of experience and to expand his knowledge – he himself had done much the same during his time in Berlin.

After returning to Braun in 1933, Calzaferri became a foreman in the engine construction department at the Idsteiner Strasse factory. Thanks to the experience he gained during his travelling years, he succeeded in making significant improvements to the design of record player motors and their production process, which soon led to his appointment as head of the phonographic and radio department. Calzaferri was considered a pragmatic man with a knack for organization. After the war, he played a key role in the reconstruction and

Claus Cobarg

Bracken P. Darrell

Fritz Eichler

planning of the new factory in Rüsselsheimer Strasse and in 1953 became manager there. He also took on responsibilities in Kronberg. Calzaferri's career began in a fairly lowly position and saw him rise to great heights: in this respect, his path is not unlike that of Max Braun.

Claus Cobarg
born 1921 in Charlottenburg, Berlin
died 2016 in Mammolshain, Königstein

Although Claus Christian Cobarg's studies in mechanical engineering and later in physics were interrupted by temporary military service, he completed them successfully in March 1945. After working in various jobs, he was hired by Braun on 1 January 1958 to work on new technical developments, including the HT 1 toaster and numerous other appliances. He set up a central laboratory in the company with state-of-the-art equipment, including a scanning electron microscope, to conduct basic research for Braun technology. Attached to the laboratory was a corresponding information and documentation centre. He was also active as a technical reviewer in various design competitions.

Even after his retirement at the end of 1986, Cobarg continued to work as a consultant for Braun and as an advisor in the field of medical physics for Erwin Braun in Switzerland, keeping him up to date on what was happening at the company. He also wrote a book, Sonnenkraft im Zentrum alternativer Energiequellen (Solar Power at the Centre of Alternative Energy Sources), which went through several editions, as well as numerous articles in technical journals. His extensive archive is now housed in the Braun Collection in Kronberg.

Bracken P. Darrell
born 1963 in Owensboro, Kentucky, US, lives in San Francisco, US and Lausanne, Switzerland

After graduating with an MBA from Harvard Business School, Darrell began his management career first at PepsiCo and Arthur Andersen, then from 1991 to 1997 as brand manager for Procter & Gamble. From 1997 to 2002, he was general manager at General Electric and then president of Braun, a subsidiary of Gillette, until 2008.

Darrell is probably the American manager after Vincent Ziegler who best understood the Braun 'DNA'. While in Kronberg, he successfully launched a reorientation of the company's designs, which sought to continue and further evolve Braun's traditions of the 1960s to 1980s. Products were once again developed to convince customers to buy at first sight, thereby creating a bond with the brand.

From May 2008 to May 2012, Darrell was executive vice president at Whirlpool, a manufacturer of household appliances. Since 1 January 2013, he has been the head of Logitech, a Swiss manufacturer of computer accessories, whose appeal he is re-establishing not least through innovative product design and because of his experience at Braun.

Fritz Eichler
born 1911 in Differdingen, Luxembourg
died 1991 in Bad Soden, Taunus

Fritz Eichler studied German, art history and theatre in Berlin and Munich from 1931 to 1936. From 1937 to 1947 – apart from his service during the war – he worked as a dramatist, director and occasional actor at the Städtische Bühnen Frankfurt, and from 1936 to 1939 as a drama teacher at the Frankfurt University of Music and Performing Arts. In 1937, he was awarded a doctorate in Munich for a dissertation on hand puppetry and marionettes.[6] In 1939 he married Franziska Selmer. The couple had two sons: Stefan Florian, who became an architect, and Christof, who became a famous photographer. In 1939, in the the midst of World War II, Eichler was selected to work in the radio company of a tank intelligence regiment in Weimar. It was there that he met Erwin Braun and the two became friends. Through Eichler, Braun was introduced to modern art, including the work of the Bauhaus.

After the war, Eichler lived and worked as a director, actor, puppeteer and set designer in Munich and Frankfurt. In 1943, he wrote and illustrated the children's book Vom König der nicht mehr lachen konnte (The King Who Could Not Laugh).[7] Together with Erich Kästner, he founded a children's theatre group in Munich's International Youth Library, for which he also wrote, directed and performed plays with hand puppets. From 1950 to 1953, he directed or co-directed five German cinema films, including Kästner's Das doppelte Lottchen (The Double Lottie). He also directed the film Christina during this period.

In 1953, Eichler was hired by Erwin Braun to produce cinema commercials, and on 1 July 1954, he was permanently employed by the company as Representative for Overall Design. He became a major driving force for new design at Braun. As part of the effort to develop the in-house product design and advertising section, he was made executive board member with overall responsibility for this department. Eichler acted as a catalyst of sorts, since he was talented in bringing together the various parties required for constructive processes. At the same time, he was a knowledgeable, culturally sensitive conversationalist, who engaged not only with Erwin Braun but the entire management team. Eichler's conversational culture, which was both open and fundamental, became an important cornerstone of the company's success.

Eichler can also be regarded as a kind of early 'chief theoretician' of Braun Design, communicating Braun's design ethos both within the company and to the public through essays and lectures. In Braun's 1972/3 annual report, he published principles of design that Dieter Rams later developed and supplemented as generally valid theses for good design. From 1968, Eichler oversaw a new

Richard Fischer

Bodo Fütterer

Oliver Grabes

format for popularizing exemplary industrial design in the guise of the BraunPrize, established by Erwin Braun in 1967. On 4 May 1973, he resigned from the executive board but remained a member of the supervisory board until 1991.

Richard Fischer
born 1935 in Neumarkt in der Oberpfalz
died 2010 in Offenbach

Richard Fischer first studied mechanical engineering in Neumarkt from 1952 to 1955 and then enrolled in a course in product design under Hans Gugelot at the HfG Ulm. He graduated in 1959, having designed a machine tool. Artur Braun was one of his two examiners. Fischer had two children with his first wife Zäcilia Ochsenkühn, whom he married in 1960. His second marriage in 1997 was to Monika Streek.

Fischer began his work as a designer in 1959 at BASF in Ludwigshafen. Then, in 1960 he joined Braun as the company's fifth product designer. Together with Dieter Rams and Robert Oberheim, he was involved in the design of the first Nizo film camera, the FA 3. He subsequently designed electric shavers and household appliances, sometimes in close collaboration with Rams.

In 1968, Fischer left Braun and worked as a freelancer and lecturer at the Werkkunstschule Offenbach. There he founded a course in industrial design and encouraged numerous HfG members from Ulm to join Offenbach for lectures and workshops. Rams was also invited to give lectures and participate in juries. Together with Lore Kramer, Fischer was instrumental in the following two years in transforming the Werkkunstschule Offenbach to HfG Offenbach, a state art and design academy. From 1975 until his retirement in 1999, he held a professorship in Offenbach and was involved in extensive research work on the indication functions of a product, publishing the findings together with Gerda Mikosch in 1984.

Fischer was a committed and persuasive teacher but with the same conviction always remained a design practitioner and freelance designer. For example, he designed the highly successful Minox-EL-35 35 mm camera, products for loudspeaker manufacturer Canton, computer-controlled exercise bikes for Keiper and equipment for Demolux, a manufacturer of overhead projectors.

In 1965, together with Rams, Oberheim and Reinhold Weiss, he won the Berlin art prize Junge Generation (Young Generation). Later, in 1980, he was awarded the Design Center Stuttgart's prize for product design.

Bodo Fütterer
born 1927 in Wächtersbach, Hesse
died 2018 in Lucerne, Switzerland

Bodo Fütterer graduated from high school in Neu-Isenburg during World War II. After a year of active service in the war and completing a technical internship from 1947 to 1950, he studied mechanical engineering at the Frankfurt School of Engineering. He began working as a designer for Braun on 2 October 1950 and was assigned to the electric shaver project. He repeatedly succeeded in making important product improvements, especially to the shaver blades, for which numerous patents were obtained. He also became responsible for the technical development of household appliances. Fütterer was regarded as a passionate, imaginative and highly successful engineer. In 1960, he was given power of attorney for the company.

A close friendship developed between Erwin Braun and Fütterer. In Braun's eyes, Fütterer was his best designer and inventor. As early as 1 July 1963, Braun made him managing director of Interelectric AG, founded in 1961 in Sachseln, Switzerland. This spin-off, which belonged solely to the Braun brothers and not to Braun AG, was involved in the manufacture of the blades for Braun shavers. Following the sale of the Kronberg-based company, Gillette acquired the patents of Interelectric AG, and the company turned its attention to miniaturized devices for the autarc system, creating, for example, an electric shaver the size of a cigarette pack.[8]

In 1969 and 1970, Fütterer also designed miniaturized high-performance motors, which were successfully launched on the market under the Maxon brand from 1970. Under his leadership the company quickly developed into a dominant supplier, even fitting the American Mars Rover with Maxon motors in 1997, 2004 and 2012. In 2020, the Perseverance rover and the Ingenuity helicopter drone landed on Mars, both with Maxon motors.

Fütterer retired on 30 June 2004 but continued to work on innovative technology, such as high-tech ceramics for gearheads.

Oliver Grabes
born 1967 in Mannheim, lives in Frankfurt/Main

After graduating from high school, Professor Oliver Grabes studied product design at the University of Art and Design in Offenbach from 1986 to 1993. He was significantly influenced by the Offenbach approach to product language and by professors Dieter Mankau and Richard Fischer – both experienced industrial designers with a strong focus on utility. During his studies, he completed internships at home and abroad at such companies as yellow design, Technology Design and npk, thus gaining extensive experience in design practice. In 1993, he co-founded Grabes & Klüber, a Design Studio in Frankfurt. His interest in high-tech products drew him to the United States' West Coast in 1999 to work as a senior designer at Teague, to London in 2001 as a design manager at the PDD Group and finally back to the States in 2004 as the creative director at Seattle's Carbon Design. He acquired expertise in different design processes through his work for numerous international clients including Microsoft, Sony, Hewlett-Packard, Panasonic, General Electric, Boeing, Johnson & Johnson, Merck, Dräger, Nike and Reebok. In 2006, he was appointed professor for 'Industrial Design/Conception and Design of Consumer Goods' at the University of Wuppertal.

In September 2009, Grabes became the new head of the Braun Design Team in Kronberg but remained a visiting professor in Wuppertal.

Oliver Grabes's Design Team

Jürgen Greubel

Hagen Gross

At Braun, he succeeded in taking up the design tradition associated with Dieter Rams. With the design approach 'Past Forward', Braun sought to reconnect with its core identity and develop it further, focusing on practical value, durability and an attractive purism of form.

Grabes's areas of responsibility within Procter & Gamble continued to expand. Since 2015, he has also been responsible for Industrial Design at Gillette, and since 2020, he has been the Vice President of Design, P&G Grooming, which includes the Braun and Venus brands.

Grabes has received numerous international design awards individually, including the IDEA Gold Award from the International Design Society of America, and together with the Braun Design Team. Moreover, his products have been granted numerous patents. In 2020, after receiving many special awards, the new Braun audio products that Grabes oversaw attracted considerable attention. Since 2012, Grabes has been the Chairman of the International BraunPrize.

Oliver Grabes's Design Team 2023

Braun's current Design Team has expanded under Oliver Grabes into the areas of graphic and UX/UI (User Experience/User Interface) design. This Design Team has taken on more tasks internally, including in the CAD design area, in order to send final data sets to internal and external toolmakers in close coordination with the product designers and technicians.

In 2023, the Team's members were:

Vice President Design, P&G Grooming: Oliver Grabes
Studio Planning & Assistance: Hannele Helminen
Design Directors: Björn Kling, Wolfgang Stegmann, Renke Thye
Product Designers: Gregor Ptok, Philipp Goeder, Franziska Schmid, Kong Linxuan
Graphic & UX/UI Designers: Nina Sagra, Irini Weiser, Maria Chiara Casarotto, Anne Hartmann, Gerlinde Brand, Claudia Steinmeyer
CAD Designers: Bernhard Sikora, Robert Schmidt, Rüdiger Boss
Model Makers: Oliver Michl (Leader, Design Modelshop), Alexander Dorn, Nikolas Pohlen, Florian Weck, Josua Messany

Jürgen Greubel
born 1938 in Wiesbaden, lives in Heidenrod, Taunus

Jürgen Greubel's father was a carpenter and glazier in Wiesbaden, which meant that from an early age he often encountered the painters for whom his father made picture frames and was thus exposed to fine arts early in life. After attending school, he began an apprenticeship as a carpenter and after five years as a journeyman passed the master craftsman's examination. This was followed by studies at the Wiesbaden Werkkunstschule. After graduating, he began an internship at Braun in 1967 and was employed first as an assistant in the design department, becoming a product designer after only a few weeks.

Greubel worked mainly in the field of appliances for household and personal care. He also collaborated with Robert Oberheim, Ludwig Littmann and Dieter Rams. In 1970, his HTG 1 table grill and MP 50 juicer were launched, followed in 1974 by the centric pocket lighter and several hair dryers.

In 1972, the MPZ 2/21/22 citromatic citrus juicer Greubel designed in collaboration with Rams was launched after having initially been developed for the Spanish market. Still in production today, it has a special significance in Braun's product history. Together with Rams, Greubel also worked on the Lectron electronics experimentation system, developed from 1967 to 1969 as a teaching tool to be used in schools and universities to help young people learn how to create their own operational systems. In particular, the Buchlabor, which combined electronic components and a print medium, was distributed internationally and won many awards. Greubel developed the HLH 1/3 Astronette dryer hood in just two weeks and it was launched in 1971. He still has a reputation at Braun for his speed and ability to work quickly.

In 1973, Greubel moved to London to join Design Research Unit, one of the first British design consultancies that brought together expertise in architecture and graphic and industrial design to produce corporate design. Back in Germany, he freelanced from 1975 onwards, again working extensively for Braun. In another collaboration with Rams, he developed a new type of iron without a handle, for which models were made by Rams and Greubel in great secrecy. Two former NASA engineers also worked on the complicated electronics. However, the iron remained only an idea, one that could not be translated into a product.

From 1995 onwards, Greubel designed the office furniture systems Axis 369° and NetWork for the school and office furniture manufacturer VS-Möbel; other clients include the lighting and sensor manufacturer Steinel, the computer manufacturer Nixdorf and Wella hair cosmetics.

Hagen Gross
born 1921 in Kattowitz
died 1989 in Bad Homburg van der Höhe, Hesse

After Hagen Gross graduated from high school in Berlin, he did a six-month stint for the Reich Labour Service before becoming a World War II fighter pilot. Badly wounded when he was shot down, he met Erwin Braun in a convalescent hospital in Kitzbühel, and the two became close friends. After the war he studied law and started work in Hanover. On 1 November 1955, he took up a position at Braun, where he was initially responsible for wholesale purchasing. In 1961, he became a member of the executive board of Braun Aktiengesellschaft and head of the household products division. From 1966, he was also director of the new products division. Gross was particularly interested in identifying potential product areas. From his numerous travels he brought back innovations from other manufacturers, which in some instances were

Hans Gugelot

Thomas Guttandin

Peter Hartwein

taken up and further developed by the engineers and product designers at Braun. A well-known example is the pocket calculator from the Japanese company OMRON, which Braun initially had manufactured unchanged with its own logo (ET 11 from 1975) and which finally, after a redesign, became an iconic Braun product – the ET 66 model – in 1987.

Hans Gugelot
born 1920 in Makassar, Celebes, Indonesia
died 1965 in Ulm

Born in the Dutch colony of Indonesia to a family of doctors and raised in the Netherlands and Switzerland, Hans (originally Johan) Gugelot finished secondary school in 1940 and went on to study architecture at the Swiss Federal Institute of Technology (ETH) in Zürich. He financed part of his studies as a jazz guitarist and continued to perform at concerts throughout his life. In 1947, he married Marie-Hélène Chuard (b. 1927 in Zürich), known as Malke. The couple had two sons, Guus Daniel (b. 1950) and David (b. 1963).

After completing his studies, Gugelot started working independently in 1950 and later also as a freelancer for Max Bill's office in Zürich. At this time, he was already designing his modular furniture system M 125, which was manufactured after 1956 on a large scale by Wilhelm Bofinger.

In 1954, Bill, then the director of the HfG Ulm, appointed Gugelot as head of the college's interior design department. Gugelot also taught in the product form and building departments. He soon became the university's most important product designer and was quick to put his ideas into practice. In 1957, he set up Development Group 2 at the HfG, which worked on corporate commissions. Braun would later become the university's largest client and in 1955, Gugelot, either alone or with his students, started designing radios and phonographs in light, almost white, maple. This marked the advent of a new style of design at Braun that was, from the very start, conceived as modular. Gugelot and Dieter Rams were jointly responsible for the SK 4, the archetypal combination of radio and phonograph in a single Hi-Fi set.

Numerous other clients, including Kodak, Agfa, Pfaff, Sonor, Girard, Weishaupt, Lumoprint and Nordmende, came along later. Gugelot's largest single project, realized together with his student Herbert Lindinger, was the design for the trains of the Hamburg Metro, which they worked on from 1959 to 1962.

In 1961, Gugelot was involved in setting up a design institute in Ahmedabad, India. He then founded the institut für produktentwicklung und design the following year with thirteen employees in Neu-Ulm. Gugelot presented his views on product design in many lectures, including at the World Design Conference in Tokyo in 1960, the Slade School of Fine Art in London in 1962 and the Grafiska Institutet in Stockholm in 1963. He published several works on design and design education. After his early death in September 1965, his wife Malke Gugelot managed the institute and, with the support of Erwin Braun, transformed it into gugelot design gmbh. The studio was one of the largest design offices in Europe until 1974 when the oil crisis, combined with the insolvency of a major client, forced it to file for bankruptcy.

Thomas Guttandin
born 1958 in Dieburg, Hesse, lives in Hattersheim am Main, Taunus

Thomas Guttandin studied social work at Frankfurt University of Applied Sciences from 1977 to 1980. However, the title of his diploma thesis 'Warenästhetik und produktkulturelle Identität' (Commodity Aesthetics and Product Cultural Identity) already indicated his interest in design, which was also reflected in his decision to pursue design professionally by starting to study it at the Offenbach University of Art and Design in 1984 and graduating in 1991 with a degree in the subject. From 1978 to 1990, he was a member of a carpentry collective and from 1984 to 2001, he worked as a project manager for the ELMA Messe- und Ausstellungsbau company in Offenbach.

As a designer and curator, he has been responsible for special and permanent exhibitions for many museums since 1986 and has been involved in the conceptualization and design of trade fair stands and exhibition systems for renowned clients, including adidas, Air France, De'Longhi, Minox and, since 1991 in a variety of ways, Braun.

Since 2011, Guttandin has been managing director of the Förderkreis BraunSammlung, an association that looks after the Braun Collection. He took charge of the Braun Archive in 2015 as successor to Horst Kaupp.

Peter Hartwein
born 1942 in Weilheim in Upper Bavaria
died 2023 in Kronberg im Taunus, Hesse

Peter Hartwein initially planned to be a carpenter, completing an apprenticeship after he finished his schooling. He then started a traineeship in an architectural office and went on to study architecture and interior design at the Werkkunstschule in Kaiserslautern, graduating in 1966.

After short periods of employment in various offices in Munich, he spent three and a half formative years working for the Munich office of the renowned architect Peter C. von Seidlein, who had been a student and collaborator of Mies van der Rohe in Chicago. For Hartwein, this was tantamount to taking a second degree.

Inspired by the designs he had seen in the Braun showroom in Munich and a meeting with Otl Aicher, Hartwein took the initiative to apply for a position working under Dieter Rams in 1970 and immediately joined the Braun Design Team. He first worked as Robert Oberheim's assistant on the Nizo film cameras and then, in 1974, developed the complex Tandem professional slide projection system. From 1978 onwards Hartwein collaborated with Rams on

Wolfgang Hasselbach

Herbert Hirche

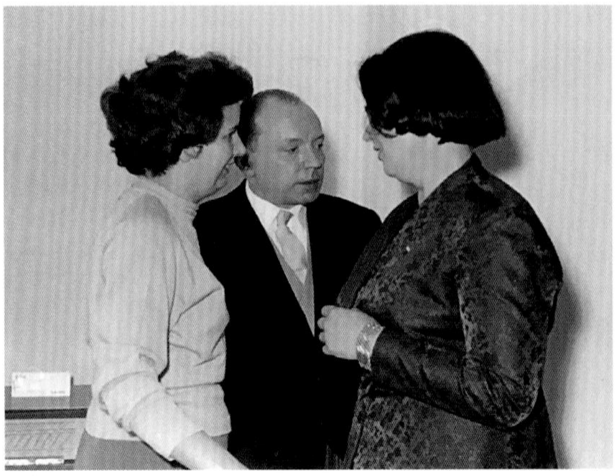

Georg Hohm

Hi-Fi components. His activities focused on designs for the atelier systems of the 1980s and, starting in 1982, all loudspeaker units.

After 1978, Hartwein also became responsible for the development of Braun oral care devices. The important breakthrough here came in 1991 with the oscillating brush, which was the result of much development work and close cooperation with the technical department. After Dietrich Lubs left the company in 2001, Hartwein took over the development of the watch lines and the management of product artwork. His last job at Braun was to oversee the construction of the new Kronberg headquarters for the architects schneider+schumacher until he retired from Braun in 2006.

Wolfgang Hasselbach
born 1924 in Falkenstein im Taunus
died 2011 in Kronberg im Taunus, Hesse

After graduating from high school in Frankfurt in 1942, Wolfgang Hasselbach served as a soldier until the end of the war. From 1946 to 1950, he studied physics at the Goethe University in Frankfurt and graduated with a diploma. From 1951 to 1957, he worked as a development engineer at Assmann GmbH, a manufacturer of tape systems for the office sector in Bad Homburg. On 1 October 1957, he moved to Braun, where he was initially responsible for the technical development of record players.

On 1 September 1960, he was appointed head of development for the electroacoustics division, especially for Hi-Fi equipment, which he played a major role in developing.

From 1 October 1964, he oversaw development for the electronics department, a field in which Braun was a market leader in Germany. After the Hi-Fi division was spun off to a/d/s/ in 1981, Hasselbach left the company on 31 March 1982.

Herbert Hirche
born 1910 in Görlitz
died 2002 in Heidelberg

After beginning an apprenticeship as a carpenter in 1924, followed by a period as a journeyman, Herbert Hirche studied at the Bauhaus from 1930 to 1933 and then worked in Mies van der Rohe's architectural practice in Berlin from 1934 to 1938. From 1939 to 1945, he worked for Egon Eiermann, from 1945 for Hans Scharoun and from 1948 as a freelance architect and designer. Unquestionably influenced by these champions of modernism, Hirche was a busy architect and designer who taught at various universities, designed exhibitions and was actively involved in professional associations. He made a name for himself primarily as a furniture designer. In 1952, he accepted a professorship for interior design and furniture construction at the State Academy of Fine Arts in Stuttgart, where he was also principal from 1969 to 1971.

In 1955, Erwin Braun contacted him through Dieter G. Mundas, his recruitment consultant, with an offer to collaborate. Hirche created numerous designs for music cabinets and TV sets, from the HFK TV music cabinet of 1956 and the FS 6 TV set of 1963 to the spectacular HF 1 TV set of 1958, with its one-button operation system and grey plastic front panel. In addition, Hirche was instrumental in the successful implementation of the company's new design approach in the mid-1950s.

In Königstein, Hirche designed a rigorously modern residential complex for Erwin Braun from 1958 to 1960, as well as three additional residential units and Werner Kuprian's physiotherapy surgery. He was also involved in the design of the buildings on the new factory site in Kronberg. Hirche acted in an advisory capacity during the planning of Erwin Braun's institute for preventive medicine in Engelberg, Switzerland. From 1978 to 1980, he represented Braun and was largely responsible for the interior design. Many of his designs were showcased at national and international exhibitions, including the Milan Triennale in 1957 and Expo 58 in Brussels, and have won numerous prizes and awards.

Georg Hohm
born 1920 in Frankfurt
died 2015 probably in Munich

Georg Hohm was a friend of Erwin Braun from early childhood. The two graduated from high school the same year and remained in close contact as officers during the war. After completing a commercial apprenticeship, Hohm was brought into the Braun company by Erwin in 1950, where he took on the roles of sales manager and, later, sales director of shavers and household appliances. From 1966, he was managing director of Braun Elektromechanik GmbH and Braun Intertechnik Commercial.

After the sale of Braun AG to Gillette, Erwin appointed him managing director of Ritterwerk GmbH near Munich in 1970, where Hohm lived and worked from then on. He subsequently took on numerous other positions in Erwin's companies and foundations in Germany and Switzerland; among other things, he was a founding member and first chairman of the Max Braun Circle in 1988 and, after 1992, managing director of the Max Braun Stiftung GmbH. Hohm was also active in numerous associations and trade fairs. He was a recipient of the Federal Order of Merit First Class Grand Cross.

Hohm was one of Erwin's closest confidants and a central figure in his attempts to build up a parallel corporate conglomerate in Switzerland – a kind of 'Braun 2.0'.

Hartwig Kahlcke
born 1942 in Szczecin, Poland, lives in Kronberg im Taunus, Hesse

After completing an apprenticeship as a woodturner, Hartwig Kahlcke studied product design (later industrial design) at the Folkwang School of Design in Essen from 1965 to 1970. He then joined the development team at Vitsœ, the

Hartwig Kahlcke

Horst Kaupp

Björn Kling

manufacturer of Dieter Rams's furniture designs, for six months. This was followed in 1970 by a position in the design department of Braun AG, where he worked mainly in the household appliances division. This also included the design for products of the subsidiary Braun Española in Barcelona, which are produced exclusively in Spain and for the Spanish market.

During Kahlcke's time at Braun, numerous new kitchen appliances were created. These include the ZK 1, a new form of blender, and the KM 210, a new concept for a food processor. The HT 40 electronic toaster from 1989, with its intelligent features, established a new quality of user experience, as did the KGZ 3 meat grinder and the Multipress MP 80 juicer, which ejects juice and pulp separately. His design of the KF 40 coffee maker in 1984, which is still in production today, is of particular significance for the small domestic appliance product segment. The design of the KF 40 became the dominant one on the coffee maker market because it was a further development of the tower design of the KF 20 but also innovative in form and inexpensive to produce. This resulted in numerous variants of the product, which has become an archetype of the coffee maker.

In 1988, Kahlcke left Braun and went into business together with Florian Seiffert, establishing the company Seiffert + Kahlcke Design.

Horst Kaupp
born 1934 in Frankfurt/Main
died 2021 in Heusenstamm

Horst Bernhard Kaupp, son of a master bookbinder in Frankfurt, encountered the Braun company when he had already finished school and was an apprentice cabinetmaker in the carpentry firm Ludwig Egger. He was involved in various interior fittings and furnishings for the reconstruction of Frankfurt's Goethehaus and Römer town hall, and at the same time was already working for Braun on its trade fair and exhibition construction, including the trade fair stand at the 19th Grand German Radio, Television and Phonographic Exhibition in Düsseldorf in 1955. The stand design for the HfG Ulm, which was created for this trade fair, was subsequently used for thirty-two years at exhibitions, with Kaupp handling the technical and organizational aspects.

Without Kaupp, the Braun Collection and the company's exhibition in Kronberg would probably not exist, or at least it would not be as complete as it is today. As early as 1952, Kaupp persuaded Braun management not to sell the appliances presented at the trade fairs or destroy the faulty devices that were still perfect on the outside. Because of this, Braun's archive is one of the most complete design collections of any German company.

From September 1968 to December 1981, Kaupp was responsible for trade fair and exhibition construction and for special exhibitions at Braun. This involved extensive global travel and meant that, in some years, he spent almost ten months on the road. From January 1982 to December 2005, he held the same position as plant manager at Elma Messe- und Ausstellungsbau GmbH.

When the Braun Collection opened in Kronberg in August 2005, Kaupp served as Braun's archivist and collection manager for a further decade.

Björn Kling
born 1965 in Ludwigshafen, lives in Glashütten-Oberems

Björn Kling grew up in a family where almost everyone was involved in some form of painting, design or sculpture. Through his Swedish mother, the family home was also influenced by Scandinavian design.

After graduating from high school in 1985 and doing his community service, Kling studied industrial design at the University of Fine Arts in Hamburg from 1986 to 1992. His professors included Dieter Rams and Karl-Achim Czemper, a graduate of the HfG Ulm. During his studies, he completed various internships, including one at Volvo's design department in Gothenburg and in product design at Braun. In 1992, he came second in the competition for the BraunPrize. This led to various freelance commissions while he was still at university.

In 1993, Kling first freelanced as a product designer before taking on a permanent position at Braun in Kronberg at a time when the firm was undergoing major personnel upheaval. At first, Kling met several designers from the early days but then he also got to know those of the next generation. He worked with Jürgen Greubel, Robert Oberheim, Dietrich Lubs, Roland Ullmann and Peter Schneider, initially still under Dieter Rams as head of the design department. This makes Kling the only designer to have created products under all three Braun Design heads and in the various phases of Braun Design.

He has been active in all of Braun and Procter & Gamble's product categories: Oral-B, hair care, shavers, trimmers, ThermoScan, household, epilators, IPL devices and Gillette Venus. Since 2011, he has also been involved in redesigning the content as well as the concept of the Braun Collection.

Kling repeatedly supervises semester projects at design colleges, including the Offenbach University of Art and Design, the Coburg University of Applied Sciences and the Burg Giebichenstein University of Art and Design in Halle. In addition, he has given numerous lectures on the subject of design. Today he is design director of female grooming at Braun.

Ingeborg Kracht-Rams
born 1931 in Bochum
died 2022 in Kronberg im Taunus, Hesse

Ingeborg Emmi Kracht, the daughter of a family of entrepreneurs from Bochum, completed her school education in 1948 and began training as a photographer for the studio of theatre, architecture and advertising photographer Grete Hamer (also in Bochum) that same year. She worked for Hamer until 30 April 1954, and towards the end took on assignments in place of her boss. From 1955 to 1957, she studied for four semesters at

Ingeborg Kracht-Rams

Ursula Kull-Sehring

Werner Kuprian

the Folkwang-Werkkunstschule in Essen in the photography class taught by Bauhaus graduate Werner Graeff.

With a little help from her friend Marlene Schneyder, she got a job as a photographer for Mediacolor in Darmstadt from February to June 1957. While taking pictures at the IBA (International Building Exhibition) in Berlin, she started to work for Braun together with Schneyder, standing in for a photographer who had dropped out and, on 1 September 1957, Albrecht Schultz hired her permanently for the company's advertising design department. A year later, Kracht and Schneyder, who remained close friends throughout their lives, travelled to Expo 58 in Brussels on a commission from Braun.

After Schneyder left the company in 1959, Kracht continued to pursue advertising photography with great sensitivity, together with Wilfried Indinger, later a professor at the Offenbach University of Art and Design, Ulf Beckert, brother of ABB partner Hans Georg Beckert, and Pit Schumann. In 1960, Kracht passed her master's examination in photography.

On 30 June 1963, Kracht ended her employment with Braun and stayed in South Africa for six months. In 1964, she returned to work as a freelance photographer for Braun and took on all product photography for the patents department. She also worked in the field of interior and object photography for Vitsœ and for the exhibition documentation of Dieter Rams, whom she had married in December 1967. On top of this she produced non-commissioned photographic works.

Ursula Kull-Sehring
born 1935 in Frankfurt
died 2023 in Kronberg im Taunus, Hesse

After graduating from high school, Ursula Sehring completed an apprenticeship as a photo lab assistant and photographer at the Frankfurt-based Dr C. Schleussner Fotowerke (ADOX), a pioneer in photographic films during the 'New Frankfurt' era of the 1920s. In 1954, at the age of nineteen she became the first woman to be hired by Braun for product photography. Between 1955 and 1956, Marlene Schneyder and Ingeborg Kracht also joined the group. The three photographers developed an objective style of product and interior photography that did not resort to dramatic lighting effects and only hinted at backgrounds. Thus, along with the company's new graphic design, a new attitude to photography emerged that avoided retouching and the use of flash lighting. Sehring also specialized in photographing production operations at Braun.

In 1958, she left the Braun company. She worked as the principal photographer at KULL & WILLING, an advertising agency founded by her first and second husbands, until the end of the 1980s. Director Robert Siodmak wanted her to be an actress in one of his films, but she declined his invitation, remaining content behind the camera instead of in front of it.

Werner Kuprian
born 1929 in Hanau
died 2003 in Königstein

After studying gymnastics and sports at the University of Frankfurt, training as a state-certified physiotherapist at Frankfurt University Hospital and graduating as a professional riding instructor, Werner Kuprian joined Braun in 1954. His task: to set up and manage a company health service. Before that, Dr Fritz Bode, who served as the company doctor from 1951 onwards, had already laid some foundations for health provision. Kuprian expanded these considerably and became Erwin Braun's 'personal trainer'. His most important employee was the legendary 'Nurse Margot', who took care of everyone and became the soul of the health service.

When he took office in 1954, Kuprian wrote in the company newsletter Werknachrichten: '"Prevention is better than cure" is therefore the precept under which we have placed a series of measures that will soon be started – we want to employ the age-old but always effective natural healing forces such as heat, air, water, earth, exercise and breathing.' This approach clearly went beyond the usual work of a factory doctor and was very much in line with Erwin Braun's philosophy. Kuprian continued: 'Real health! This is a state of perfect physical, spiritual and mental harmony.'[9]

The Braun health centre was available to all company employees and focused on offering physical education, compensatory workplace gymnastics, sports and ergonomics, respiratory treatment, connective tissue massage, sunlamp treatment and chiropratic therapy – all free of charge. In 1962, he established the Kuprian Institute in the buildings designed by Herbert Hirche in Königstein. The Austrian conductor Herbert von Karajan was among the institute's prominent patients.

During the 1972 Summer Olympics in Munich, after three years of preparation, Kuprian took over the management of the team of physiotherapists looking after athletes during the games, a team formed on his initiative. Kuprian published essays and books on therapeutic riding, sports and physiotherapy[10] and was actively involved in several specialist associations. Every two years, the German Board of Trustees for Therapeutic Riding now awards a Werner Kuprian Prize for outstanding scientific or literary work. Kuprian supervised Braun's occupational health service until 1995.

Gerhard Lander
born 1922 in Frankfurt/Main
died 2002 place unknown

Gerhard Lander had been friends with Erwin Braun since early childhood – a friendship that would last until Braun's death. Both attended the Liebig Ober-Realschule senior high school in Frankfurt. After graduating from the school in 1940, Lander was drafted into service with the Frankfurt electrical company Hartmann & Braun and began studying mathematics, physics and chemistry in 1943 at Goethe University. After the war, he became a lecturer there and in 1950

Gerhard Lander

Ludwig Littmann

Archibald Livis

earned his doctorate at the Institute for Applied Physics at Goethe University. Nine years later he received his postdoctoral degree.

After taking over the management of the company, Erwin Braun initially recruited him on a freelance basis to deliver technical training. In the course of this work, Lander developed a novel concept for an electronic flash unit under the name of Hobby, which, from 1952 onwards, contributed significantly to the company's post-war success. He also played a key role in the development of stereo and Hi-Fi equipment and the company's first slide projector. Lander was both an engineer and a successful manager. In 1962, he was appointed to the board of Kodak AG and was involved in the carousel projector designed by Hans Gugelot. From 1972 to 1983, he was a member of the board and head of research for accumulator, battery and fuel-cell development at VARTA AG in Kelkheim. From 1969 to 1987, he held a professorship at the Institute of Applied Physics at Goethe University and was elected president of the Physikalischer Verein.

In October 2011, he and his wife Margo established the Professor Gerhard Lander Foundation for Research at the Institute of Applied Physics, which he endowed with assets of one million euros.

Ludwig Littmann
born 1946 in Mülheim, lives in Kronberg im Taunus, Hesse

Ludwig Littmann[11] is a graduate of the Folkwangschule in Essen. With a design for an 'automated litter collection system', he won first place at the BraunPrize in 1972, together with Peter Schneider, Lothar Brendgens and Wolfgang Rebentisch. From 1973 onwards, he was employed as a product designer at Braun. Littmann quickly specialized in household appliances, a field in which he was very active. For example, after 1981, he was responsible for hand mixers and hand blenders and, after 1984, steam irons. In 1991, his E 400 T espresso machine with its distinctive angler design, was launched. In 2004, his organically shaped metal HT 600 toaster and WK 600 electric kettle appeared.

His Multiquick MR 550 hand blender received an iF Award in 1996 and a successor model won the Red Dot Award in 2002. Littmann got through an enormous design workload at Braun. During his time under Peter Schneider, he transitioned the objective, geometric design ethos of the Rams era into a more emotional, organic approach. From 2005 onwards, together with four other designers, he headed the Group Design 4, Design Project.

Archibald Livis
born 1938 in Montreal, Canada, lives in Boston, US

In 1959, after studying business administration in Montreal, Archibald Livis began to work in the controlling department of the Canadian subsidiary of the American Gillette Group, which marked the start of his managerial career in the company. After holding various positions in finance, in 1975, he moved to the marketing department at Braun Española, SA, in Barcelona, where he was appointed managing director in 1978. In 1982, he joined Braun in Kronberg. In 1984, he became a member of the executive board and, in October 1993, chairman of the board. Livis wrote in a Design Centre Munich publication in 1997:

> We want to concentrate our resources on core product areas. These core areas include electric shavers, hair removal devices for women, household, hair care and oral care products and, more recently, home diagnostic devices ... Our company's objective is to achieve a leading global market position in our core product areas.

The corporate strategy also includes 'offering a wide range of different models in each product category to maximize market and sales potential'. To achieve this, new models needed to be launched on the market as quickly as possible.[12] This was a new strategy that clearly deviated from Braun's corporate and design attitude of the 1960s. On 18 May 1998, Livis stepped down from the Braun executive board and joined its parent company Gillette in Boston as executive vice president with responsibility for the Braun, Oral-B, Waterman, Parker and Papermate brands. He was succeeded as chairman of the board by Bernhard Wild. On 31 July 2000, Livis retired from Gillette.[13]

Dietrich Lubs
born 1938 in Berlin, lives in Bad Homburg vor der Höhe, Hesse

At the age of fifteen, Dietrich Lubs moved with his family from East Berlin to West Germany and began an apprenticeship as a shipbuilder at the Cologne shipyard Hansa Stahl und Schiffbau, where he was later taken on as a parts designer and technical draftsman.

He had his first formative encounter with modern design when he visited Expo 58 in Brussels, especially the West German exhibition pavilions designed by Sep Ruf and Egon Eiermann. Braun's first products embodying the company's new approach to design were also shown there. Later he visited Braun's Düsseldorf information centre and took the initiative to apply to the Kronberg-based company. In July 1962, he was hired as an assistant in the design department. Here, due to his previous training, he was initially responsible, together with Roland Weigend, for the typography of the appliances. His consistent use of the Akzidenz Grotesk typeface, especially in small point sizes, enhanced both the desired objectivity and the recognizability of the devices. Lubs subsequently became involved in the design process and, as he put it, learned 'design' at Braun.

One of Lubs's main areas of activity in his evolution as a designer was in the clock sector. With his phase 3 alarm clock, he created a device that he would vary and further develop in many subsequent models.

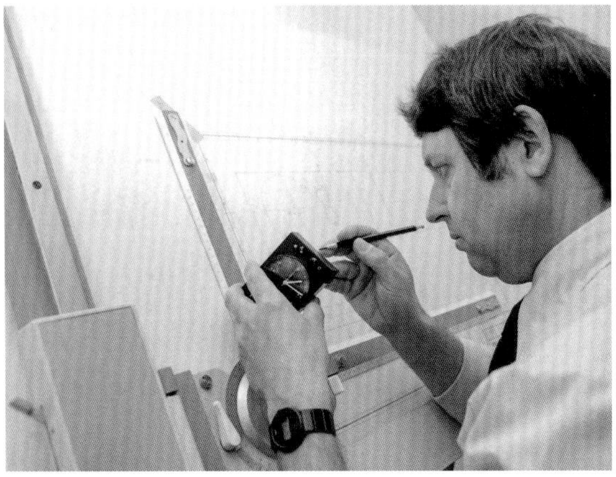

Dietrich Lubs

Traugott Malzan

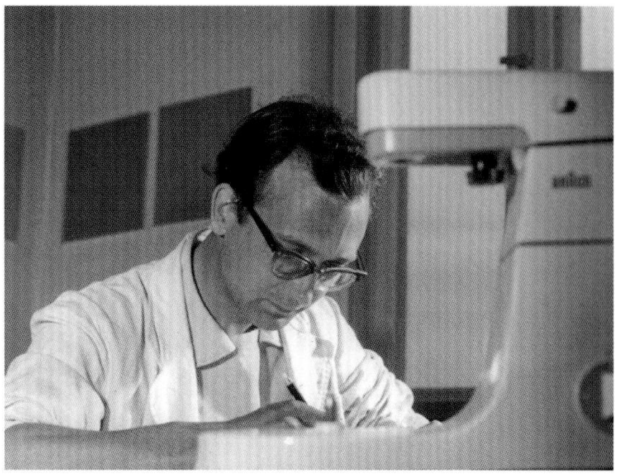

Gerd A. Müller

In 1975, Braun launched the first ET 11 pocket calculator, the technology and form of which were adopted without alteration from the Japanese manufacturer OMRON. However, from the first Braun-designed successor model, the ET 22, onwards Dieter Rams and especially Lubs gave pocket calculators a completely different look, designing them with flat, convex and high-gloss buttons and a subtle colour scheme.

Regarding design processes at Braun, Lubs recalled: 'a conversation in the corridor could be more important than one in a meeting. The whole company was like a root system in which there were diverse contacts independent of hierarchies and departments. There was an open climate of discussion.' And, despite the many prizes and awards, there were equally no egos: 'we simply made products. Each product was born out of the need to make a typical Braun product.'[14]

Traugott Malzan
born 1927 in Darmstadt
died 1990 in Port Grimaud, France

After graduating from high school, Traugott Malzan was drafted into military service for a short time. After 1945, he studied history in Mainz, where he also earned his doctorate. Starting in 1952, he worked for two years as an editor at the Darmstädter Echo before becoming a foreign correspondent for various German newspapers and magazines. After returning from abroad to Germany in 1956, Malzan initially became chief press officer at Ullstein publishing house. From 1958 to 1968, he worked for Braun, initially as advertising manager for radio/television/phono, and was granted power of attorney for the new products division in 1961. Throughout the 1960s Malzan was the worldly and experienced spokesperson 'for what we understood by industrial design', as Erwin Braun later expressed it.[15] In 1968 and 1972 respectively, Malzan became a member of the executive boards of Kahla, a porcelain manufacturer, and Hutschenreuther. After 1978, he worked as an independent management consultant.

At the end of the 1960s, Malzan and his American wife Patricia bought a house built by the architect François Spoerry in the postmodern lagoon town of Port Grimaud near Saint-Tropez on the Côte d'Azur. Here, the yachts of the well-heeled residents were moored on the doorstep, but for the Malzans it was a Chinese junk ship, imported from Hong Kong, that they used for cruises along the French Mediterranean coast. In 1987, Malzan published an extensive travel report about his voyages as a book.[16]

Gerd A. Müller
born 1932 in Frankfurt/Main
died 1991 in Wilhelmshaven

Gerd Alfred Müller, son of a family of dentists in Frankfurt, was not at all drawn to his father's profession. After finishing school, he began an apprenticeship as a carpenter and while at vocational college he met and befriended Dieter Rams. His parents did not support his choice of career, so he earned extra money as a mime, poet and singer in allotment garden clubs. Even then, there were signs of what would later become typical for him: Müller was interested in many different things – above all in books about the cultures of the world.

After completing his carpentry apprenticeship, he was encouraged by Rams to attend the Wiesbaden Werkkunstschule. Although he graduated as an interior designer, he was also interested in sculpture.

At the end of 1955, he was hired by Braun on Rams's recommendation, since Artur Braun was looking for a model-maker for the company's new kitchen appliances. Müller built the models for the KM 3 out of plaster, like a three-dimensional sculpture. The food processor would become one of Braun's most successful products. Further kitchen and shaver models were created, all of which reflected a more organic approach to design. He also designed a forerunner of the sixtant, the shape of which was later adopted by Hans Gugelot for the successful shaver with minor changes.

In 1960, Müller left the Braun company to set up his own design studio, Atelier 8. His most important client was Lamy, a company for which he not only designed the revolutionary Lamy 2000 fountain pen but also, together with Manfred Lamy, developed a design-orientated corporate strategy. Numerous other models, packaging and marketing ideas for Lamy followed. Also an ambitious amateur chef, Müller cooked for and entertained his guests while swapping ideas and brainstorming, and so the dinner party became a part of his design process.

Müller also worked for other industrial companies, such as WEGA, Buderus and the wooden toy manufacturer Naef in Switzerland. His output extended to graphic design, such as the corporate design for the city of Eschborn and the furnishings for its museum. He was active in the Deutscher Werkbund and a co-founder of the Deutscher Designer Club. He was more than critical of postmodernism and soon turned away from industrial design altogether. Instead, he embraced nature conservation and designed campaigns and exhibitions for the Bund für Umwelt und Naturschutz Deutschland.

Robert Oberheim
born 1938 in Gedern, Hesse
died 2014 in Liederbach near Frankfurt/Main

Robert Oberheim graduated from the Wiesbaden Werkkunstschule in 1960 and joined Braun AG as a designer that same year. In 1965, together with Reinhold Weiss, Richard Fischer and Dieter Rams, he received the Young Generation Berlin art award. From 1972 to 1994, he was one of the deputy heads of product design at Braun.

Oberheim became one of Braun's most productive designers. In 1962, he was already involved in the first Nizo camera, the FA 3, and from 1963 onwards designed all cameras as well as numerous slide and film projectors until the mid-1970s. Particularly noteworthy is the D 300 slide projector, which was introduced in 1970 and received an iF Award two years later. In 1964, he revised Gerd A. Müller's food processor, which was then produced

Robert Oberheim

Markus Orthey

Dieter Rams

until 1991. In collaboration with Karl-Heinz Lange, Oberheim also designed Braun's only camera, the Nizo 1000, which was launched in 1968. Until 1972, Rams designed the flash units but then he transferred the work to Oberheim, who added numerous other designs over the next ten years. In addition, from 1979, Oberheim completed an impressive amount of work designing personal care appliances.

His work exemplifies the long-standing specialization of Braun designers in specific design areas. In 1994, he left Braun but continued to work for the company in a freelance capacity, as well as for a variety of other firms.

Markus Orthey
born 1973 in Hachenburg, lives in Eltville am Rhein

Growing up in a rural area in the Westerwald, Markus Orthey joined local craftsmen in their purposeful and creative work at an early age. Attending a humanistic secondary school with an artistic focus subsequently broadened his understanding of education.

After completing his schooling, he initially began studying interior design at the University of Applied Sciences in Mainz in 1995, graduating five years later with a diploma and the accolade 'best student of the year'. Working as assistant to Professor Florian Seiffert, he became acquainted with product design and started freelancing for the office of Seiffert+Kahlcke. In addition, he took part in an exchange programme with Swinburne University in Melbourne, Australia, and became a trainee at Matsushita Electric Works in Osaka after completing his studies. This was followed by a permanent position as a product designer at Braun in Kronberg in 2001.

At Braun, Orthey initially worked with Ludwig Littmann and designed a steam-ironing station as his first independent product. Together with Littmann, he then designed numerous products in the field of household appliances. Through the senior designers of the Rams era, he also became familiar with the design DNA of the company's earlier years. When this era of designers retired, Orthey quickly assumed a management role. In 2007, he became senior product designer at Braun. In 2013, he joined De'Longhi Braun Household as senior design manager, where he has been design director since 2019 and is responsible for the production of the company's household appliances. He also holds teaching positions and guest lectureships in Italy at Politecnico Milano and ISIA Roma, and in the Netherlands at Delft University of Technology.

Orthey always begins new design projects with observation, an in-depth analysis of the challenge being posed and a library of images that give form to his ideas: 'I need an environment that does not stimulate me visually.' He often finds inspiration for new projects while in nature. 'I like to walk through the nearby vineyards of my home with my sketchbook, let my thoughts wander and try to capture them in sketches. Ultimately, my goal is to create innovative products that are useful to people, that improve their lives, that represent meaning. For me, the purpose of Braun is to make life more manageable and to give people freedom.'

Dieter Rams
born 1932 in Wiesbaden, lives in Kronberg im Taunus, Hesse

As a child, Dieter Rams spent a lot of time in his grandfather's carpentry workshop. In 1947, at the age of fifteen, he became a student at a school for arts and crafts, the Handwerker und Kunstgewerbeschule Wiesbaden, but interrupted his studies after one semester to complete a carpentry apprenticeship. When he resumed his studies, the school had a new director, Dr Hans Soeder, and a new name, the Werkkunstschule (School of Arts and Crafts). Soeder also had plans to change the title of the interior design programme to Gestaltung für Raum und Gerät (Room and Appliance Design). Rams passed his interior design exams with distinction in 1953 and went on to work for two years in the office of Frankfurt architect Otto Apel. He joined Braun in July 1955 initially as an interior designer but soon started to receive commissions for product designs. After 1956, he also worked on furniture designs for Otto Zapf in Eschborn, who produced his system furniture in his father's workshop. Erwin Braun granted Rams permission to pursue a side career, which gave him a certain independence as a designer. In 1959, Danish furniture dealer Niels Wiese Vitsoe partnered with Otto Zapf and further expanded the company as Vitsoe & Zapf. Rams's 606 universal shelving system and his 620 chair programme, one of the very few design products to have acquired an artistic copyright, are still manufactured today by Vitsœ in England and their popularity has remained undiminished for the past sixty years.

In 1961, Rams became head of the now institutionalized design department at Braun. The department was expanding and becoming more productive with a steadily increasing portfolio. New designers were often initially employed by Vitsœ 'on a trial basis'. In 1968, Rams was appointed director of product design at Braun AG and became general manager in 1988. The new design focus under his leadership resulted in numerous prizes and exhibitions. In 1965, together with Reinhold Weiss, Richard Fischer and Robert Oberheim, Rams received the Young Generation art prize in Berlin. In 1968, he was made an Honorary Royal Designer for Industry (RDI) by the Royal Society of Arts and, in 1991, received an honorary doctorate from the Royal College of Arts. There were numerous other awards, memberships and presidencies between these two dates and in later years. From 1987 to 1998, for example, he was president of the German Design Council and, from 1981 to 1997, he held a professorship for industrial design at the University of Fine Arts, Hamburg.

In the forty years he worked for Braun, his Design Team created almost 1,000 products and product variants, more than 300 of which were designed by him or in close collaboration with a team member.[18] In the process, he created milestones of Braun Design: the SK 4 radio-phono combination in 1956, the TP 1 phono transistor

Nina Sagra

Wolfgang Schmittel

in 1959, the T 1000 world receiver in 1963, the wall-mounted Hi-Fi stereo system in 1965, the F1 Mactron pocket lighter in 1971, the DW 30 digital wristwatch in 1978, the ET 66 pocket calculator in 1987 and, last but not least, the aforementioned 606 universal shelving system and 620 chair programme system for Vitsœ.

In 1995, Rams handed over the management of Braun's design department to Peter Schneider and retired from Braun in 1997. One highly noteworthy exhibition (one of very many), Less and More: The Design Ethos of Dieter Rams, attracted more than 400,000 visitors and travelled to major museums in Osaka, Tokyo, London, Frankfurt, Seoul and San Francisco between 2008 and 2012. In November 2018, the documentary film Rams by Gary Hustwit premiered in New York and was subsequently released in cinemas worldwide proving immensely successful.

In 1992, Rams and his wife Ingeborg founded the Dieter and Ingeborg Rams Foundation, which is dedicated to promoting design. In cooperation with the Museum Angewandte Kunst in Frankfurt, it is responsible for the Dieter Rams Archive, which looks after the designer's personal estate and makes it accessible to the public.

Nina Sagra
born 1977 in Hamburg, lives in Frankfurt/Main

After studying Communication Design at the University of Applied Sciences in Potsdam, Nina Sagra moved from Berlin to Italy, where she had spent some time during her studies. Starting in 2005, she worked in various agencies in Milan, including Syneo Design Consultancy, Studio C Milano and Continuum, where she oversaw brand strategy, exhibitions and print and packaging design for brands like Whirlpool (kitchen appliances), Persol (sunglasses) and Prodir (pens). During her time in the north Italian metropolis, she found the collaboration with designers from various disciplines to be especially inspiring. In 2010, she left Milan to freelance in London for a year. Sagra's passion for high-quality design led her in 2011 to the Braun brand, which she regards as a design role model and associates with many childhood memories – her father, who was something of a technology aficionado, had possessed not only a Nizo film camera but also a micron shaver and Braun wristwatch.

Now acting as an intermediary between various functions in marketing and product research at Braun, she is responsible for the company's coherent and contemporary brand presence with a focus on visual communication and packaging design, as well as the fields of user interface and user experience design.

Wolfgang Schmittel
born 1930 in Frankfurt/Main
died 2013 in Kronberg im Taunus, Hesse

Wolfgang Schmittel studied at the Städelschule in Frankfurt in the late 1940s and early 1950s under Theo Garve, who in turn was a student of Max Beckmann in the 1920s. Although Schmittel certainly considered himself an artist, he also completed an internship producing commercial art alongside his fine art studies. In August 1952, he became a freelancer at Braun. One of his first jobs was to rework Will Münch's Braun logo lettering in an exact grid pattern, which would remain unchanged until the 1990s. Today's logo is also somewhat rounded but not fundamentally different. One of his early advertisements for the Hobby flash unit is presented as a pre-war modernist-style collage.

Schmittel, who initially worked with typography in a conventional manner, using Antiqua fonts and a central axis, adopted the ideas of the HfG Ulm for visual communications, as did Dieter Rams in product design. However, both soon emancipated themselves from the HfG model and continued to develop their approaches in a way that was appropriate to the design.

After the move to Kronberg, the communications department became lavishly equipped, produced cinema advertising and boasted a large photo studio and numerous employees. Eckard Tischer, who was an advertising manager in Schmittel's department from 1961 to 1968, later recalled the way things worked. He described the atmosphere as 'unique', something he put down not only to the influence of Erwin Braun and Fritz Eichler, but also to an approach to work within the department that was strongly based on open discourse. Themes were debated and discussed, not simply directed by someone from the top. For example, the advertising department discussed why erotic photos should not be used. Over time, these conversations led to a common-sense outlook that everyone accepted and practiced consistently. Once something had been determined, it was not questioned again and again.[19]

In his spare time, Schmittel became a respected photographer of jazz concerts and even received a stage permit, probably through his connection to concert organizer Fritz Rau via graphic artist Günther Kieser, who was a mutual friend.

In 1980, Schmittel parted company with Braun. Subsequently, he ran his own consultancy firm for corporate identity and visual communications in Kronberg. Braun remained one of his clients for a while, but Schmittel also worked for the photo manufacturer Rollei, the mail-order company Quelle, the Austrian Creditanstalt and Volkswagen AG. On 1 September 1982, he was appointed a professor of graphic design at the University of Design Schwäbisch-Gmünd. He published several books on corporate design. Numerous prizes, awards and exhibition participations honour his work as a graphic designer and photographer.

Peter Schneider
born 1945 in Kitzbühel, Austria, lives in Sinzheim near Baden-Baden

Peter Schneider, the grandson of Blendax entrepreneur Rudolf Schneider from Mainz, lived briefly with his family in Peru from the end of 1950 until May 1958. After the family returned to Germany, Schneider attended secondary school in Neubeuern, Upper Bavaria, and graduated

Peter Schneider

Peter Schneider's Design Team

Marlene Schnelle-Schneyder

from there in 1967. From 1968 to 1972, he studied product design at the Folkwangschule in Essen. In 1972, he won the BraunPrize.

Initially, Dieter Rams hired the young designer for Vitsœ and shortly afterwards for the Braun Design Team. Thus, in 1972, Schneider was the second-to-last person to join Rams's Design Team. In 1976, Schneider took over the development of the new sound film cameras, the penultimate chapter of Braun photographic technology. Continuing the designs of Robert Oberheim, Schneider created two new product families, with special emphasis on the Nizo integral camera.

At Braun, Schneider was also responsible for design commissions from the Gillette Group, including its JAFRA cosmetics brand in 1989, the Silk·épil epilator in 1992 and external commissions from clients, including Siemens and Lufthansa. For Gillette's Oral-B brand, he developed the area of manual and electric toothbrushes as a new field for Braun and the Design Team.

For more than twenty years, Schneider worked with Rams before taking over from him as manager in 1995. In this role, he developed a design mindset that focused more on the internationalization of Braun's markets, while fully supporting the new strategies in product development (the introduction of CAD/CAM, short 'time to market' or product lead time) and business management (market research including design testing).

From 1995 to 2009, he was responsible for the design and implementation of the BraunPrize, and from 1998 to 2009, he served as the jury chairman. In 2005, Schneider, together with Bernhard Wild and Horst Kaupp, was instrumental in setting up the Braun Collection in Kronberg, which brings together the company's collection and archive and makes them accessible to the public. Schneider retired in January 2009.

Peter Schneider's Design Team

Under Peter Schneider, the new head of design after Dieter Rams, the Design Team at Braun increased steadily and by 2005 included several female members.[20]

In 2005, its members were:

Director Corporate Design: Peter Schneider
Assistant: Josefa Gonzales
Group Designers: Roland Ullmann, Ludwig Littmann, Peter Hartwein, Björn Kling
Product Designers: Till Winkler, Duy Phong Vu, Fritz Schubert (computer designer), Markus Orthey (junior designer), Benjamin Wilson (junior designer), Rory McGarry (junior designer), Wolfgang Stegmann (intern)
Graphic Designers: Ella Korell, Catharina Gasseling and Claudia Hempel
Corporate Identity Designers: Sven Wuttig
Model Makers: Klaus Zimmermann (head), Christoph Marianek, Oliver Michl, Alexander Dorn, Karl-Heinz Wuttge, Udo Bady

The designers were supported by two specialists from the Pro-E/CAD Department, Bernhard Sikora and Joachim Nickel.

Marlene Schnelle-Schneyder
born in 1931 in Insterburg, lives in Berlin

Marlene Schneyder completed her photographic apprenticeship in Grete Hamer's studio in Bochum from 1949 to 1952, having won a regional photography competition in North Rhine-Westphalia. She then began studies at the first state college for photography in Cologne in 1954 (Staatliche Höhere Fachschule für Photographie), graduating in 1955.

On 1 June 1955, she started work as an advertising photographer at Braun in Frankfurt, at a time when Fritz Eichler had initiated a new advertising concept that introduced customers to the newly designed electrical appliances. He believed that the buyer should be informed of Braun products objectively, and to this end, he worked with the HfG Ulm to come up with an advertising style that dispensed with photographic effects. Products were placed in front of a bright monochrome background; indirect light ensured neutral, uniform illumination. So-called milieu shots, which showed corresponding settings and were supposed to convey the 'style of our time', became pivotal to Braun's advertising strategy. Accordingly, several companies, including Rosenthal, Knoll International, WMF and Braun, joined forces to create an advertising campaign that conveyed the idea of modern interiors. Producing the photographs for the campaign was a challenge, however, since any clichés introduced through added effects or retouching were frowned upon.

To document Braun's presentations at Interbau Berlin in 1957, Schneyder brought on board a former colleague from Bochum, photographer Ingeborg Kracht. The two friends also worked together successfully for Braun at Expo 58 in Brussels.

After leaving Braun in 1958, Schneyder worked as a freelance photographer in Cologne, Bonn, Bochum and Berlin. From 1975 to 1981 she studied history of art, journalism, philosophy and English and received a doctorate in 1988 for her dissertation on the theme of the depiction of movement in nineteenth-century photography. Afterwards she taught at various German universities, published widely and designed several exhibitions.

Schneyder was appointed a member of the German Photographic Society in 1987, served on its board for ten years and was made an honorary member in 2002. Since 1986 she has been an appointed member of the German Photographic Academy, where she likewise became an honorary member in 2021.

Rudolf Schönwandt

Albrecht Schultz

Florian Seiffert

Rudolf Schönwandt
born 1935 in Nordeck-Rabenau near Giessen
died 1997 in Hofheim

Rudolf Schönwandt came from a family that embraced the Lebensreform Bewegung (reform of life movement) of the 1920s. His father was an engineer and silversmith. Both his parents were involved in establishing the anthroposophical reform settlement Loheland in the Röhn. From 1927 to 1928, they worked at the Folkwangschule in Essen. Schönwandt also initially learned the trade of silversmith but after finishing secondary school began to read philosophy, German and art history alongside other subjects. From 1955 to 1965, he studied in Heidelberg and Göttingen under the philosopher Hans-Georg Gadamer. He had a broad and well-founded range of interests, and his dream job was to be a writer.

Initially, however, he worked as a copywriter for several advertising agencies and then, from 1973 to 1981, at Braun in the advertising department. Here he developed a friendship with Dieter Rams, for whom he became a partner in intellectual discussions. He also maintained close contact with the progressive Frankfurt publisher KD Wolff, as well as with Michael Schneider from the Institute for New Technical Form (INTeF) in Darmstadt. At Braun, Schönwandt excelled in his work; an employee with a wealth of knowledge, he became increasingly expert in complex strategic corporate identity (CI).

After leaving Braun, Schönwandt was in great demand as a CI specialist. Among other things, he developed the major trade fair presentations for Mercedes-Benz and together with Atelier Markgraph wrote scripts and designed the DASA (German Occupational Safety and Health Exhibition), a permanent exhibition of the Federal Institute for Occupational Safety and Health in Dortmund on the world of work.

In addition, he worked as a freelance communications consultant and independent publicist specializing in product design, including for the FAZ, Handelsblatt, md magazine and the Werkbund magazine werk und zeit. He published several essays on Dieter Rams. He also continued to work for Braun on a freelance basis, for example, as the author of the brochure 40 Years of Braun Design in 1995.

After Schönwandt's untimely death, Ulrich Kern, head of Braun's advertising department at the time, characterized him and his most important texts in a publication published by the company as follows:

> Superficially, his texts are concerned with specific, matter-of-fact questions: What is the status of German design in the world? What distinguishes outstanding designers? ... If you take a closer look ... Rudolf Schönwandt was concerned with more: For him, design was a paradigm. In the things of this world, our values and world views become visible and comprehensible, but also vulnerable. And that was what really concerned him.[21]

Albrecht Schultz
born 1919 in Naunhof near Moritzburg
died 2007 in Honolulu, Hawaii, US

After serving in World War II as a fighter pilot, Albrecht Schultz studied business administration in Munich from 1949 to 1953. After 1953, he worked for Braun in market research and sales. In 1960, he joined the management team and in 1963 became a member of the Braun AG executive board. He became head of sales and the shaver division, which he developed significantly and which later became Braun's core field of business. From 1976 until his retirement in 1984, he was vice chairman of the executive board, then a member of the advisory council.

Schultz was one of Braun's longest-serving managers, and his focus was the highly profitable expansion of the company's shaver business. At the same time, with his strong educational background and interest in culture, he was the company's businessman who was most interested in and informed about design, and he presented the company from a design perspective in numerous lectures to business organizations. In addition to his work for Braun, Schultz was substantially involved in associations and institutions, especially in the field of trademarks and on the board of the International Design Center Berlin (IDZ).

In 1994, when he was seventy-five, his wife became seriously ill and Schultz emigrated with her to her native Honolulu, Hawaii, where he spent the rest of his life.

Florian Seiffert
born 1943 in Jamlitz in Lower Lusatia, lives in Bierstadt, Wiesbaden

Florian Seiffert comes from a family of artists: his father was an interior designer, painter and engraver, his mother a weaver. In 1960, the family fled from East to West Germany. Initially, Seiffert had an apprenticeship in West Germany as a precision mechanic and then an internship as a technical draftsman and designer. In 1965, he studied industrial design at the Folkwangschule in Essen and, in 1968, he and the Japanese designer Masanori Umeda won the first BraunPrize for their idea for a new type of 16-mm film camera. The prize money of 15,000 Deutschmarks was presented to him by Fritz Eichler at the prestigious Grand Hotel Frankfurter Hof. It was immediately followed by an offer to become a member of the Braun Design Team.

Though Seiffert spent only four years working at Braun – from 1968 to 1972 – he made his mark with a total of fourteen products that he either designed or was involved in designing, including the mach 2 pocket lighter, the cassett travel shaver, the sixtant 8008 shaver and, above all, his Aromaster KF 20 coffee maker with its tower-like design, which was launched in 1972. Not only did the KF 20 establish a new type of coffee maker but it was added to the collection of the Museum of Modern Art (MoMA) in New York.

After leaving Braun, Seiffert worked for Tomás Maldonado's office in Italy for ten months in 1972 before opening his own design studio, which he ran in Monza for five years before

Dieter Skerutsch

Wolfgang Stegmann

Renke Thye

settling in Munich for six years and finally in Wiesbaden. From 1988 to 2005, he ran the Seiffert+Kahlcke studio in partnership with Braun designer Hartwig Kahlcke. They worked for Krups, Loewe, Schulte-Ufer, Thomas vacuum cleaners and Wella in Germany and for Black+Decker in the United States. For more than thirty-two years Seiffert has received commissions from the Japanese company Matsushita Electric Industrial, which includes the Panasonic, Technics and JVC brands.

Additionally, Seiffert taught at Pforzheim University and was a professor of product design at Mainz University of Applied Sciences from 1987 to 2007, as well as a visiting professor at the Istituto superiore per le industrie artistiche (ISIA), the first highly renowned university for industrial design in Rome, from March 2016 to June 2019.

Dieter Skerutsch
born 1926 in Berlin
died in Kronberg im Taunus, Hesse

Dieter Skerutsch studied physics and mathematics in the 1940s and 1950s and was an assistant to the professor Max Bense in Stuttgart. After his studies, he worked as an editor for various newspapers and magazines. From 1958 to 1978, he was employed by Braun, where he headed the electronics advertising section from 1963 to 1968, was head of general advertising from 1968 to 1972 and, finally, creative director for promotional activities in the field of design. In addition, he conceived the design exhibitions Was nützt die gute Form? (What is the Use of Good Form?) and Form nicht konform (Non-Conforming Form) in Essen and Frankfurt, respectively.

Even after becoming self-employed in 1972, with commissions from Rosenthal, Canton, Letraset, the Hi-Fi Institute, the Association of German Interior Designers and Ciba-Geigy, among others, he continued to work on the basic rules for communication at Braun. Much of his freelance work was done in collaboration with Braun graphic designer Bernd Ziepluch. Skerutsch was largely responsible for Braun's original communication style of lightness, subliminal humour and factual text. From 1962 to 1969, he conceived and wrote a total of nine comprehensive brochures on the Hi-Fi sector. These were highly informative and easy to understand, had an attractive, objective graphic design and were illustrated with photographs by Darmstadt graphic designer Christoph Gassner.[22]

In addition to scientific essays on control technology in the chemical industry, for example, he also wrote about nude photography and compiled cookbooks for Braun. He even tried out all the recipes himself using Braun kitchen gadgets and food processors. Skerutsch was an early exponent of crossing boundaries, favouring the combination of high-end and low-end culture and design.

Wolfgang Stegmann
born 1978 in Illertissen, lives in Mörfelden-Walldorf

From 2000 to 2003, Wolfgang Stegmann studied industrial design at the State Academy of Fine Arts in Stuttgart. During his studies, he freelanced at the UP Designstudio in the same city.

From 2003 to 2005, he continued his studies at the Aalto University School of Arts, Design and Architecture in Helsinki, which cooperated closely with the Nokia company. As part of a collaborative project between the university and Braun, Stegmann's submission received an award, which prompted an invitation from Braun to undertake an internship in the company's design department. He graduated in 2005 and initially worked at the Milan-based design studio Syneo for companies such as Motorola, LG and Whirpool, as well as the eyewear manufacturers Persol and Rayban. In 2007 Stegmann returned to Braun, now employed on a permanent basis, initially working for the Oral-B brand.

Since the appointment of Oliver Grabes as the new head of product design at Braun, Stegmann has been significantly involved in the development of a new design language and the realignment of Braun's product design, with the aim to forge links between the past and the future. Since 2012, as design director for the dry shaver, hair care and healthcare divisions he has also been responsible for the development of new product concepts.

Stegmann is active as an external study supervisor at various design colleges and has on several occasions sat on the jury for the German Design Council's German Design Award.

Renke Thye
born 1983 in Westerstede, lives in Düsseldorf

After growing up in a rural environment in north Germany, Renke Joris Thye studied industrial design at the University of Wuppertal from 2004 to 2011, where one of his instructors was Oliver Grabes. During his studies, he interned at IDEO in Chicago and freelanced from Germany for Seattle's Carbon Design.

As part of his postgraduate studies after his bachelor's degree, which he passed with distinction, he worked with three other fellow students at the university as a consultant for the redesign of the Wuppertal suspension railway. In 2011, Thye graduated from Wuppertal with a master's degree on Braun's brand strategy and a practical section on Braun Audio.

Even before graduating, he began working as a designer at Braun, starting in 2009 in the dry shaver, hair care and household segments. He was involved in the development of new product concepts, such as the relaunch of Braun Audio, and in the opening of the new Braun Collection. From the very beginning, Thye has been instrumental in the reorientation of Braun's product design and the development of its new design language, one that seeks to forge links between the past and future to reposition Braun as a design brand in a changing world.

Since 2015, his main task has been to apply these changes in product design to product communications, digital media and graphics using a holistic design approach. After 2017, Thye headed the visual communications and packaging team as design director, where he

Roland Ullmann

Duy Phong Vu

was responsible for developing a consistent and confident image to match the product design. During this time, the new Braun Linear house typeface was created under his leadership.

Since 2022, he has been responsible as design director for the Incubator and Ventures divisions, which deal with the design and innovations of future product generations.

Roland Ullmann
born 1948 in Offenbach am Main, Hesse, lives in Rumpenheim, Offenbach

After completing an electrical engineering apprenticeship at Siemens and gaining his vocational college certificate, Roland Ullmann began studying industrial design at the Offenbach University of Art and Design, a course that was still in its infancy in 1968.

Then, in 1972, he joined Braun as a designer and over the course of four decades he contributed to all of Braun's product categories. His experience in electrical engineering made him the ideal candidate to produce a shaver design as his first project for the institute established by Artur Braun after the sale of the company. Subsequently, Ullmann specialized in the shaver sector and, during his time at the company, shaped the development of shavers at Braun. Thanks to Braun's typical approach of developing products in interdisciplinary teams, he was able to advance many technical innovations. Numerous patents granted to his ideas bear witness to his successful work.

As coordinator for advanced design and head of the CAD group, he was responsible for products destined for the more distant future. Ullmann has received numerous international awards for his designs, including a prize from the American Industrial Design magazine and the prestigious Japanese G Mark Award. When Procter & Gamble acquired the Braun-Gillette Group, Ullmann became one of the few research fellows of this large corporation in Europe.

One of Ullmann's design milestones was the micron plus shaver, with its hard-soft technology, that was on the market from 1979 to 1982. Soft plastic, which is used inside the device to dampen the sound and vibration, is also bound to the harder plastic of the housing and can be seen and felt on the outside as a non-slip dimpled structure. The micron plus would also be the first shaver with a shaving foil cartridge.

In 1985, the micron vario 3 was developed, manufactured for the first time using a sleeve construction for easier assembly. The product was produced in large numbers for a period of twenty years. The year 1999 saw the launch of Ullmann's Syncro system, a shaver with an automatic cleaning system that also featured simultaneous battery charging. Then 1986 marked the birth of Dual and, in 1990, Dual Aqua launched as a standalone shaver programme with 100-volt technology designed exclusively for the Asian market. Dual Aqua was the first waterproof Braun shaver and was developed in cooperation with Japanese manufacturers.

Before the Syncro shaver, exact universal had already been created in Japan in 1986 and for the first time included a memory function for personal beard hair length. In 1994, Ullmann designed the highly successful Aroma Select coffee maker programme, which is the world's best-selling model and is still sold today.

In 1999, Ullmann's toothbrush project started out life as an 'advanced project', and in 2000, preliminary work began on the technically challenging Oral-B Triumph D 25 electric toothbrush, which was initially launched in the United States in 2005. The shape of the casing was developed through a series of experiments using highly complex, multi-dimensional software (CAD). For the first time, the calculated formal computer solutions were incorporated into the toothbrush's design. Pulsonic, the world's first shaver on the market with sonic technology, followed in 2006. To this day, a slim handle and the free-moving shaving head continue to be central to the design of Braun's shaver range.

Duy Phong Vu
born 1972 in Biên Hòa, Vietnam, lives in Schwalbach am Taunus, Hesse

After the end of the Vietnam War in 1975 and the withdrawal of US forces, more than 1.6 million South Vietnamese people risked their lives to escape the inhumane conditions under the new communist government, leaving the country by boat. Duy Phong Vu was one of these so-called 'boat people'. He and his older brother arrived in West Germany as children in 1979. After leaving school, he began an apprenticeship as a model maker for the automotive industry at the Max Weis company in Dreieichenhain, Hesse. After successfully completing his training, Vu studied industrial design at Darmstadt University of Applied Sciences and joined Braun in 1997, initially for two years as a designer in the model-making department and then as a member of the Design Team. After several career moves, he finally became group design manager of Gillette/Procter & Gamble/Braun GmbH.

After Peter Schneider's retirement, Vu became interim head of design at Braun in 2007. At that time, the motto 'Strength of Pure' was already being developed. When De'Longhi took over Braun Household, he switched to the Italian family-owned company and set up his own six-person design team in February 2013 in Neu-Isenburg for the household appliances that now trade under the Braun brand.

Vu remained in close contact with the head of design, Oliver Grabes, and the Braun team in Kronberg. In addition, he also actively nurtured contacts with design schools in Coburg, Delft, Paris, Tokyo and the Politecnico Milano. He has been a professor at Donghua University in Shanghai since May 2016 and has taught at Politecnico Milano since 2018.

Today, as Group Director Global Design & IoT Ecosystem, Vu is responsible for all design teams of the De'Longhi Group, which includes the brands Braun Household and De'Longhi, as well as Kenwood. With his teams in Italy, England, China and Germany, he has received

Wilhelm Wagenfeld

Roland Weigend

Reinhold Weiss

several hundred design prizes and awards, including the Good Design Award in the United States, the iF Design Award in Hanover, the Plus X Award – Design Team of the Year 2016 in Bonn, the Red Dot Award in Essen and the German Design Award from the German Design Council in Frankfurt.

Wilhelm Wagenfeld
Born 15 April 1900, in Bremen
died 28 May 1990, in Stuttgart

After training as a silversmith, Wilhelm Wagenfeld studied at the Bremen School of Applied Arts and at the Hanau Drawing Academy. In 1923, he came to the Bauhaus in Weimar, where he studied primarily under Lázló Moholy-Nagy at the school's metal workshop and designed the famous so-called 'Bauhaus lamp' together with Carl Jacob Jucker. In 1925, he became an assistant at the Weimar Staatliche Bauhochschule, and the following year he joined the Deutscher Werkbund. After 1930, Wagenfeld worked independently, then, in 1935, he became creative director at the Lausitzer Glaswerke, a glassworks factory in Weisswasser. It was there in 1938 that he created the modular glass Kubus-Geschirr, a food container system that anticipated America's Tupperware. He became a professor at the Berlin Academy of Arts in 1947 and then from 1954 to 1978 he ran his own workshop, Werkstatt Wagenfeld, in Stuttgart with a few employees, working for numerous important clients such as WMF, Rosenthal and Braun. More than 600 products were created for these clients.

Wagenfeld resolutely advocated an organic design approach without corners and edges, and the products above all had to be usable and suitable for mass production. For his designs and approach, he received numerous national and international prizes and awards starting as early as 1937.

His 1948 collection of essays Wesen und Gestalt der Dinge um uns (Essence and Design of the Things around Us)[23] definitely had an influence on Erwin Braun's new corporate concept. Erwin's popular phrase that Braun appliances should be as reserved as an English butler can be traced back to Wagenfeld's book.

Wagenfeld designed only one device for Braun, the combi portable radio and phono combination in 1955. Though it was not very successful, his influence on the initial phase of the new design was considerable. In Wagenfeld, Braun dealt with a professional and experienced designer for the first time.

Roland Weigend
born 1933 in Mosern (formerly Czechoslovakia),
lives in Kronberg im Taunus, Hesse

After attending primary school, Roland Weigend was a cabinetmaker's apprentice in Idstein, which mainly produced furniture. In 1952, he began studying interior design at the Wiesbaden Werkkunstschule, graduating in 1956.

In March 1956, immediately after his diploma examination, Dieter Rams recommended him to Braun and he joined the small in-house design group, which consisted of two other members: Rams and Gerd A. Müller, also graduates of Wiesbaden Werkkunstschule. Weigend was mainly responsible for the technical drawings and model construction. The three designers from the Wiesbaden school were assigned their own design room in Rüsselsheimer Strasse, although it was still sparsely equipped: 'there was a workbench, a drawing board and a lathe next to each of us. That's all we had – except for the plaster,'[24] commented Rams.

Initially, Weigend, along with Rams, was involved with furnishing offices, creating trade-fair stands and designing furniture for the health service and the company's own premises. In 1957, together with Müller and Rams, he attended Interbau Berlin and, the following year, the World Expo 58 in Brussels.

After founding the Braun design department in 1962, Weigend was head of model making from 1965 to 1987. At Braun, model makers used to be called 'model designers', an indication of their importance in the design process. Weigend always collaborated closely with Rams and supervised all his audio projects in the model-making department. Weigend worked for Braun until his retirement in 1996.

Reinhold Weiss
born 1934 in Stuttgart,
died 2022 in Tucson, Arizona, US

After graduating from the Freie Waldorfschule in Stuttgart, Reinhold Weiss[25] trained as a cabinetmaker at the renowned Stuttgart furniture manufacturer Erwin Behr (now Behr International). He passed his journeyman's examination with the grade 'very good' and received the first prize from the Nürtingen Chamber of Industry and Commerce. He then completed a two-year internship at the architectural firm Steim und Mühleisen in Stuttgart.

In 1955, he began studying under Hans Gugelot at the HfG Ulm, where he designed alternatives to the Hobby camera flash unit and a grill pan for Braun. In the summer of 1958, Weiss worked at Braun during the semester break and Artur Braun offered him the opportunity to join the company as another product designer after he had graduated. His diploma project was a new type of iron with an angled handle, which was purchased by the Braun company but not realized, since the first steam irons were starting to come onto the market in the early 1960s.

In the winter semester of 1959, he graduated from the HfG and began working for Braun. He worked on basic conceptual research for new appliances and designed his first household appliances. He subsequently created numerous iconic appliances that helped to establish Braun design, including the HE 1 express kettle, the HF 1 toaster, the HG 1 combination grill in 1962, the KM 1 and KSM 1 coffee grinders, the M 12 and M 140 hand mixers, the HLD 2 and HLD 3 hair dryers in 1964, the Manulux NC flashlight in 1967, the KH 1000 headphones, the zb 1 electric toothbrush and the H 7 fan heater. When the

Wilhelm Wiegand

Bernhard Wild

Benjamin Wilson

in-house Braun design department was officially established in 1961, Weiss became its deputy head. In 1965, together with Rams, Richard Fischer and Robert Oberheim, he received the Berlin Young Generation art prize.

In 1967, Weiss accepted a job offer as executive designer at the US design company Unimark International. He settled in Chicago and worked as vice president of the product design department until 1970. In 1971, he set up his own design studio, reinhold weiss, inc., in Evanston, Chicago, and worked for the Hi-Fi equipment manufacturer NAD, among others. From 1989, he was also vice president for design and corporate identity at the Taiwanese radio and Hi-Fi manufacturer Proton Electronic Industrial for two years.[26]

Weiss's design work is characterized by a precise, well-thought-out understanding of the requirements of model making, something he continued to develop over the years at Braun, as well as a rational, yet ultimately elegant, aesthetic.

In 1972, he received the Design Award of the American Institute of Graphic Arts for packaging design and in 1982 a gold medal at the Industrial Fair in Brno, Czech Republic.[27] The products he designed can be found in the collections of MoMA in New York, the Indianapolis Museum of Art, the Israel Museum, Jerusalem, and the Museum Angewandte Kunst in Frankfurt, among others.

Wilhelm Wiegand
born 1904 in Frankfurt/Main
died 1961 in Frankfurt/Main

Wilhelm Wiegand, a trained banker, met Max Braun in 1927 and joined his fledgling company on 20 November that same year. As head of the foreign section, he was involved in establishing the branches in Brussels, Paris and London. On 1 January 1935, Wiegand was given power of attorney and became a close confidant of the entrepreneur. After Max Braun's death, his sons Erwin and Artur Braun also relied on Wiegand's commercial expertise. Among other things, he negotiated the licence and supply contract for $10 million with the American lighter manufacturer Ronson. In 1960, he moved from the operational side of the company to the supervisory board.

Bernhard Wild
born 1943 in Achern, Western Baden-Württemberg, lives in Königstein

After an apprenticeship as a precision mechanic from 1965 to 1968, Bernhard Wild began studying precision engineering at the State School of Engineering in Aalen, Germany, where he graduated with a diploma. In 1968, he joined Ritter AG/Sybron Corporation and worked for the technical manufacturer for dental surgeries, first as a test engineer and later in production planning. In 1973, Wild joined Braun in Kronberg as head of materials management and shortly thereafter was appointed head of corporate manufacturing engineering. In 1979, he became plant manager of the Braun factory in Marktheidenfeld.

After two years as senior vice president of technical operations at Oral-B in Redwood City, California, Wild was appointed general manager responsible for production, materials, maintenance and manufacturing at Braun on 1 September 1988. This included the plants in Barcelona and Mexico City. From 1996, he was also in charge of business management and became a member of the management board in 1990. In 1998, he became chairman of the management board of Braun AG. With the organizational restructuring of the global technical and manufacturing division at Gillette, his technical area of responsibility was extended to all product lines in May 2004. He then became managing director of Braun GmbH and senior vice president of technical development at Gillette. In 2005, Wild, together with Peter Schneider, became the pioneer of the Braun Collection, with public showrooms in Kronberg. On 1 January 2006, he retired as chairman of the Braun executive board and joined the supervisory board as chairman until 2012. From 2005 to 2022, Wild was chairman of the Friends of the Braun Collection.

Benjamin Wilson
born 1979 in Melbourne, Australia, lives in Schwalbach am Taunus

Benjamin Wilson grew up in Melbourne, Australia, and completed his Bachelor of Industrial Design in 2002 with first class honors from Swinburne University. He was motivated to work in the field of design field by his enthusiasm for the design process, which often begins with a frustration, or with something that can be improved upon technologically, in its use, production or end-of-life cycle. After travelling the world and gaining experience through several traineeships, including one with Tassilo von Grolman, he eventually settled in Europe. In 2003, Wilson joined the Braun Design Team in Kronberg. At Braun, he found a home to leverage his collaborative, curious nature and industrial design skills, working with incredible technical teams and familiarizing himself with German design methods.

His passion for making the future tangible, strategically challenging the status quo and employing material- and process-driven design led to key innovations in various categories, including Braun's return to premium materials and its first ceramic watch. His 'Aussie design foundations', passion for German design and understanding of the brand has led him to become an advocate for the Braun brand and Good Design.

Since 2017, as Director of Communications at Braun, his mission has been to share with the next generation the idea of the Braun mindset, namely that good design is useful, simple and built to last. He believes that mindset reflects a visionary approach to design that leads to a particular corporate culture, identity and extraordinary products.

Vincent C. Ziegler
born 1910, in Greenfield, Michigan, US
died 1979, in Boston, US

Vincent C. Ziegler grew up in the United States to a German American family. His German mother raised him bilingually. In 1946, he joined Gillette, becoming president in 1958, executive vice

Vincent C. Ziegler

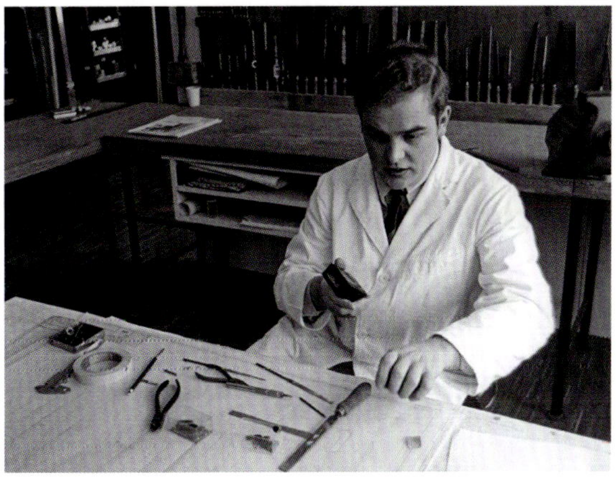

Klaus Zimmermann

president in 1964 and chairman of the board and CEO in 1966. Under Ziegler's leadership, Gillette experienced the greatest growth in its seventy-three-year history, with sales increasing from $340 million to more than $1 billion between 1966 and 1973. This expansion was driven by extensive product diversification with corresponding corporate acquisitions.

When the Braun company was taken over by Gillette in 1967, Ziegler assured Braun's management and staff that the company would be run independently and that all product areas would be retained, including the loss-making Hi-Fi division. This remained the case until his departure.

Ziegler attempted to transpose Braun's design attitude, which he appreciated and supported, to Gillette and other subsidiaries – but he was not always successful in this endeavour. Dieter Rams and Jürgen Greubel were able to design the very successful Sensor shaver for Gillette, but the subsidiary Paper Mate rejected designs from Kronberg.

Klaus Zimmermann
born 1945 in Frankfurt/Main, lives in Eppstein

After an apprenticeship as a model maker, Klaus Zimmermann joined Braun in 1968 as the first trained model maker. In 1974, he passed his master craftsman's exams. From 1987, he was head of the 'model shop', where he worked with six other model makers and helped expand the colour database and colour-mixing system so that there was a palette of 14,000 possible colour options. In a long series of tests conducted with Roland Ullmann, he developed the hard-soft technology for shaver housing that was later used in other product areas. From 1987 onwards, all models manufactured up to that time were documented.

From 1991 to 1994, Zimmermann was instrumental in Braun's switch to using CAD/CAM technology in design and model making, which, along with up to five-axis simultaneous CNC-milling machines and other high-tech equipment, enabled a whole new level of representation fidelity for models. Model makers were increasingly evolving from being precision mechanics to computer specialists. The design process began with the production of rough 'rapid models', followed by preliminary and then final models. The constant exchange between designers and model makers – a process in which far-reaching design decisions are often made and enabled because of the particular proximity of the two departments was still extremely important. Although, like the designers, the individual modellers tended to have their specialist areas, there was also an emphasis on task rotation, so that it was possible for everyone to try different jobs. Test models were also increasingly being made for business management to assess. After thirty-eight years at Braun, Zimmermann retired in 2008.

Braun Management from its Beginnings to the Present Day

After more than one hundred years, Braun's management today naturally looks very different from Braun's management in 1921. As a backyard factory with just four employees, it had only one boss – Max Braun. He was responsible for everything. Today's corporation, comprising about 2,000 direct 'Braunians', requires more management, especially since it is part of the much larger Procter & Gamble group, which boasts a payroll of 100,000 people worldwide. Today, Braun is primarily a Procter & Gamble brand. Since 1 November 2021, the company's chief executive has been David S. Taylor, whose titles include Chairman of the Board, President and Chief Executive Officer. The other members of the board are John R. Moeller, President and Chief Executive Officer, and Shailesh G. Jejurikar, Chief Operating Officer. At the next level, there are six chief executive officers (or presidents), who head individual product areas, horizontally for all the group's brands. Since 2018, Gary Coombe has been the CEO of P&G One Grooming. One of his vice presidents is Alessandra Dolfini in Geneva, who oversees P&G One Grooming and is the direct superior of Braun's chief designer Oliver Grabes. Victor Leal Negre is responsible for Braun's marketing in Geneva. On the management board of Braun GmbH in Kronberg, Stefan Schamberg, who is also vice president of R&D (Research and Development) at Braun, has rendered outstanding services to the company for years, brought Oral-B to Kronberg and is also responsible for the Braun Collection.

CEOs Braun AG/GmbH (1961–1999)
1961 to 1963: Artur Braun
1962 to 1970: Dr Rudolf Gros
1970 to 1974: Hans-Wilhelm Herrmann
1974 to 1977: Alfred M. Zeien
1977 to 1979: Dr Paul Stern
1979 to 1986: Lorne R. Waxlax
1986 to 1990: Robert R. Murray
1990 to 1993: Jaques Lagarde
1993 to 1997: Archibald Livis
1998: Bernhard Wild (Managing Director in 1999)

Chairmen of the Supervisory Board (1961–2005)[28]
1961 to 1966: Erwin Braun
1966 to 1968: Artur Braun
1968 to 1977: Vincent C. Ziegler
1977 to 1983: Colman M. Mockler
1983 to 1986: Joseph F. Turley
1986 to 1993: Lorne R. Waxlax
1993 to 1997: Jaques Lagarde
1998 to 2000: Archibald Livis
2001 to 2005: Edward F. DeGraan

Presidents Responsible for Braun (2002–2018)
2002 to 2008: Bracken Darrell
2008 to 2009: Juan Pedro Hernández
2009 to 2012: John Goodwin
2012 to 2014: Austin Lally
2014 to 2017: Ihab Rafik

P&G Leadership Responsible for Braun (2018–2023)
Gary Coombe, Chief Executive Officer, P&G One Grooming
Alessandra Dolfini, Senior Vice President, Brand Franchise Leader, P&G One Grooming
Jean Ibrahim, Senior Vice President, R&D, P&G One Grooming
Phil Duncan, Chief Design Officer, P&G (since 2009)

Index

Underlined page numbers refer to illustrations

Notes

The Braun Story

1 Cited from Die Deutsche Wirtschaft, https://die-deutsche-wirtschaft.de/auslands-unternehmen/braun-gmbh-kronberg-im-taunus.
2 Dieter Rams 'Im Mittelpunk: Wachsame Nachdenklichkeit' (1995b), 6f.
3 Saint-Exupéry, Wind, Sand and Stars (1989).
4 Demetrios Eames, Essential Eames (2017), 54.
5 DRA: Erwin Braun, 'Zum Gestalten bei Braun' (1961).

01 A Start-Up in Frankfurt

1 BA: Betriebsspiegel, vol. 7/65, 2.
2 Klaus Klemp, Annika Sellmann, Matthias Wagner K and Grit Weber, Moderne am Main 1919–1933 (2019), 62.
3 BA: Manuel Aicher, 'Biographie Max Braun' (1990), 61.
4 Ibid., 70.
5 Ibid.
6 Günter F. Abele, 'Eine Chronik in Wort und Bild', http://historische-radios.info/firmen/braun/firmengeschichte.htm
7 BA: Aicher, 'Biographie' (1990), 83f.
8 Ibid., 67.

Notes for 01 Contexts

1 Klemp et al., Moderne am Main (2019), 62.
2 Ibid.
3 Ludwig Landmann, 'Zum Geleit' (1926), 1f.

02 Dark Times, New Hope

1 The following remarks are based for the most part on BA: Aicher, 'Biographie' (1990).
2 Ibid., 94f.
3 Hans Wichmann, Mut zum Aufbruch (1988), 260.
4 BA: Company anniversary FWL (1957).
5 Ibid., 123f.
6 While in 1932 there were about 4 million paying radio listeners, by 1939 there were already 12 million.
7 See BA: interviews file and Sonja Meldau-Stagge, Britta Stücker and Braun GmbH, 90 Jahre Braun (2011), 32.
8 BA: 'Cosmophon 333' (1933).
9 BA: Aicher, 'Biographie' (1990), 119.
10 Quoted from ibid., 128.
11 DRA: Artur Braun interview (23 July 2010).
12 Artur Braun, Max Brauns Rasierer (2007), 48.
13 BA: Aicher, 'Biographie' (1990), 134.
14 Patent specification no. 802 052 at the German Patent Office; ibid., 143.
15 TBA: Max Braun business card, 1949.

Notes for 02 Contexts

1 See Winifried Nerdinger, Walter Gropius (2019).

03 A New Company

1 BA: Cobarg estate, Erwin Braun file.
2 Quoted from BA: 'Gesagt!' (1973), 68.
3 Theodor Heuss, 'Bewährung in der Gewöhnung' (1952). Heuss (1884–1963) was managing director and board member of the German Werkbund from 1918 to 1933 and the first President of the Federal Republic of Germany from 1949 to 1959.
4 New York 1951, Düsseldorf 1953.
5 Christopher Oestereich, 'Gute Form' (2000a), 224.
6 Wilhelm Wagenfeld, 'Künstlerische Zusammenarbeit mit der Industrie' (1998), 178–80.
7 Wichmann, Mut zum Aufbruch (1998), 181.
8 Wagenfeld, Wesen und Gestalt der Dinge (1948).
9 Oestereich, 'Gute Form' (2000a), 129.
10 In 1960, West Germany provided for a tenth of the total worldwide exports; after ibid., 169.
11 Ibid., 168.
12 According to the German American designer Carl Otto; after ibid., 174.
13 BA: Betriebsspiegel, vols 6/57, 7/57 and 10/57.
14 BA: Betriebsspiegel, vol. 7/57.
15 Speech, held on 17 October 1979 on '25 years of Braun health service', after Wichmann, Mut zum Aufbruch (1998), 103.
16 On the development of the S 50, see Artur Braun, Max Brauns Rasierer (2007), 30–57.
17 Wichmann, Mut zum Aufbruch (1998), 58.
18 Albrecht Schultz, 'Problem zu Ende gedacht', in Max Braun Stiftung, Mehra Is Erinnerungen (2000), 27–33.
19 Max Braun Stiftung, Mehr als Erinnerungen (2000), 13f.
20 Werner Apel-Dube in Max Braun Stiftung, Mehr als Erinnergungen (2000), 21f.
21 Max Braun Stiftung, Mehr als Erinnergungen (2000), 14.
22 BA: Betriebsspiegel, vol. 7/52.
23 Max Braun Stiftung, Mehr als Erinnergungen (2000), 35f.
24 BA: Aicher, 'Biographie' (1990), 143f.
25 Louis Lasagna, The Doctors' Dilemmas (1963), 284f.
26 BA: Aicher, 'Biographie' (1990), 142.
27 See Ulrich Voelkel, 'Operation "combi"', Max Braun Stiftung, Mehr als Erinnergungen (2000), 57–9.
28 Advert in Max Braun Stiftung, Mehr als Erinnerungen (2000), 58.
29 Erwin Braun, 'Öffentlicher brief Erwin Brauns an Fritz Eichler' (1980), 105.
30 Oestereich, 'Gute Form' (2000a), 228.
31 See Wichmann, Mut zum Aufbruch (1998), 68f.

Notes for 03 Contexts

1 This furnishing style has entered the German vocabulary colloquially as 'Gelsenkirchener Barock' (Gelsenkirchen Baroque), which is somewhat unfair to Gelsenkirchen, the working-class city in the Ruhr area, as 'Gelsenkirchener Barock' could be found everywhere.
2 Oestereich, 'Gute Form' (2000a), 175.

04 Braun Design

1 DRA: Artur Braun interview (23 July 2010).
2 René Spitz, HfG Ulm (2002), 32.
3 Letter cited in Wichmann, Mut zum Aufbruch (1998), 181.
4 Jo Klatt, 'Leute um Hans Gugelot' (1987), 13.
5 BA: 'Gesagt!' (1973), 15f.
6 See Wichman, Mut zum Aufbruch (1998), 72f.
7 See Katharina Kurz and Christiane Wachsmann, 'Seitwärts der Avantgarde' (2020), 116–131.
8 DRA: Marlene Schnelle-Schneyder interview (30 August 2019).
9 Under the name 'Deutsche Werkstätten für Wohnkunst' a network of eleven German furniture houses was founded that fostered modern design. Today, exclusive furniture is sold under the label now called 'WK Wohnen'.
10 DRA: Schnelle-Schneyder (2019).
11 DRA: Erwin Braun, 'Zum Gestalten bei Braun' (18 December 1961).
12 Artur Braun, 'Wie das Braun Design entstand' (2012), 25.
13 HfGUA: Otl Aicher holdings (n.y.a).
14 Ibid.
15 HfGUA: Aicher holdings (n.y.a and n.y.b).
16 HfGUA: Aicher holdings (n.y.c).
17 The term, relating to prefab buildings manufactured to be as cheap as possible and devoid of architectural ambition, was coined in the 1960s by Heinrich Klotz, the architectural historian teaching in Marburg who later went on to establish the German Architecture Museum in Frankfurt and the ZKM in Karlsruhe.
18 Hartmut Jatzke-Wigand, 'Dieter Rams' (1995), 8.
19 All quotes from DRA: 'Braun. Funktionsbeschreibung Produktgestaltung' (1977).
20 After Königsdorfer Medienhaus, 'Designlexikon International', https://designlexikon.net.
21 Expanded slightly from Klaus Klemp and Keiko Ueki-Polet, Less and More (2009), 484.
22 DRA: Rams, 'The Designer's Contribution to Company Success' (22 September 1975), 5.
23 See Jo Klatt, 'Nicht produzierte Braun-Geräte aus dem Unterhaltungselektronik-Bereich' (2004), 16.
24 BA: Betriebsspiegel, vol. 9/57.
25 Oestereich, 'Umstrittene Selbstdarstellung' (2000b), 149; where it is extensively documented through sources.

Notes for 04 Contexts

1 Otl Aicher, Innenseiten des Kriegs (1985), 250.

05 Kronberg and the World

1 All data according to BA: Braun Annual Reports 1959 and 1960.
2 BA: Braun Annual Report, 1961/2.
3 BA: Braun Annual Report 1963/4.
4 BA: 'Presse Information A 804 der Braun AG v. März 1966' (1966).
5 BA: 'AT und VO an Führungskräfte' (14 March 1966).
6 HfGUA: Otl Aicher holdings (n.y.a), Manuscript.
7 BA: Betriebsspiegel, vol. 1/62.
8 Ibid.
9 BA: Braun Annual Report 1963/64.
10 BA: Betriebsspiegel, vols. 8/64, 9/64 and 10/64.
11 'Kosmische Skulpturen im Braun-Studio', in BA: Betriebsspiegel, vol. 4/63.
12 BA: Betriebsspiegel, vol. 10/63.
13 BA: Betriebsspiegel, vol. 4/67.
14 According to inventory MA 2084, Municipal Archive of Frankfurt/Main.
15 BA: Betriebsspiegel, vol. 6/67.
16 BA: Betriebsspiegel, vol. 7/67.
17 Frankfurter Allgemeine Zeitung of 7 September 1968.
18 The International Council of Societies of Industrial Design, the umbrella organization of all national design associations today calls itself the World Design Organization (WDO).
19 Rudolf Schönwandt, 'Wie arbeitet Dieter Rams und sein Team bei Braun?' (1980a).
20 BA: 'Leitlinien für die Gestaltung der Kommunikation von Braun' (1976).
21 Artur Braun, Max Brauns Rasierer (2007), 53.
22 BA: 'Leitlinien für die Gestaltung von Braun Verpackungen' (1975).
23 BA: 'Leitlinien für die Gestaltung der Kommunikation von Braun' (1976).
24 Ibid.
25 BA: 'Braun Design – die Verwirklichung eines Unternehmenskonzepts' (1975).
26 Ibid.
27 DRA: Horst Czipluch interview (20 February 2014).

Notes for 05 Contexts

1 Dorothy Todd Hénaut, 'Weltausstellung in Montreal' (1967), 346.

06 'The Times They Are a-Changin'

1 Wichmann, Mut zum Aufbruch (1998), 137.
2 BA: 'Presseinformation A 852 der Braun AG' (18 October 1967).
3 BA: Vincent C. Ziegler (19 December 1967).
4 BA Cobarg estate. Braun AG, Design Branch (GB-G), Vorschlag February 1968, Braun Archive, with the note 'Co von Dr. Ei, 2/68'.
5 The participants were Dr Rudolf Gros, Fritz Eichler, Ferdinand Simon, Hagen Gross, Ernst Krull, Werner Greutert, Karl Busch, Albrecht Schultz, Hans Steybe and Dr Traugott Malzan.
6 Original: 'Dear Vin ... Turbulent years followed. Many mistakes were made, the process of learning was painful on both sides

of the Atlantic. But in spite of all these difficulties you kept your promise. You protected Braun design from too much interference from the managers that came and went.' BA: Erwin Braun file (trans. KK).
7 BA: 'Organisationsverfügung AB-S' (18 October 1968).
8 BA: Traugott Malzan interview (1968).
9 DRA: Rams, 'Treffen mit Mr. Ziegler in Bonn' (7 December 1970).
10 Ibid.
11 All quotes in ibid.
12 Gunhild Frese, 'Exodus aus Kronberg' (1975).
13 Ibid.
14 The Max-Braun-Kreis was founded in 1988 by Erwin Braun and Georg Hohm together with thirty other former Braun employees as a registered association. The articles of association state: 'The purpose of the "Max-Braun-Kreis" is to preserve and pass on the industrial culture embodied by the life's work of Max Braun and by his family business.' According to the articles of association, the association should 'in the future, if necessary, attempt to exert appropriate influence on the company's reputation and corporate sponsorship.'
15 Sales Division – Vertriebsabteilung
16 Erwin Braun to the Max-Braun-Kreis on Max Braun's 100th birthday on 25 October 1990; transcript in Wichmann, Mut zum Aufbruch (1998), 238.
17 HfGUA: Concept book 'Braun Design' (1989).
18 BA: Albrecht Schultz, 'Referat Albrech Schultz' (19 September 1984).
19 Letter from Erwin Braun to Dr J.O. Gebbers, Pathological Institute of the Cantonal Hospital in Lucerne, of 17 February 1992, in Wichmann, Mut zum Aufbruch (1998), 163.
20 See Wichmann, Mut zum Aufbruch (1998), 146.
21 Zeitschrift Braun + Wohnen (1967).
22 'One of the interesting developments in design education in the last few years is a renewed interest among students today in the traditional crafts ... It seems that we all agree that our quality of life, economic strength, and our ability to solve the world's most challenging problems in the 21st century will be in large part shaped by design ... One of the reasons that Apple has done so well is because Dieter Rams trained our sensibilities – he taught the world how good design should look and function ... the BraunPrize is the best organized, most rigorous, most prestigious design competition for young designers in the world', from DRA: Mark Breitenberg, 'The Braun Prize 2009 Awards Ceremony' (2009) (trans. K.K.).
23 Oliver Grabes, 'Why Do We Promote Design?' https://braunprize.org/en/the_idea.html.
24 According to Sales Economics, issue 5 (1991), Düsseldorf.
25 Ibid.
26 Hamilton had already made Braun appliances the subject of two still lifes multiple years earlier. Perhaps this referred back to these.
27 Concept and design: Jo Klatt, Ivica Maksimovic, Thomas Sauer, Michael Schneider; cover design: Dieter Rams and Peter Hartwein; photos: Jo Klatt, copyright a/d/s/ Braun, Kronberg 1990.
28 The article's German title is 'Stilvoller Abgang'. Der Spiegel, issue 25 (1990).
29 All Peter Schneider quotes from: BA: 'Braun Design' (1995), 20–26.
30 Roland Barthes, Mythen des Alltags (1964).
31 BA: 'Braun Design' (2002).
32 Schneider+Schumacher, 'Braun Hauptverwaltung 1998–2000', https://schneider-schumacher.de/projekte/project-details/90-braun-hauptverwaltung.

Notes for 06 Contexts

1 The British Arts and Crafts Movement in the later nineteenth century was a reaction to design historicism and machine production. Instead of historical designs, floral motifs were used and, instead of machine production, a return to craftmanship was promoted.
2 See Paul Feyerabend, Against Method (1975).

07 Braun Today and Tomorrow

1 'Grooming' means 'body care' at P&G. This term covers Gillette, Venus, Braun and The Art of Shaving, which are all of the brands that create products that deal in hair removal.
2 Grabes, 'Design Today', https://www.braun.de/de-de/world-of-braun/braun-design/design-today
3 Products with a fast turnaround, or objects used every day, such as foodstuffs, body care products or cleaning agents that swiftly move off store shelves and are purchased regularly.
4 The name of Braun's 2021 report was 'Good Design for Better Future – Gen Z Study'.

Core Identities

1 Le Corbusier, Ausgabe Feststellungen zu Architektur und Städtebau (1964), 84.
2 BA: Aicher, 'Biographie' (1990), 156.
3 The 'Super' in the name of the portable radio (Kofferradio) stands in radio technology for 'superheterodyne transmitter' and is an electronic circuit for receiving and processing high-frequency electromagnetic radio signals. It is a technology used in radios and TVs but also in modern GPS devices. Back then it was cutting-edge technology.
4 Jatzke-Wigand, 'Dieter Rams' (1995), 12.
5 'In the practice of design, function also becomes aesthetic.' This is the key theory proposed by Daniel Martin Feige, see Design (2018).
6 Kenya Hara, Weiss (2009), 68.
7 Herbert Marshall McLuhan, Understanding Media (1964).
8 Richard Fischer and Gerda Mikosch, 'Grundlagen einer Theorie der Produktsprache' (1984), 93.
9 BA: 'Gesagt!' (1973), 64f.
10 Rams, Weniger, aber besser (1995), 6.
11 Ibid., 7.
12 Rams, 'Ramsifikationen' (2010), 100.
13 BA: Albrecht Schultz, 'Referat Albrecht Schultz' (19 September 1984).
14 The concept of a 'circular economy' was first set out in 1990 by British environmental economists David W. Pearce and R. Kerry Turner.
15 In the late 1990s, German chemist Michael Braungart and US architect William McDonough advocated a 'cradle-to-cradle' principle as an approach for a comprehensive and consistent circular economy, see Michael Braungart and William McDonough, Cradle to Cradle (2014).
16 DRA: Erwin Braun, 'Zum Gestalten bei Braun' (1961).
17 Rams, Weniger, aber besser (1995), 25–7.
18 BA: 'Gesagt!' (1973), 35.
19 DRA: Rams, 'Die Rolle des Designers im Industrieunternehmen' (14 October 1977).
20 Wolfgang Welsch, Transkulturalität. Realität, Geschichte, Aufgabe (2017).
21 All quotes come from brochures in BA.

Design Is Made By People

1 Aicher, Innenseiten des Kriegs (1985), 250.
2 BA: Aicher, 'Biographie' (1990).
3 Cited from Wichmann, Mut zum Aufbruch (1998), 243.
4 BA: Hans Horst Semmel (2001), 57.
5 The biographical details are based largely on BA: Aicher, 'Biographie' (1990).
6 Fritz Eichler, Das Wesen des Handpuppen- und Marionettenspiels (1937).
7 Eichler, Vom König der nicht mehr lachen konnte (2016).
8 The autarc project, probably initiated by Erwin Braun, was intended to produce miniaturized, modular and battery-powered devices that could also be used outdoors.
9 BA: Betriebsspiegel, vol. 4/54.
10 Amongst others Werner Kuprian, Sport Physiotherapie (1981).
11 As Ludwig Littmann has withdrawn from the public eye and could not be contacted as an interviewee, it was not possible to consult him on his biography.
12 Archibald Livis, 'Success by Design' (1997), 105.
13 MA: 'Livis, Archibald', http://www.munzinger.de/document/00000021544.
14 DRA: Dietrich Lubs interview (8 May 2008).
15 Wichmann, Mut zum Aufbruch (1998), 261.
16 Traugott Malzan, Blaue Küste Rote Segel (1987).
17 See Lucia Hornfischer, Gerd A. Müller (2021).
18 See Klemp, Dieter Rams (2020).
19 DRA: Eckard Tischer interview (6 February 2014).
20 Cited from Design+Design, vol. 69/70 (2004–5), 31.
21 Rudolf Schönwandt, Die unmissverständliche Sprache der Vernunft (1998).
22 Cited from Deutsches HiFi-Museum. http://www.hifimuseum.de/
23 Wagenfeld, Wesen und Gestalt der Dinge (1948).
24 Rams, 'Die Verantwortung des Designs' (2004), 18f.
25 As a rule, the biographics of persons who are still alive were produced in consultation with them, and personal emphases were taken into account. The biography of Reinhold Weiss is an exception. It was compiled on the basis of extensive source material that he had left to the author; nevertheless, no consensus could be reached on his biography, so the version presented here reflects solely the author's view to the best of his knowledge and belief.
26 Cited from Günter Staeffler, 'Reinhold Weiss Design-Studien für Braun' (2000b).
27 Charlotte Fiell and Peter Fiell, 'Reinhold Weiss', in Industriedesign A-Z (2006), 520.
28 Assembled 1961 to 2005: Thomas Guttandin and Britte Siepenkothen, Kronberg.

Archival Sources and Bibliography

Artur Braun Archive, Königstein im Taunus (ABA)

ABA: Max Braun on the trip to the United States [business card] (1949)

Braun Archive, Kronberg im Taunus (BA)

Files

BA: Aicher, Manuel, Grünbuch [unpublished manuscript], 'Biographie Max Braun – Version lang, no. 1–27.3.90 (m.a.)' (1990)

BA: 'AT und VO an Führungskräfte' (AT and VO to Executives) [circular] (14 March 1966)

BA: 'Betr: Zukünftige Zusammenarbeit mit Gugelot Design GmbH (Future Collaboration with Gugelot Design GmbH), 24.5.1974 G-PGL-Ra/Ut'. Signature Dieter Rams (24 May 1974)

BA: 'Entwicklung des Unternehmens' (Development of the Company) with additions by Erwin Braun (18 February 1953–23 September 1957)

BA: Company anniversary FWL, in Mappe Firmenjubiläum FWL (1957)

BA: Braun Annual Reports (1959–2005)

BA: Gros, Rudolf, 'An unsere Aktionäre' (To our shareholders) [circular], in Sammlung Cobarg (16 October 1967)

BA: 'Organisations- und Personalverfügung AB-S' (Organizational and HR allocations), in Sammlung Cobarg (15 October 1968)

BA: 'Organisationsverfügung AB-S' (Organizational allocations), in Sammlung Cobarg (18 October 1968)

BA: 'Presse Information A 804 der Braun AG v. März 1966' (press release; 1966)

BA: Circular in Ordner Cobarg (19 October 1967)

BA: Schultz, Albrecht, 'Referat Albrecht Schultz beim "Kamingespräch" der Bundesarbeitsgemeinschaft der Mittel- und Grossbetriebe des Einzelhandels (BAG) 19. September 1984. Thema: Unternehmensstrategie eines Herstellers technischer Markenartikel – Beispiel Braun', in Ordner Albrecht Schultz (19 September 1984)

BA: Schwarzbuch, typed excerpt for a biography on Max Braun, author unknown, n.d.

BA: Semmel, Hans Horst, Collected Works, vol. 2, pt 2, 'Ein Leben als Roman' (type manuscript; 2001)

BA: Ziegler, Vincent C., 'Vincent C. Ziegler, Chairman of the Board Gillette, an die Mitarbeiterinnen und Mitarbeiter von Braun' (letter; 19 December 1967)

BA: Cobarg estate

Interviews

BA: Malzan, Traugott, interview by the Head of Braun Information Department Dr Traugott Malzan with Vincent C. Ziegler, in Ordner Historie 67/Verkauf an Gillette (1968)

Company Publications

BA: 'Ansichten zum Design' (Views on Design; 1974)

BA: Betriebsspiegel. Unsere Werksnachrichten, Hauszeitschrift für Mitarbeiter und Mitarbeiterinnen des Unternehmens Braun (Braun in-house magazine; 1951–96)

BA: 'Braun Aspekte' (Braun Aspects) (1995 and 1999)

BA: 'Braun – Designed to Perform Better' (1995)

BA: 'Braun Design' (1995 and 2002)

BA: 'Braun Design – die Verwirklichung eines Unternehmenskonzepts' (Braun Design – the Realization of a Corporate Concept; 1975)

BA: 'Braun Design', catalogue for the travelling exhibition, concept Dietrich Lubs, Peter Hartwein, Peter Schneider, text Rudolf Schönwandt (1995)

BA: 'Braun Hi-Fi. Zukunft als Konzept' (Braun Hi-Fi. Future as Concept; 1983)

BA: 'Braun im Rückblick' (Braun in Retrospect; 1980)

BA: 'Braun im Rückblick, 1921–1987' (Braun in Retrospect, 1921–1987; 1988)

BA: 'Braun in Retrospect. Dates and Facts 1921–1979' (1980)

BA: 'Braun Markenprofil' (Braun Brand Profile; 2002)

BA: BraunPrize documentation: 'BraunPreis. Dokumentation der 10 Ausschreibungen des BraunPreises mit diversen Materialien' (Documentation of the 10 Braun Prize Competitions with Various Materials; 1995). 'BraunPreis 1999. Kooperation, Technik und Design' (Cooperation, Technology and Design). 'BraunPreis 2003'. 'BraunPreis 2005'. 'BraunPreis 2007'. 'BraunPreis 2009. Design Förderung' (Design Promotion).

BA: 'Canada Commerce', The Topic: The role of design in industry; The example: Braun, Reprint from 'Canada Commerce' (1976)

BA: 'Cosmophon 333' (facsimile of the brochure) in Mappe Unterhaltungselektronik (1933)

BA: 'Darum. Gute Form hat guten Grund. Bilder zum Thema "Industrial Design"' (That's Why. There's No Need to Justify Good Design. Images on the Topic of Industrial Design; 1986)

BA: 'Das Gesicht einer Firma' (The Face of a Company; c.1970)

BA: De Gasperi, Magdalena, 'Die Braun Brüder' (The Braun Brothers; 2015)

BA: 'Der 10. Braun Preis. Dokumentation des internationalen Wettbewerbs für Nachwuchsdesigner' (The 10th Braun Prize. Documentation of the International Competition for Young Designers; 1994)

BA: 'Dieter Rams' (c. 1988)

BA: 'Form – nicht konform. 20 Jahre Braun Design' (Form – Non-Conforming. 20 Years of Braun Design; 1975)

BA: 'Gesagt! von Dr Fritz Eichler 1963 ... 1972' (Said! By Dr Fritz Eichler 1963 ... 1972; 1973)

BA: 'Hallo Simplicity' (2015)

BA: Kress, Gerlinde, ed., Braun Design (1999)

BA: 'Leitlinien für die Gestaltung von Braun Verpackungen' (Guidelines for the Design of Braun Packaging; 1975)

BA: 'Leitlinien für die Gestaltung der Kommunikation von Braun' (Communication Guidelines for Braun; 1976)

BA: 'Profit Opportunities from Technological Advances – Speech by Dr Paul G. Stern Chairman of Braun AG to the 1979 Top-Management Forum Organized by Management Centre Europe in Monte Carlo on 17 and 18 September 1979' (Kronberg, 1979)

BA: 'Tatsachen über Braun' (Facts About Braun; n.d.)

Dieter Rams Archive (DRA)

Files

DRA: 'Braun. Funktionsbeschreibung Produktgestaltung', no. 77.333–16, valid from 1.1.1977, agreed between A. M. Zeien and D. Rams, archive no. 3.9.4. (1 January 1977)

DRA: Braun, Erwin; 'Zum Gestalten bei Braun', Frankfurt/Main (position paper), archive no. 3.5.2.1. (18 December 1961)

DRA: Breitenberg, Mark, 'The Braun Prize 2009 Awards Ceremony – Keynote Address' (lecture manuscript), archive no. 1.3.1.19. (2009)

DRA: Rams, Dieter, 'Elektrorasierer' (concept paper), archive no. 3.2.5.1., n.d.

DRA: ———, Braun AG Abt. PGF 15.8.1968, presented on 26.8.1970 by Dieter Rams, archive no. 8.3.2. (26 August 1970)

DRA: ———, 'Treffen mit Mr. Ziegler in Bonn am 7.12.1970, Produktvorschläge Ra-PG', archive no. 8.3.1. (7 December 1970)

DRA: ———. 'The Designer's Contribution to Company Success' (manuscript), archive no. 1.1.2.1. German version (October 1975), archive no. 1.1.2.1.2 (22 September 1975)

DRA: ———, 'Die Rolle des Designers im Industrieunternehmen', talk at the I. Deutscher Designertag in Karlsruhe, archive no. 1.1.2.9. (14 October 1977)

DRA: ———, 'Boston USA Oct. 1984' (lecture manuscript), archive no. 1.1.5.1. (1984)

DRA: ———, Opening speech for the exhibition Designed in Germany, Mexico, archive no. 1.1.2.23.1. (1991)

Interviews

DRA: Braun, Artur, interview with Klaus Klemp in Königstein, archive no. 1.2.4.4. (23 July 2010)

DRA: Czipluch, Horst, interview with Klaus Klemp in Museum Angewandte Kunst, Frankfurt/Main, archive no. 1.2.4.12. (20 February 2014)

DRA: Guttandin, Thomas, and Roland Weigend, interview with Klaus Klemp in the Braun Collection, Kronberg, archive no. 1.2.4.11. (10 September 2019)

DRA: Hartwein, Peter, interview with Klaus Klemp in Museum Angewandte Kunst, Frankfurt/Main, archive no. 1.2.4.2. (5 April 2012)

DRA: Lubs, Dietrich, interview with Klaus Klemp in Museum Angewandte Kunst, Frankfurt/Main, archive no. 1.2.4.13. (8 May 2008)

DRA: Rams, Dieter, interview with Klaus Klemp in Kronberg, archive no. 1.2.1.49. (audio; 2010)

DRA: Schnelle-Schneyder, Marlene, interview with Klaus Klemp in the Museum Angewandte Kunst, Frankfurt/Main, archive no. 1.2.4.9. (30 August 2019)

DRA: Tischer, Eckard, interview with Klaus Klemp in Museum Angewandte Kunst, Frankfurt/Main, archive no. 1.2.4.14. (6 February 2014)

DRA: Ullmann, Roland, interview with Klaus Klemp in Museum Angewandte Kunst, Frankfurt/Main, archive No. 1.2.4.10. (16 March 2021)

Magazines

DRA: Klatt, Jo, Design+Design, ed., Zeitschriftenreihe, nos 1–94. Klaus Rudolph, ed. original title 'Der Braun Sammler' (1984–6), Jo Klatt, ed., 'Braun+Design', nos 8–21 (1987) 'Design+Design' (1984–2011)

HfG-Ulm Archive (HfGUA)

HfGUA: Aicher memo. About a meeting with Dr Seifert and Dr Eichler on 3.3.1959, in Signature HfG Archive AI AZ 3167 (3 March 1959)

HfGUA: Aicher memo, in Signature HfG Archive AI AZ 3167 (18 December 1958)

HfGUA: Aicher memo, in Signature HfG Archive AI AZ 3167 (13 March 1959)

HfGUA: Artur Braun to Herbert Ohl, in Signature HfG Archive AI AZ 648.3 (13 September 1960)

HfGUA: Concept book 'Braun Design', in Signature HfG Archive AI AZ. 922 (1989) HfGUA: Otl Aicher holdings, in Signature HfG Archive AI AZ 2212, AI AZ 2214 and AI AZ 3167

HfGUA: 'haus 62' development programme, in Signature HfG Archive AI AZ 31 (1959)

HfGUA: Project sketch Aicher, in Signature HfG Archive AI AZ 31, n.d.

Institute for the History of Frankfurt, Frankfurt/Main (IfS)

IfS: Sakowski, Norbert, To the Head of the Municipal Press and Information Office of the City of Frankfurt Joachim Peter (letter), in: MA 2084 (8 May 1963)

Munzinger Archive, Ravensburg (MA)

MA: 'Livis, Archibald' (entry) in Munzinger Online/Personen – Internationales Biographisches Archiv (accessed 28 July 2021), http://munzinger.de/document/00000021544

Literature

Abromeit, Günter, Erzeugnisplanung und Produktionsprogramm im Lichte der Produktions-, Absatz- und Wettbewerbspolitik (Wiesbaden, 1995)

Aicher, Otl, Innenseiten des Kriegs (Frankfurt/Main, 1985)
———, die welt als entwurf (Berlin, 1991)
Appel Design Gallery, MöBel by Dieter Rams (Berlin, 2012)
Axis Gallery, Braun Exhibition – Not Just a Matter of Form (Tokyo, 2005)
Baptista, Miguel, Dieter Rams Haus (Lisbon: Museu do Design Lisbon, 2001)
Barthes, Roland, Mythen des Alltags (Mythologies) (Frankfurt/Main, 1964)
Bergische Universität Gesamthochschule Wuppertal, ed., Braun Gegenstand. 200 Produkte aus vier Jahrzehnten (Wuppertal, 1992)
Braun, Artur, ed., Max Brauns Rasierer. Erinnerungen von Artur Braun (Hamburg, 1996)
———, Max Brauns Rasierer. Die Geschichte einer erfolgreichen Erfindung mitgeschrieben von Artur Braun (Hamburg, 2007)
Braungart, Michael, and William McDonough, Cradle to Cradle: Einfach intelligent produzieren (Munich, 2014)
Breuer, Gerda, Andrea Peters and Kerstin Plüm, eds, Die 60er. Positionen des Designs (Cologne, 1999)
Breuer, Romana, ed., Radio Zeit. Röhrengeräte, Design-Ikonen, Internetradio (Bielefeld, 2016)
Bürdeck, Bernhard E., Design. Geschichte, Theorie und Praxis der Produktgestaltung (Cologne, 1994)
Burkhardt, Francois, and Inez Franksen, Design: Dieter Rams & (Berlin, 1980)
Crowley, David, and Jane Pavitt, eds, Cold War Modern: Design 1945–1970 (London: Victoria & Albert Museum, 2008)
de Jong, Cees W. Zehn, Thesen für gutes Design: Dieter Rams. Die Sammlung Jorrit Maan (Munich, 2017)
documenta III, Industrial Design. Grafik (Cologne, 1964)
Dormer, Peter, Design Since 1945 (London, 1993)
Eames, Demetrios, Essential Eames. Zitate und Bilder (Weil am Rhein, 2017)
Edelmann, Thomas, and Gerrit Terstiege, Gestaltung denken. Grundlagentexte zu Design und Architektur (Basel, 2010)
Eichler, Fritz, Das Wesen des Handpuppen- und Marionettenspiels, dissertation (Emsdetten, 1937)
———, Vom König der nicht mehr lachen konnte [written 1943] (Berlin, 2016)
Erlhoff, Michael, ed., Designed in Germany. Since 1949 (Munich, 1990a)
———, Deutsches Design 1950–1990 (Munich, 1990b)
Feige, Daniel Martin, Design. Eine philosophische Analyse (Berlin, 2018)
Feyerabend, Paul, Against Method: Outline of an Anarchistic Theory of Knowledge (New York, 1975)
Fiell, Charlotte, and Peter Fiell, Design des 20. Jahrhunderts (Design of the 20th Century) (Cologne, 2001)
———, Industriedesign A–Z (Industrial Design A–Z) (Cologne, 2006)
Fischer, Volker, ed., Design heute. Masstäbe: Formgebung zwischen Industrie und Kunst-Stück (Frankfurt/Main, 1988)
———, and Anne Hamilton, Theorie der Gestaltung. Grundlagentexte zum Design, vol. 1 (Frankfurt/Main, 1999)
Förderkreis Braun Collection e.V.: The Director / The Man Who Directed (ed.), with texts by. Magdalena De Gasperi, Mascha Ebelding, Cornelia Heymann and others, Kronberg i.Ts. and Cologne (2021)
Frye, Annika, Design und Improvisation. Produkte, Prozesse und Methoden, dissertation (Bielefeld, 2017)
Fukasawa, Naoto, and Jasper Morrison, Super Normal – Sensations of the Ordinary (Tokyo, 2007)
Gassner, Christof, and Ute Schauer, eds, Designgeschichte und -geschichten. Das Institut für Neue Technische Form (Darmstadt, 2018)
Giedeon, Sigfried, Mechanization Takes Command. A Contribution to Anonymous History (New York, 1948)
———, Die Herrschaft der Mechanisierung (Frankfurt/Main, 1982)
göppinger galerie, Jazz hören, Jazz sehen (Frankfurt/Main, 1961)
Günther, Sonja, and Siegfried Maser, eds, Braun Gegenstand (Wuppertal, 1992)
Hara, Kenya, Weiss (Baden, 2009)
Hauffe, Thomas, Geschichte des Designs (Cologne, 2014)
Heine, Achim, Michael Lenz and Peter Zizka, eds, Ansichten von und zu Dieter Rams (Frankfurt/Main, 2002)
Hesse, Petra, and René Spitz, System Design. Über 100 Jahre Chaos im Alltag/Over 100 Years of Chaos in Everyday Life (Cologne, 2015)
Hirche, Herbert, and Dieter Gobel, herbert hirche architektur innenraum. Design 1945–1978 (Stuttgart, 1978)
Hochschule für Künste Bremen, ed., Der weisse Kittel. Eine Hommage an Wilhelm Wagenfeld (Bremen, 2000)
Höhne, Günter, Die geteilte Form. Deutsch-deutsche Designaffären 1949–1989 (Hanover, 2009)
Hornfischer, Lucia, Gerd A. Müller. Der unbekannte Designer (Stuttgart, 2021)
iF Industrie Forum Design Hannover and Uta Brandes, eds, Dieter Rams, Designer. Die leise Ordnung der Dinge (Göttingen, 1990)
Institute for New Technical Form, Dieter Rams Design. Die Faszination des Einfachen. Produkte, Prototypen, Skizzen, Dokumente (Darmstadt, 2004)
Jatzke-Wigand, Hartmut, Fritz Eichler (Stuttgart, 2021)
Klatt, Jo, ed., Möbel-Systeme von Dieter Rams (Hamburg, 2002)
Klatt, Jo, and Günter Staeffler, eds, Braun+Design Collection. Braun Produkte von 1955 bis heute (Hamburg, 1990)
———, Braun+Design Collection. 40 Jahre Braun Design. 1955–1995 (Hamburg, 1995)
Klemp, Klaus, Pure Design. Deutschland und benachbarte Länder in der ersten Hälfte des 20. Jahrhunderts (Amsterdam, Berlin and Stuttgart, 2011)
———, Design in Frankfurt 1920–1990. Mit einem Essay von Dieter Rams (Frankfurt/Main: Museum Angewandte Kunst, 2014)
———, Dieter Rams: The Complete Works (London, 2020)
Klemp, Klaus, Annika Sellmann, Matthias Wagner K and Grit Weber, Moderne am Main 1919–1933 (Frankfurt/Main: Museum Angewandte Kunst, 2019)
Klemp, Klaus, and Keiko Ueki-Polet, Less and More: The Design Ethos of Dieter Rams (Japanese/English edition, Osaka, 2008; German/English edition, Berlin, 2009)
Kloos, Werner, Braun. Das Gesicht einer Firma (Bremen: Hefte des Focke-Museums, 1966)
Kuprian, Werner, ed., Sport Physiotherapie (Stuttgart, 1981)
Lasagna, Louis, The Doctors' Dilemmas (New York, 1963)
Le Corbusier, Ausgabe Feststellungen zu Architektur und Städtebau (Frankfurt/Main, 1964)
Lindinger, Herbert, ed., Ulm Design School. Die Moral der Gegenstände (Berlin, 1987)
Loewy, Raymond, Never Leave Well Enough Alone: The Personal Record of an Industrial Designer (New York, 1951)
———, Hässlichkeit verkauft sich schlecht. Die Erlebnisse des erfolgreichsten Formgestalters unserer Zeit (Düsseldorf, 1953)
Lovell, Sophie, Dieter Rams: As Little Design as Possible (London, 2013)
Malzan, Traugott, Blaue Küste Rote Segel. Mit der Dschunke im Mittelmeer (Bielefeld, 1987)
Max Braun Stiftung, Mehr als Erinnerungen. Eine Dokumentation der Max Braun Stiftung (Munich, 2000)
McLuhan, Herbert Marshall, Understanding Media: The Extensions of Man (Boston, MA, 1964)
Meldau-Stagge, Sonja, Britta Stücker and Braun GmbH, 90 Jahre Braun (Cologne, 2011)
Müller, Dieter, Superacht mit Pfiff und einer Nizo (Winterthur, 1966)
———, Alles über die Nizo (Düsseldorf, 1973)
———, Mehr über die Nizo, Nizo S 56 und Nizo S 80 (Düsseldorf, n.d.)
Mukai, Shutaro, and Shukuro Habara, Origin of Design: Braun's Design Concept and Background (Tokyo, 1978)
Museum für Kunst und Gewerbe Hamburg, Mehr oder Weniger. Braun Design im Vergleich (Hamburg, 1990)
Nerdinger, Winfried, Walter Gropius. Architekt der Moderne (Munich, 2019)
Oestereich, Christopher, 'Gute Form', im Wiederaufbau. Zur Geschichte der Produktgestaltung in Westdeutschland nach 1945 (Berlin, 2000a)
Papanek, Victor, Design für eine reale Welt. Anleitungen für eine humane Ökologie und sozialen Wandel, 2nd ed. (Vienna, 2009)
Pfaender, Heinz G., Meine Zeit in der Werkstatt Wagenfeld. Tagebuch 1954–1957 (Hamburg, 1998)
Phaidon Editors, Phaidon Design Classics (London, 2006)
Polster, Bernd, Braun. 50 Jahre Produktinnovationen (Cologne, 2005)
Rams, Dieter, Weniger, aber besser – Less but better (Hamburg, 1995)
German Design Council, ed., Designed in Germany. Die Anschaulichkeit des Unsichtbaren (Frankfurt/Main, 1988)
———, Designing for product success. Essays und Fallstudien der Ausstellung des TRIAD Design Project (Düsseldorf, 1990)
———, Designkultur 1953–1993. Philosophie, Strategie, Prozess (Frankfurt/Main, 1993)
———, design deutschland. 80 years of 'made in germany' (Tokyo, 2006)
Ricard, André, Hitos del diseño. 100 diseños que hicieron éoca (Barcelona, 2009)
Saint-Exupéry, Antoine de, Wind, Sand and Stars, 18th ed. (Dessau, 1989)
Schepers, Wolfgang, and Peter Schmitt, Das Jahrhundert des Design. Geschichte und Zukunft der Dinge (Frankfurt/Main, 2000)
Schmittel, Wolfgang, Design Concept Realisation. Braun, Citroen, Miller, Olivetti, Sony, Swissair (Zürich, 1975)
———, process visual. Entwicklung eines Firmenprofils (Zürich, 1978)
Schneider, Beat, Design. Eine Einführung (Basel, 2005)
Schneider, Peter, live. Ein High Fidelity Report von Braun. Sonderausgabe zur fünften Wechselausstellung der Braun Sammlung 'Braun HiFi: Ursprung einer Designkultur' (Kronberg, 2008)
Schönwandt, Rudolf, Die moderne Trockenrasur. Technik – Tatsachen – Tipps (Munich, 1984)
———, German Design. Neue Dimensionen für Form und Funktion (Dortmund, 1990)
———, Die unmissverständliche Sprache der Vernunft. Aufsätze und Essays zum Design (Klein-Auheim, 1998)
Schwer, Thilo, Produktsprachen. Design zwischen Unikat und Industrieprodukt (Bielefeld, 2014)
Seckendorff, Eva von, Die Hochschule für Gestaltung in Ulm. Gründung (1949–1953) und Ära Max Bill (1953–1957) (Marburg, 1989)
Selle, Gerd, Geschichte des Design in Deutschland (Frankfurt/Main, 1994)
Spitz, René, HfG Ulm. Der Blick hinter den Vordergrund. Die politische Geschichte der Hochschule für Gestaltung 1953–1968 (Stuttgart, 2002)
Terstiege, Gerrit, The Making of Design. Vom Modell zum fertigen Produkt (Basel, 2010)
Themenkreis im IDZ, ed., design? Umwelt wird in Frage gestellt (Berlin, 1970)
Wachsmann, Christiane, ed., Hans Gugelot. Die Architektur des Design (Stuttgart, 2020)
Wagenfeld, Wilhelm, Wesen und Gestalt der Dinge (Potsdam, 1948)
Walker, John A., Design History and the History of Design (London, 1989)
Welsch, Wolfgang, Transkulturalität. Realität, Geschichte, Aufgabe (Vienna, 2017)
Wichmann, Hans, ed. System-Design Bahnbrecher: Hans Gugelot 1920–65 (Munich: Die Neue Sammlung Staatliches Museum für Angewandte Kunst, 1984)
———, Mut zum Aufbruch. Erwin Braun 1921–1992 (Munich, 1988)

Essays and Articles

'A Braun Study: An Interview with Dieter Rams, Chief Designer of Braun A.G.', Gillette International Quarterly, no. 1 (1973), 16–18
Abendroth, Uta, Christian Pixis and Karin B. Phillips, 'Dieter Rams', in Das Designbuch (Augsburg, 1999), 304f
Albus, Volker, 'Braun Phonosuper SK4. Dieter Rams und Hans Gugelot', in Design! Das 20. Jahrhundert, ed. Volker Albus et al. (Munich, 2000), 114f
Aldersey-Williams, Hugh, 'Modern Masters: Dieter Rams', I.D. Magazine of International Design, vol. 36, no. 5 (1989), 32–5
Angress, Gina, and Inez Franksen, 'Dieter Rams 25 Years in Business', mobilia, no. 298 (1981), 33–43
Apel-Dube, Werner, untitled article in Mehr als Erinnerungen. Eine Dokumentation der Max Braun Stiftung, Max Braun Stiftung (2000), 21f
Bertoni, Franco, 'Dieter Rams', in Minimalistisches Design (Basel, 2004), 118–21
Beuker, Ralf, 'Braun Design. Dream Real Products, Make Real Products', Design Management Journal, vol. 12, no. 4 (2001), 15–23
Böhme, Gernot, 'Atmosphären wahrnehmen, Atmosphären gestalten, mit Atmosphären leben. Ein neues Konzept ästhetischer Bildung', in Atmosphäre(n), ed. Rainer Goetz and Stefan Graupner (Munich, 2008), 38
Braun, untitled article in Design Journal, no. 44 (1991), 20f
Braun, untitled article in AZIMUTS, no. 26 (2005), 34–9
Braun, Artur, 'Wie das Braun Design entstand', Design+Design zero (2012), 25
Braun, Erwin, 'Öffentlicher Brief Erwin Brauns an Fritz Eichler', form, no. 90 (1980), 105
'Braun: Eine Lehre der industriellen Formgebung', Der Braun Sammler, no. 7 (1987), 8–12
'Braun a/d/s/ verabschiedet sich ...', form, no. 132 (1990), 32–43
Carter, Erica, 'The Aesthetics of Rationality: Braun in 1950s West Germany', Things, no. 3 (1995), 57–69
Choi, Dae-Hyuk, 'Mr Braun, He's Here: Dieter Rams', Design, no. 392 (2011), 114–23
Cobarg, Claus. C., 'Max Braun, Unternehmensgründer und Innovator', Design+Design, no. 17 (1990), 6–12
———, 'Braun Produkte auf der Expo 1958 und eine kleine Vorschau auf die Expo 2000', Design+Design, no. 52 (2000), 3–6
Cobarg, Claus. C., and Jo Klatt, 'Der 'ungewöhnliche' Braun-Messestand auf der Düsseldorfer Funkausstellung 1955', Design+Design, no. 72 (2005), 4–6
Cobarg, Claus. C., Michael Unger and Peter Ziegler, 'Die Diaprojektoren von Braun', Design+Design, no. 27 (1994), 4–11
De Lucchi, Michele, ed., 'Dieter Rams', in International Design Yearbook 2001 (New York, 2001), 71, 232
Derichs, Claus Uwe, 'Anfangen neu zu denken – werkundzeit-gespräch mit Dieter Rams', werkundzeit, no. 2 (2002), 8–11
'Dieter Rams: An Interview With Dieter Rams, Chief Designer of Braun AG of Frankfurt – Ideals Know No Compromise', Building Design (1973), 10f
'Dieter Rams', in Türklinken. Workshop in Brakel, ed. Franz Schneider Brakel (Cologne, 1987), 36f
'Dieter Rams. Lucky Strike Award für ein Lebenswerk', in Designers Digest, no. 102 (2007), 22–8
Doblin, Jay, 'Dieter Rams', in One Hundred Great Products Designs (New York, 1970), 84, 104f
Duhalde, Benedicte, 'Concours Braun: Visions Futures, the Braun Competition, the Way Forward', Intramuros, no. 63 (1996), 24f
Edelmann, Thomas, 'Zeitlos praktisch', design report, no. 9 (1994), 58–65
———, 'Identitet for andra tider', Form, vol. 95, no. 725 (1999a), 40–45
———, 'Braun-Design auf neuen Wegen', design report, no. 7/8 (1999b), 46–51, 108f
———, 'Fragebogen: Dieter Rams', design report, no. 5 (2008), 12
Eichler, Fritz, 'Eichler an Wagenfeld', form, vol. 23, no. 23 (1963), 11
Fairs, Marcus, 'Dieter Rams', Icon, no. 19 (2004), 60–68
Fischer Richard, and Gerda Mikosch, 'Grundlagen einer Theorie der Produktsprache', Anzeichenfuntionen, issue 3 (1984)
Fischer, Wend, 'Gestaltung unter dem Primat der Intelligenz', werkundzeit, vol. 6, no. 11 (1957), special supplement Werkbericht Braun, 2f
Fogarty, Kate, 'The Hi-Fi Heard Around the World: 50 years of Braun Design', Modernism Magazine, issue 9 (2006), 140
Franksen, Inez, 'Montageprogramm 571/72 Vitsoe', Bauwelt, no. 44 (1974), 1460f
———, 'Regalsystem 606 Vitsoe Kollektion', Bauwelt, no. 23 (1976), 709
Frese, Gunhild, 'Exodus aus Kronberg. Warum bei Braun die Manager weglaufen', Die Zeit, no. 23 (30 May 1975)
Fukasawa, Keita, 'Braun Exhibition – Not Just a Matter of Form', Axis, no. 118 (2005), 144–7
Günther, Sonja, 'Dieter Rams', Design+Design, no. 33 (1995), 18–20
Hansen, Astrid, 'Natürlich: Braun Design. Konkretes Design', Die 60er: Positionen des Designs, ed. Gerda Breuer (Cologne, 1999), 32–43
Henkels, Jochen, 'Sachen und Menschen. Erkenntnisse aus dem Studium bei Dieter Rams', Design+Design, no. 33 (1995), 21
Heuss, Theodor, 'Bewährung in der Gewöhnung', werkundzeit, no. 4 (1952)
Horsham, Michael, 'Brains und Braun', Wallpaper, no. 15 (1998), 119f
Irrek, Hans, 'Wir haben damals eine Menge über die Zusammenstellung der Dinge von ihm gelernt', Design+Design, no. 33 (1995b), 24–9
Jatzke-Wigand, Hartmut, 'Dieter Rams. Zu diesen Entwürfen stehe ich noch heute, I am still sticking to my design ...' Design+Design, no. 33 (1995), 6–17
Jockel, Nils, and Ivo Schelle, 'Mehr oder weniger, Braun Design im Vergleich', Braun+Design, no. 18 (1990), 4–9
Justice, Lorraine, and Dieter Rams, 'Dieter Rams: 1996 World-wide Design Award Winner', Innovation, vol. 15, no. 4 (1996), 14f
Kalb, Joachim, 'Braun Plakate, ein neues Genre?', Der Braun Sammler, no. 7 (1987), 21–3
Klatt, Jo, 'Leute um Hans Gugelot. Gespräch mit Helmut Müller-Kühn', Der Braun Sammler, no. 7 (1987), 13f
———, 'Die Radio-Phono-Kombination von Braun', Design+Design, no. 15 (1990), 4–14
———, 'Nicht produzierte Braun-Geräte aus dem Unterhaltungselektronik-Bereich', Braun+Design, no. 21 (1991), 4–9
———, '40 Jahre Braun Gesundheitsdienst', Design+Design, no. 30 (1994), 33f
———, 'Das besondere Einzelstück. Der Braun Kofferempfänger exporter 2', Design+Design, no. 68 (2004), 4–6
———, '"Algo de Braun". Eine kleine Auswahl von Braun Produkten, die im Werk Barcelona hergestellt wurden', Design+Design, no. 88 (2009), 4–7
Kleefisch-Jobst, Ursula, 'Dieter Rams rgs-Serie', in Architektur zum Anfassen. FSB Greifen und Griffe, ed. Gernot Böhme, Gera Breuer, Oliver Elser and Ursula Keelfisch-Jobst (Frankfurt/Main, 2002), 124–7
Klemp, Klaus, 'Design Dieter Rams', in Dieter Rams meets Nikolai Petrovich Sheremetev (Moscow: State Museum of Ceramics and Kuskovo, 2007), 2–29
———, 'Dieter Rams: Ethics and Modern Philosophy – What Legacy Today?', docomomo Journal, no. 46 (2012a), 68–75
———, 'Metamorphosen der Haltbarkeit im Design', in Metamorphosen. Design zwischen Erneuerung und Anpassung, ed. Deutscher Designer Club e.V, DDC Design Bibliothek – History 2 (Frankfurt/Main, 2012b), 8–27
Klemp, Klaus, and Fabian Wurm, 'Apostel der Coolness', Beef (2010), 34–7
Komachi, Hanae, 'Dieter Rams', Axis, no. 105 (2003), 81–5
Kramer, Lore, 'Braun Design und Vorläufer industrieller Formgestaltung – Pioniere und Leitbilder', Der Braun Sammler, no. 6 (1986), 15–23
Landmann, Ludwig, 'Zum Geleit', Das Neue Frankfurt, issue 1 (1926), 1f
Livis, Archibald, 'Success by Design', in Szenenwechsel. German Design goes Rocky Mountain High, ed. Hans Hermann Wetcke (Frankfurt/Main: Design Zentrum München, 1997), 105
Lossau, Jürgen, 'Braun Nizo lebt', Schmalfilm, no. 4 (2010a), 9
———, 'Sagenhafte Silberlinge', Schmalfilm, no. 4 (2010b), 30–36
Lovell, Sophie, 'Dieter Rams. Master and Commandments', Wallpaper, no. 103 (2007), 317–39
Moller, Svend Erik, 'Dieter Rams', Mobilia, no. 145, n.d., 45–53
Moss, Richard, 'Braun Style' Industrial Design, no. 11 (1962), 36–47
Moyano, Neus, 'Braun. Industrial Construction System und Technical Drawings', in Otl Aicher. Metro Bilbao. Architecture and Landscape, ed. Gillermo Zuaznabar (Bilbao: Bilbao Fine Arts Museum, 2020), 37–79
Naylor, Colin, 'Dieter Rams', in Contemporary Designers, 2nd ed. (Chicago, IL, 1990), 477f
Neumann, Eckhard, 'Dieter Rams. Seit 25 Jahren "vernünftiges Design"', design report, no. 3 (1981), 1–3
Oestereich, Christopher, 'Umstrittene Selbstdarstellung. Der deutsche Beitrag zur Weltausstellung in Brüssel 1958', Vierteljahreshefte für Zeitgeschichte, vol. 48, no. 1 (2000b), 127–53
Peterek, Michael, 'Wohnstadt Limes. Mehr als eine Siedlung', in Wohnen in der Nachkriegsmoderne. Siedlungen in der Region Rhein-Main, ed. Maren Harnack, Matthias Brunner and Natalie Heger (Berlin, 2020), 26–33
Plonka, Kay Alexander, 'Less. But Better', stilwerk (2019), 14–19
Polster, Bernd, and Peter Schneider, 'Design-Evolution. Die aktuelle Produktentwicklung seit 1995 bei Braun', Design+Design, no. 69/70 (2004), 30f
Quadejacob, Lars, 'Die Einführung des neuen Braun Designs im Spiegel der zeitgenössischen Presse', Design+Design, no. 42 (1997), 21–6
Rams, Dieter, 'Hören und Sehen im System', form, no. 29 (1965a), 40–45
———, 'Produktdesign hos Braun', form, vol. 61, no. 1 (1965b), 26–9
———, 'Kann Design zum Erfolg eines Unternehmens beitragen?', werk archithese. Zeitschrift und Schriftenreihe für Architektur und Kunst, no. 4 (1977), 9–15
———, 'And That's How Simple It Is To Be a Good Designer', Designer (1978), 12f
———, 'Io disegno la funzionalità', Auto & Design, no. 11 (1981a), 6f
———, 'Die Rolle des Designers im Industrieunternehmen', in Helmuth Gsöllpointner, Design ist unsichtbar, ed. Helmuth Gsöllpointner (Vienna: Österreichisches Institut für Visuelle Gestaltung, 1981b), 493–506
———, Funktionales Design ist eine Zukunftsaufgabe', in Design Dasein. Ausgewählte Objekte zum Sitzen, Stellen und Leben (Hamburg: Museum für Kunst und Gewerbe Hamburg, 1987a), 155–9
———, 'Funktionales Design ist eine Zukunftsaufgabe', in Donation Braun, Hans Wichmann (Munich: Die Neue Sammlung Staatliches Museum für angewandte Kunst, 1987b), 270–330
———, 'Ramsification', Designer (April 1987c), 24f
———, 'Erinnerungen an die ersten Jahre bei Braun', in Johannes Potente, Brakel. Design der 50er Jahr, ed. Otl Aicher et al. (Cologne, 1989b), 70–76
———, 'Omit the Unimportant', in Design Discourse: History, Theory, Criticism, ed. Victor Margolin (Chicago, IL, 1989b), 111–14
———, 'Market Performance with "Technology Design"', in The Seventh Arango International Design Exhibition: Good and the Dieter Rams Retrospective Exhibition (Fort Lauderdale, FL, 1989c), 14–17
———, 'Die Kultur der Geräte für den Alltag', in German Design. Neue Dimensionen für Form und Funktion, Rudolf Schönwandt (Dortmund, 1990), 7–9
———, 'Beyond the Logic of Consumerism', Ottagono, issue 102 (1992), 22–9

———, 'Ten Principles for a Good Design', Domus, no. 748 (1993), 21–8
———, 'Design als Verantwortung – neue Konzeption der Gestaltung', in Der Stoff der Dinge. Material und Design, ed. Arnica Verena Langenmeier (Munich, 1994), 97–105
———, 'Fremtidens Design', Design, no. 5 (1995a), 37–44
———, 'Im Mittelpunkt: Wachsame Nachdenklichkeit', in Design = unsichtbar, ed. Hans Höger. Published on the occasion of Lucius Burckhardt receiving the Federal Award for Promoting Design (Bundespreis für Förderer des Designs) (Ostfildern, 1995b), 6f
———, 'Cosi si acquisita', Ottagono, no. 114 (1995c), 52–5
———, 'Weniger, aber besser', in Szenenwechsel. German Design goes Rocky Mountain High, ed. Hans Hermann Wetcke (Frankfurt/Main, 1997), 97–9
——— 'Less is More', Design Week, vol. 14, no. 44 (1999), 18f
———, 'Die Verantwortung des Designs', in Im Designerpark. Leben in künstlichen Welten, ed. Kai Bucholz and Klaus Wolbert (Darmstadt, 2004), 106–13
———, 'Ramsifikationen. Die zehn Prinzipien für gute Gestaltung von Dieter Rams', in The Making of Design. Vom Modell zum fertigen Produkt, ed. Gerrit Terstiege (Basel, 2010), 97–101
Sano, Keiko, 'Braun Design Today', Axis, no. 117 (2005), 108–11
Scheuer, Winfried 'Rams Gives the Green Light in LA', Blueprint, no. 69 (1990), 7
Schlüter, Ralf, 'Lang lebe die Kaffeemaschine', art Kunstmagazin, no. 5 (2002), 24–31
———, 'Die zehn größten Designer der Gegenwart', art, no. 6 (2008), 18–33
Schock mit Sarg, 'Der Mann, der den Braun-Stil schuf', Der Spiegel, no. 25 (1975), 122f
Schönwandt, Rudolf, 'Wie arbeitet Dieter Rams und sein Team bei Braun?', in Dieter Rams &, ed. Francois Burkhardt and Inez Franksen (Berlin, 1980a), 31–46
———, '25 Jahre Dieter Rams & Braun-Design', form, no. 91 (1980b), 20–23
———, 'Bilanz nach 35 Jahren. Design Portrait: Braun AG', in Jahrbuch 1989 (Haus Industrieform Essen, 1989)
———, 'Porträt Dieter Rams', Lufthansa Bordbuch, no. 4 (1993), 22–9
Schultz, Albrecht, 'Braun. Von der Design-Idee zum System der Produktgestaltung und Vertriebspolitik', in Marktfaktor Design. Grundlagen für die Marketingpraxis, ed. Ludwig G. and Gudrun S. Poth (Landsberg am Lech, 1986)
———, 'Designphilosophie bei Braun', in Die stillen Designer – Manager des Designs, ed. Norbert Hammer (Essen, 1994), 177–87
Sparke, Penny, 'Dieter Rams: The New Modernism', in A Century of Design. Design Pioneers of the 20th Century, ed. Penny Sparke (Hauppauge, NY, 1998), 184–7
Staeffler, Günter, 'Was ist Design? Eine zusammenfassende Betrachtung zur Braun Design-Philosophie', Der Braun Sammler, no. 4 (1985), 22f
———, 'Braun und Herbert Hirche', Braun+Design, no. 16 (1990), 16–23
———, 'Braun Kleinsuper SK 1, SK 2, SK 3, SK 2/2 und SK 25', Design+Design, no. 19 (1991), 4–13
———, 'Braun Design by Reinhold Weiss: Part 1', Design+Design, no. 48 (1999a), 7–11
———, 'Braun Design by Reinhold Weiss: Part 2', Design+Design, no. 49 (1999b), 19–23
———, 'Braun Design by Reinhold Weiss: Part 3', Design+Design, no. 51 (2000a), 11–15
———, 'Reinhold Weiss Design-Studien für Braun', Design+Design, no. 52 (2000b), 12–15
———, 'Braun Design-Studien. Produkt Ideen, die nicht verwirklicht wurden', Design+Design, no. 78 (2007a), 8f
———, 'Braun Design-Studien. Innovative Produkt-Ideen, die nicht in Serie gingen. Part 1', Design+Design, no. 79 (2007b), 3–12
———, 'Braun Design-Studien. Innovative Produkt-Ideen, die nicht in Serie gingen. Part 2', Design+Design, no. 81 (2007c), 3–7
———, 'Vorbildliche Braun Produktpräsentation in den 1950er und 1960er Jahren. Das Ausstellungs- und Display-System', Design+Design, no. 86 (2008), 7–11
———, 'Design Studien für Braun. Elektronenblitzgeräte von Reinhold Weiss', Design+Design, no. 94 (2010), 6
Stam, Mart, 'Das Masse, das richtige Masse, das Minimum-Masse', Das Neue Frankfurt, 2nd ed., issue 1 (1928), 29f
Stürzebecher, Jörg, 'Braun blickt zurück', form, no. 205 (2005), 108f
———, 'Systematisches Understatement', design report spezial imm cologne (2008), 52f
Sudjic, Deyan, 'Dieter Rams. Ripensamenti', Domus, no. 849 (2002), 130f
Terstiege, Gerrit, 'Der Systematiker', form, no. 183 (2002), 76–81
———, 'Gripping stuff. Interview with Alfredo Häberli, Matali Crasset, Davin Stowell, Ole Bobjer, Dieter Rams and Ron Arad', form, no. 203 (2005), 42–7
———, 'Schneewittchen und die sieben Designer/Snow White and the Seven Designers', form, no. 207 (2006), 34
———, 'Dieter Rams. "Der Designprozess hat noch keine Form", Interview von Gerrit Terstiege mit Dieter Rams', in The Making of Design. Vom Modell zum fertigen Produkt (Basel, 2009), 8–15
'To the Bauhaus With Love From Dieter Rams', Interiors (2007), 45f
Todd Hénaut, Dorothy, 'Weltausstellung in Montreal', Graphis, no. 132 (1967), 346
Trappschuh, Elke, 'Verfechter des Vernünftigen. Ein Besuch bei Dieter Rams', in Designpreis der Bundesrepublik Deutschland 2007, ed. German Design Council (Basel, 2007), 4–21
Wagenfeld, Wilhelm, 'Künstlerische Zusammenarbeit mit der Industrie, Vortrag gehalten am 18. September 1954 in der TH Darmstadt', in Mut zum Aufbruch. Erwin Braun 1921–1992, Hans Wichmann (Munich, 1998), 178–80
Werk, Ingo, 'Rams in LA', md Interior Design Architecture (November 2019), 66f
Werkbericht Braun, werkundzeit, no. 11 (1957), special supplement
Witke, Boris, 'Braun "SK 4" – die erste Kompaktanlage', Funkgeschichte, no. 241 (2018), 228–39
Woudhuysen, James, 'The Apostle of Cool', Blueprint, no. 39 (1987), 28–30
Zec, Peter, 'Interview mit Dieter Rams', in German Design Standards (Essen, 2005), 122–9
Zehntbauer, Markus, 'Das Werkzeug [Designprozess von Roland Ullmann]', in The Making of Design. Vom Modell zum fertigen Produkt, ed. Gerrit Terstiege (Basel, 2010), 92–9
Zeitschrift Bauen + Wohnen, issue 9 (1967)

Online Publications

Abele, Günter F, 'Eine Chronik in Wort und Bild. Braun, Firmengeschichte', Historische Radios (accessed 21 January 2021), http://historische-radios.info/firmen/braun/firmengeschichte.htm
'design wissen' (accessed 31 March 2021), https://designwissen.net
Grabes, Oliver, 'Why Do We Promote Design?' (accessed 14 July 2021), https://braunprize.org/en/the_idea.html
———, 'Design Today' (accessed 14 July 2021), https://braun.de/de-de/world-of-braun/braun-design/design-today
'IG Metall kündigt Streikbeginn an (21. August 1951)', Zeitgeschichte in Hessen (accessed 11 March 2021), https://lagis-hessen.de/de/subjects/idrec/sn/edb/id/4705
'Beendigung des Streiks der hessischen Metallarbeiter (22. September 1951)', Zeitgeschichte in Hessen (accessed 27 March 2019), https://lagis-hessen.de/de/subjects/idrec/sn/edb/id/3447
Jatzke-Wigand, Hartmut, 'Design + Text' (accessed 31 March 2023), https://designundtext.com/de/index.php
Königsdorfer Medienhaus, 'Designlexikon International' (accessed 27 March 2021), https://designlexikon.net
'The Global E-Waste Monitor 2020', Jeder Deutsche produziert 19,4 kg Elektroschrott, zitiert nach de.statista.com (accessed 1 February 2021), https://de.statista.com/infografik/12272/die-zehn-laender-mit-dem-groessen-elektroschrott-aufkommen
schneider+schumacher, 'Braun Hauptverwaltung 1998–2000' (accessed 5 January 2021), https://schneider-schumacher.de/projekte/project-details/90-braun-hauptverwaltung

Glossary of Product Names

1000 The number 1000 denotes a high-quality and pioneering 'Grand Design' product series. Well-known examples include the T 1000, studio 1000, CSV 1000, CE 1000, PS 1000, L 1000, TG 1000.
4547 First post-war radio, the numbers stand for 4 tubes, 5 circuits, year (19)47. Other product examples include 4648, 4549, 4650.

A Amplifier
AB Analogue display Battery
ABW Analogue display Battery Wall Clock
AC Amplifier Control
AP Amplifier Powera
atelier Braun product name for Hi-Fi stereo system in the 1990s
Atelier Product name for compact device (1957–62)
audio Product name for compact system (1962–75)
AW Arm Watch (wristwatch)
BSK Batterie Super Koffer – portable battery-operated radio

C Cassette deck
CC Control Centre
CD Compact Disk Player; Console Demodulator (for PSQ)
CE Consolen Empfänger (console receiver/tuner)
CEV Consolen Empfänger Verstärker (radio receiver + amplifier)
CES Consolen Empfänger Steuerverstärker (radio receiver + control amplifier)
CET Consolen Empfänger Tuner (radio receiver and Hi-Fi tuner)
CM Compact Monitor (speaker with artificial stone housing)
combi Product name for portable radio-phono unit by W. Wagenfeld
combiscope Product name for combined slide projector and viewing device
CS Consolen Steuergerät (console control device/ preamplifier)
CSQ Consolen Steuer Quadroverstärker (console control Quadro amplifier)
CSV Consolen Stereo Verstärker (console stereo amplifier)
CT Consolen Tuner (console tuner)
CV Consolen Verstärker (console amplifier/ power amplifier)

D Diaprojektor (slide projector)
DB Digital display Battery
'Langer Heinrich' (Long Henry) nickname for PK-G by H. Gugelot in reference to the length of the unit
DW Digital Watch

E Espresso machine
ET Elektronischer Taschenrechner (electronic calculator)

functional Product name for digital alarm clock (1975)
FS Fern Sehgerät (television set)
FS-G Fern Sehgerät (television set) by H. Gugelot
FZ Foto Zelle (photo cell)

G 11 Table Radio Superhet Series (RC 60, 61 A, 62-2, 7) by H. Gugelot
G 11 Table Stereo Superhet in two versions (RC 8, 81B) by H. Gugelot
G 12 Record Player (PC 3-chassis 3-speed turntable) by H. Gugelot
G 12 V Record Player (PC 3-chassis 4-speed turntable) by H. Gugelot
GS Geräte Schrank (unit cabinet)
GT Geräte Turm (unit tower)

ic in concert (loudspeaker)

H Heizlüfter (fan heater)
HF 1 Hirche Fernsehgerät (Hirche television)
HFK Hirche Fernseh Musikschrank (combined Hirche television and music cabinet)
HFS Hirche Fernseh Schrank (Hirche television cabinet)
HLH Heiss Luft Haube (hot-air hood/hair-dryer hood)
HM Hirche Möbel (Hirche furniture)
Hobby Product name for electronic flash units (1952–78)
HT Haushalt Toaster (household toaster)

'Kaninchenstall' (Rabbit hutch) The nickname for G11
'Kängurufüsse' (Kangaroo feet) The nickname for the Hi-Fi system-base because of its long feet
KF Kaffee Filtermaschine (filter-based coffee machine)
KH Kopf Hörer (headphones)
KM Küchen Maschine (kitchen machine or food processor)
KMM Kaffee Mühlen Mahlwerk (coffee grinder)
'Kommissbrot' / 'Knäckebrot' (Commissary bread / crispbread) A nickname for SK 1/2

L Lautsprecher (loudspeaker)
LE Lautsprecher Elektrostat (electrostatic loudspeaker)
Lectron Name for a system for teaching electronics by Georg Franz Greger; Braun purchased the licence and revised the design in 1967
LF Lautsprecher Fuss (loudspeaker base)
linear Product name for pocket lighter (1976)
LS Laut Sprecher (loudspeaker)
LSV Laut Sprecher Video (video loudspeaker)
LTV Lautsprecher Tele Vision (television loudspeaker)
LV Lautsprecher Verstärker (loudspeaker amplifier)
LW Lautsprecher Woofer (loudspeaker woofer)

M Monitor (floor-based loudspeaker); Multiquirl (hand blender)
micron vario 3 Name for electric shaver derived from the three shaver settings (1985)
MM Musik Möbel (Hi-Fi furniture)
MP Multi Press, a fruit juicer
MPZ Multi Press Zitrus (citrus fruit juicer)
MS Musik Schrank (Hi-Fi cabinet)
MX Multi Mix

P Plattenspieler; Powerplate (record player; Powerplate amplifier)
PA Power Amplifier
PC Platten Cassetten; Plattenspieler Chassis (record and cassette player; turntable body)
PCK Plattenspieler Chassis Koffer (turntable body case)
PCS Plattenspieler Chassis Stereoverstärker (turntable body with stereo amplifier)
PCV Plattenspieler Chassis Verstärker (turntable body with amplifier)
PDS Platten Direktdrive Spieler (record player with direct drive)
Phono-Super Combined with a model no., the name for a radio receiver and record player combination made after World War II
PK Phono-Kombination (radio and record player combination)
PK-G 1-5; 51 Phono-Kombination (radio and record player comination [RC 60, 61, 62, 7, 8, 81 B, 82 B] by H. Gugelot)
PS Platten Spieler; Powerplate Stereo (record player; stereo amplifier/mixer)
PSQ Platten Spieler Quadro (Quadro record player)

QF Quadro Fernbedienung (for CSQ, Quadro remote control)

R Radio; Receiver
RA Receiver Analog (analogue receiver)
RC Radio Chassis (body for table radio, radio and record player combinations, music cabinets and compact systems with record players [RC 4, 31, 60, 61, 61-1, 61 A, 62, 62-1, 62-2, 62-3, 62-5, 7, 8, 8A, 8B, 81, 81 A, 81B, 82, 82 A, 82 B, 82 C, 82 US, 9])
RCS Radio Chassis Steuergerät (radio body with control device)

regie Product name for control unit (1968–78)
RM Regal Monitor (loudspeaker)
RR Remote Receiver
RS Radio Super; Receiver Synthesizer
RT Rundfunk Tischgerät (tabletop radio)

'Schneewittchensarg' (Snow White's Coffin) The nickname for the SK 4
sixtant Product name for electric shaver, derived from the shaving foil with hexagonal openings (1962–84)
SK Super-Het Klein (small radio using het technology)
SM Studio Monitor (loudspeaker); Studio Master (loudspeaker)
studio Product name for compact system (studio 1 by H. Gugelot/H. Lindinger, 1957) or control device with record player (studio 2 by D. Rams, 1959)
Super Until the end of World War II, the broadcast-only receivers were known by the name of 'Super' from Super-Het (short for superheterodyne receiver; the standard technology for receiving radio frequencies until the digital radio era)
SW Subwoofer or Systemwagen (Hi-Fi systems trolley)

T Transistor; Tuner; Tischfeuerzeug (table lighter)
TB Ton Band or Vollspule (tape or full reel)
TC Transistor Complement; Tonband Cassette (compact unit; tape cassette)
TG Tonband Gerät (tape recorder)
TG 60 Tonband Gerät 60 (Tape Recorder 60), the name derives from the model name of the tube amplifiers CSV 60 and the loudspeaker unit with base L 60
TGC Tonband Gerät Cassette (tape recorder cassette)
TGF Tonband Gerät Fernbedienung (tape recorder remote control)
'Toaster' Nickname for the TFG 1 table lighter owing to its rectangular form enclosed by a metal body, a shape reminiscent of the Braun toaster
TS Tisch Super (tabletop radio); Transistor Steuergerät (receiver); Tuner Synthesizer (tuner)
TS-G Tisch Super (tabletop radio by H. Gugelot)
TV Tele Vision

VC Video Cassettenrecorder (video cassette recorder)

Compiled by Hehn-Chu Ahn / Dieter Rams Archive

Awards and Exhibitions

Awards

1935, Grand Prize for Battery-Koffersuper, International Exhibition Thessaloniki
1937, Gold medal for special achievements in phonography and the making of musical appliances, World Expo Paris
1954–2021, iF Die gute Industrieform (1954–87), later renamed iF Industrieform Hannover (1988–96), iF Design Awards (1997–) – in total Braun has won 369 awards
1957 and 1960, Gran Premio for Braun's entire product range, 11th and 12th Triennale di Milano
1958, Etoile d'Or for the complete Braun range of products, World Expo Brussels, Belgium
1961, 1963 and 1985, Supreme Award for outstanding use of plastic for Phono-Transistor TP 1 (1961) and F 21 flash unit (1963); Main prize (1985), Interplas International Trade Fair, London
1961, Compasso d'Oro for good design of industrial products, Italian Embassy in Bad Godesberg (1962) / Associazione per il Disegno Industriale (ADI), Milan, Italy
1962–92, Permanent show at the State Trade Office (LGA) Zentrum Form (1962–64), afterwards known by the name Permanent Exhibition of International Product Design (1967–70), Annual Selection (1971–78), Design Selection (1979–92) – in total 286 awards for Braun products, exhibition in the LGA-Zentrum Form (1962–69), in the LGA Baden-Württemberg (1969–92) / Landesgewerbeamt Baden-Württemberg (1962–69), Design Center Stuttgart (1969–92)
1964, 1968, 1981 and 1984, Gold medal for D 40 projector and mentions for T 52 transistor, sixtant shaver and F 25 flash unit (1964), Gold medal for learning and experimenting system Lectron (1968), Gold medal for visotronic DN 50 Desk Clock, honorary mention for micron (1981), Prize for Braun hair dryer travelcombi PGI 1200 (1984), Biennale of Design Ljubljana
1964, Gold medal for radio phono combination audio 1, 13th Triennale Milano
1964–74, Participation in the special show Die Gute Form, Frankfurt Autumn Fair / Institute for New Technical Form (INTeF), Frankfurt/Main
1965 and 1966, Award for Braun advertising material, Art Directors Club Deutschland, Düsseldorf
1967, Award for parat BT shaver, Council of Design Promotion, Tokyo
1968, Prize for exemplary design of Braun exhibition booth, Interbytmash Municipal and Service Trade Fair, Moscow
1969 and 1970, Gute Form (Good Design) Award for studio 500 Hi-Fi system, second prize for 250 compact system audio (1969) and for sixtant 6006 shaver, T 2 /TFG 2 Table Lighter, M 140 Hand Blender, HL 1 Desk Fan (1970), Berlin Industrial Fair, award ceremony in IDC / Federal Ministry for Economic Affairs and German Design Council, Berlin
1971, 1972 and 2022, Prize for Hi-Fi system studio 500 (1971), a total of 38 prizes for Braun products (1972), Oral-B (2022), Consumer Electronics Show (CES) / Consumer Technology Association, Chicago / Las Vegas
1972, Isenkraemmers design pris for T 2 Desk Lighter, KSM 11 Coffee Grinder, HLD 4 Hair Dryer, Danmarks Isenkraemmerforening (Association of Danish Hardware Dealers)
1972, Highest distinction for KSM 1 Coffee Grinder, Center of Research of Industrial Design, Buenos Aires
1978–2005, Busse Longlife Design Award for Multipress MP 32 and mentions for KM 32 (1978), Multipress MP 50 and MPZ 2 Citromatic juicer (1984), micron plus shaver and calculator ET 33/55 (1987), atelier system (1990), micron vario 3 shaver (1993), KF 32 coffee maker (2002), MPZ 22 Citromatic juicer (2005), Longlife Award Competition / Rido Busse, Elchingen
1978–2015, Good Design Award G-Mark – in total 30 Awards to Braun, Design Selection Competition / Japanese Ministry of international Trade and Industry and Japan Institute of Design Promotion, Tokyo
1983–96, in total 22 awards to Braun in the categories Corporate Identity Trophy, Top Ten of the Year, Best of Category, Award for Good Industrial Design, High Quality of Design and Best of Group, Hannover Fair
1985 and 1986, Permanent product display with 9 (1985) and 5 (1986) Braun products, Haus Industrieform Essen, Design Zentrum Nordrhein Westfalen
1985, Prize to Braun AG and honorary award to Dieter Rams as Academico de Honor Extranjero, Academia Mexicana de Diseño, Mexico City
1986 and 1987, 1st Prize for micron vario 3 universal and 2nd Prize for AB 45 (1986), 1st Prize for the Aromaster KF 80 (1987), I.D. Magazine Annual Design Contest, New York
1987–98, Design innovations – in total 49 awards to Braun, Design Zentrum Nordrhein Westfalen, Essen
1991, mention for shaver pocket de luxe and DB 10 sl alarm clock, 1st Swiss Design Prize competition
1992, German Product Design Award for calculator World Traveller ET 88, Ambiente Trade Fair / Federal Ministry for Economic Affairs and German Design Council, Frankfurt/Main
2002–22, Red Dot Award – in total 59 awards to Braun in Product Design and Communication Design, Awards ceremony in Aalto Theatre in Essen / Design Zentrum Nordrhein Westfalen (until 2006), afterwards Red Dot GmbH & Co. KG, Essen
2003, Design Award ONDI for special contribution to industrial design and culture to Dieter Rams, Oficina Nacional de Diseño Industrial, Havana
2012–22, Good Design – in total 18 awards to Braun, Good Design Awards / Chicago Athenaeum Museum of Design and Architecture, Illinois
2013–20, CoolTec Shaver (2013), Series 9 Shaver (2014), CareStyle 5 Steam Station (2015), BN0211 Classic Slim Watch, Braun Silk Expert IPL, Braun Silk·épil 5 Wet & Dry Series 3 (2016), BT 3040 Beard Trimmer (2017), Braun CareStyle Compact (2019), Braun MultiQuick MQ7077X (2020), Good Design Award Australia / Good Design Australia Pty Ltd., Rose Bay, New South Wales
2014–23, German Design Award – in total 53 awards to Braun, GDA Awards ceremony during Ambiente Fair / German Design Council, Frankfurt/Main
2013–23, + x Award – in total 44 awards to De'Longhi / Braun Household, former Federal Parliament in Bonn / MEDIA SOCIETY NETWORKS, Grevenbroich
2020, Oral-B iO (Finalist), IDEA International Design Excellence Award / IDSA Industrial Designers Society of America, Herndon, Virginia

Exhibitions

1937, World Expo Paris – contribution with Braun products
1957, International Building Exhibition, Berlin – Braun products displayed in model flats
1957, 1960, and 1964, 11th, 12th and 13th Milan Triennial – contribution for the German Section
1958, World Expo Brussels – contribution to the German pavilion
1958–59, 20th Century Design from the Museum Collection, Museum of Modern Art (MoMA), New York
1959, The Package, MoMA, New York
1961, Hear Jazz – See Jazz, göppinger galerie, Frankfurt/Main – Braun products and photography by Wolfgang Schmittel
1964, documenta III, Kassel – works by Dieter Rams, Gerd A. Müller and Reinhold Weiss exhibited in the Industrial Design Section
1965, Japanese Design Council, Tokyo – exhibition featuring Braun's complete programme
1965–70, Braun. The Face of a Company – touring exhibition, Focke Museum, Bremen; Technisches Nationalmuseum, Prague, Brno and Bratislava; 17 universities in Japan; Die Neue Sammlung, Munich; Technical University Bilbao; Landesmuseum Oldenburg; Gillette AGM, Boston; several universities and art academies throughout the US and Canada
1967, World Expo 67, Montreal – contribution with 12 Braun exhibits
1969, What Is Design?, Louvre Museum, Paris
1970, World Expo 70, Osaka, – contribution with 7 Braun products
1973, Good Design from the Museum of Modern Art – touring exhibition), several stops in South America and then permanent exhibition in São Paulo
1974, FRG Design, Moscow
1979, Exhibition on Braun products and Gate graphic design, Centre Pompidou, Paris
1980–84, Design: Dieter Rams & – touring exhibition, IDZ, Berlin, 1980; Biennale of Industrial Design, Ljubljana, 1981; Amos Andersen Museum, Helsinki, 1981; Palazzo dell'Esposizione, Milan, 1981; Boilerhouse Project of the Victoria & Albert Museum, London, 1982; Stedelijk Museum, Amsterdam, 1982; Kunstgewerbemuseum, Cologne, 1983; museums throughout the Americas until 1984
1983–84, Design since 1945, Philadelphia Museum of Art
1987–89, ulm... The Moral of Objects, Bauhaus Archives, Berlin; Centre Pompidou, Paris; other stops in Zurich, Milan, Barcelona, London, Netherlands, US, Japan and Ulm
1988, Design heute. Masstäbe: Formgebung zwischen Industrie und Kunst-Stück, Deutsches Architekturmuseum, Frankfurt/Main, other stops in Tokyo, 1988–89; Los Angeles, 1989; London, 1990
1990, Braun Design Collection, reopening of exhibition to mark the Design Centre Hessen's opening in the Alfred-Messel-Haus of the INTeF, Darmstadt
1990–91, More or Less. Braun Design in Comparison, Museum für Kunst und Gewerbe, Hamburg
1991, Designed in Germany since 1949, sponsored by the Design Council, several cities in the US and Mexico
1998, '68 – Design and Everyday Culture between Consumption and Conflict, Kunstmuseum Düsseldorf und Galerie im Karmeliterkloster, Frankfurt/Main
2000, 4:3 – Fifty Years of Italian and German Design, Art and Exhibition Hall of the Federal Republic of Germany, Bonn
since 2000, Museum opening of the Braun Ettel Collection, Spandauer Zitadelle, Berlin, since 2013 in Berlin Moabit
2001, Dieter Rams Haus, Centro de Exposicoés Centro Cultural de Belém, Lisbon
2002–4, Dieter Rams Design – The Fascination of the Simple, INTeF, Darmstadt; Design Zentrum im Wilhelm-Wagenfeld-Haus, Bremen
2002–3, Dieter Rams. Less but Better, Museum für Angewandte Kunst, Frankfurt/Main; 2005, Kenninji Temple, Kyoto
2005–6, 50 Years of Braun Design, Braun GmbH, Kronberg; Royal College of Arts, London; Axis Gallery, Tokyo; Massachusetts College of Art and Design, Boston
2005–, permanent exhibition of the Braun Collection, Förderkreis BraunSammlung e.V., Westerbach Center Kronberg
2008, Cold War Modern, Victoria and Albert Museum, London
2008–12, Less and More. The Design Ethos of Dieter Rams, Suntory Museum, Osaka; Fuchu Art Museum, Tokyo; Design Museum, London; Museum für Angewandte Kunst, Frankfurt/Main; Daelim Contemporary Art Museum, Seoul; San Francisco Museum of Modern Art
2014, Making Music Modern: Design for Ear and Eye, MoMA
2016, Radio Era. Tube Radios, Design Icons, Internet Radio, Museum für Angewandte Kunst, Cologne
2016, Modular World, Vitra Design Museum, Weil am Rhein
2017–, Style Room. Dieter Rams, Museum für Angewandte Kunst, Frankfurt/Main
2017–18, Driven by German Design, Qatar Museums Gallery Al Riwaq, Doha, Qatar
2018–19, Dieter Rams. Principled Design, Philadelphia Museum of Art
2019, The Value of Good Design, MoMA, New York
2019–20, A Face of Second Modernity. Braun Design 1951–1967, China Design Museum, Hangzhou
2021, Braun 100, Bröhan-Museum, Berlin
2021, 1921 Braun 2021, INTeF, Darmstadt
2021–23, Dieter Rams. Looking Back and Ahead, Museum für Angewandte Kunst, Frankfurt/Main, 2021; Goethe Institute, New York and Washington DC, 2022; ADI Designmuseum, Milan, 2023; Stadtmuseum am Markt, Wiesbaden, 2023

Compiled by Hehn-Chu Ahn / Dieter Rams Archive

TEXT CREDITS
Concept: Klaus Klemp, Renke Thye, Oliver Grabes and Benjamin Wilson
Text: Klaus Klemp (Parts 1–3, Appendix), Oliver Grabes (Chapter 7, Part 2), Renke Thye (Part 2) and Hehn-Chu Ahn (Appendix)
Translation into English: Jeremy Gaines / German Editor: Daniela Böhmler

PICTURE CREDITS
Unless otherwise stated, all product photography is courtesy Gerhard Kellermann. All product icons were created by Stefan Dziallas, Iconwerk.

43, 89, 243: Niels Geisselbrecht; 102–3: Ernst Scheidegger; 104, 136–37: Marlene Schnelle-Schneyder; 119: Marlene Schnelle-Schneyder / Ingeborg Kracht-Rams; 135: Ingeborg Kracht-Rams; 139: Braun Photo Team; 161: Bernd Cziepluch; 170: Paul Ferrara; 179: Günther Becker; 304: Tobias Weinacker; 387: Jinwon Kim (Photoeye)

PUBLISHER'S ACKNOWLEDGEMENTS
The publisher would like to thank Oliver Grabes, Klaus Klemp, Nina Sagra, Renke Thye and Benjamin Wilson for their generosity and patience during the making of this book, and João Mota, Daniela Böhmler, Emma Caddy, Blanche Craig, Jeremy Gaines and Gillian Northcutt Liles for their contributions.

AUTHOR'S ACKNOWLEDGEMENTS
Klaus Klemp would like to express his gratitude to the following persons for information and research: Hehn-Chu Ahn, Helge Aszmoneit, Artur Braun, Bernd Cziepluch, Jürgen Greubel, Thomas Guttandin, Peter Hartwein, Hartwig Kahlcke, Peter Kapos, Horst Kaupp, Björn Kling, Dietrich Lubs, Markus Orthey, Cassandra Peters, Dieter Rams, Ingeborg Rams, Linus Rapp, Sara Rybakow, Marlene Schnelle-Schneyder, Florian Seiffert, Britte Siepenkothen, Eckhard Tischer, Roland Ullmann, Vu Duy Phong, Roland Weigend, Reinhold Weiss, Bernhard Wild and Klaus Zimmermann.

ABOUT THE AUTHOR
Prof. em. Dr. phil. Klaus Klemp

Born 1954, professor of design history and design theory at the Offenbach University of Design until 2020 and curator of design at the Museum of Applied Arts Frankfurt a.M., executive director of the Dieter and Ingeborg Rams Foundation since 2022, numerous exhibitions and publications on architecture, art and design.

Phaidon Press Limited
2 Cooperage Yard
London E15 2QR

Phaidon Press Inc.
111 Broadway
New York, NY 10006

phaidon.com

First published 2023
Reprinted 2024

ISBN 978 1 83866 389 6

A CIP catalogue record for this book is available from the British Library and the Library of Congress.

Commissioning Editor: Emilia Terragni
Project Editor: Olivia Clark
Production Controller: Lily Rodgers
Art Direction: Nina Sagra, Renke Thye, Oliver Grabes
Design: Cantina

Printed in China